Turfgrass Physiology and Ecology

Advanced Management Principles

This book is dedicated to my wife, Diana, who has supported me in all things for over 35 years and to my mother and father who taught me the value of hard work and determination that made this effort possible.

Turfgrass Physiology and Ecology

Advanced Management Principles

Gregory E. Bell

Oklahoma State University

CABI is a trading name of CAB International

CABI Head Office	CABI North American Office
Nosworthy Way	875 Massachusetts Avenue
Wallingford	7th Floor
Oxfordshire OX10 8DE	Cambridge, MA 02139
UK	USA
Tel: +44 (0)1491 832111	Tel: +1 617 395 4056
Fax: +44 (0)1491 833508	Fax: +1 617 354 6875
E-mail: cabi@cabi.org	E-mail: cabi-nao@cabi.org
Website: www.cabi.org	

© CAB International 2011. All rights reserved. No part of this publication may be reproduced in any form or by any means, electronically, mechanically, by photocopying, recording or otherwise, without the prior permission of the copyright owners.

A catalogue record for this book is available from the British Library, London, UK.

Library of Congress Cataloging-in-Publication Data

Bell, Gregory E.
　Turfgrass physiology and ecology : advanced management principles / Gregory E. Bell.
　　p. cm. -- (Modular text)
　ISBN 978-1-84593-648-8 (alk. paper)
1. Turfgrasses--Physiology. 2. Turfgrasses--Ecology. 3. Turf management. I. Title. II. Series: Modular texts.

SB433.B29 2011
635.9'642--dc22

　　　　　　　　　　　　　　　　　　　　　　　　2010028031

ISBN-13: 978 1 84593 648 8

Commissioning editor: Meredith Carroll/Sarah Mellor
Production editor: Fiona Chippendale

Typeset by SPi, Pondicherry, India
Printed and bound in the UK by Cambridge University Press

Contents

About the Author ... ix

1. **Diagnosing Plant Need** ... 1
   1.1 An Introduction to Using Scientific Concepts ... 1
   1.2 A Look at Terminology ... 2
   1.3 Using the Science ... 3
   1.4 Observing Turf Performance ... 6
   1.5 Photosynthesis ... 8
   1.6 Respiration ... 8
   1.7 Transpiration ... 9
   1.8 Diagnosing Plant Need ... 10
   1.9 Chapter Summary ... 11

2. **Understanding Photosynthesis** ... 12
   2.1 Encourage Photosynthesis to Encourage Turfgrass Growth ... 12
   2.2 The Importance of Light ... 13
   2.3 The Importance of Carbon Dioxide ... 17
   2.4 The Importance of Water ... 21
   2.5 Chapter Summary ... 22

3. **Why C_3 and C_4 Grasses Require Different Management** ... 24
   3.1 Use Species Adaptation to Your Advantage ... 24
   3.2 Photorespiration ... 25
   3.3 Kranz Anatomy ... 30
   3.4 Characteristics of C_3 and C_4 Plants ... 33
   3.5 Chapter Summary ... 35

4. **Respiration and Transpiration** ... 36
   4.1 Encouraging Efficient Use of Carbohydrates and Water ... 37
   4.2 Plant Respiration ... 37
   4.3 Transpiration ... 51
   4.4 Chapter Summary ... 57

5. **Why Our Management Practices Affect Our Turf** ... 58
   5.1 Understanding What We Know From a Turfgrass Perspective ... 58
   5.2 Mowing Causes Chronic Damage to Turf ... 60
   5.3 Thatch and Thatch Management ... 66
   5.4 Issues Pertaining to Soil Compaction ... 72
   5.5 Managing Wear Caused by Traffic ... 72
   5.6 Chapter Summary ... 73

6. **The Importance of Light and Managing Shade** ... 75
   6.1 The Practice of Making Adjustments to Improve Photosynthesis in Shade ... 75
   6.2 The Influence of Irradiance ... 76
   6.3 Managing Turf in Shade ... 81
   6.4 Chapter Summary ... 86

7	**Understanding and Prescribing Nutrition**	**88**
	7.1 Manage Your Fertilization to Match Your Turfgrass Needs and Local Conditions	88
	7.2 Basic Soil Attributes	89
	7.3 Plant Nutrition	93
	7.4 Chapter Summary	105
8	**Irrigation and Water Management**	**107**
	8.1 Manage Your Irrigation and Drainage for Site-specific Objectives	107
	8.2 Irrigation Management	108
	8.3 Species Adaptation to Low Water Use	123
	8.4 Managing Water Overload	124
	8.5 Chapter Summary	127
9	**Adjusting for Seasonal Conditions and Temperature Stress**	**129**
	9.1 Optimal Turfgrass Health is a Result of Flexible Management Under Differing Temperatures	130
	9.2 Seasonal Turfgrass Management	130
	9.3 High-temperature Stress in Grasses	131
	9.4 Cold-temperature Stress in Grasses	139
	9.5 Chapter Summary	144
10	**Growing Grass on Soil, Sand and Salt**	**146**
	10.1 Different Media Require Different Management	146
	10.2 Problems With Soil	147
	10.3 Problems With Sand	150
	10.4 Sand System Design and Construction	152
	10.5 Managing Turfgrass on Soil and Sand	154
	10.6 Managing Turf on Soils High in Salts	156
	10.7 Chapter Summary	161
11	**The Ecology of Turfgrass Management**	**163**
	11.1 How to Make Nature Work For You	164
	11.2 Introduction to Ecological Theory	164
	11.3 Turfgrass Ecology	170
	11.4 Chapter Summary	174
12	**Managing Competition Among Plant Species**	**175**
	12.1 Manage Your Sites to Favor Turfgrass and Discourage Other Plant Species	175
	12.2 Competition Among Turfgrass Species	176
	12.3 Competition Between Turfgrass and Other Plants	177
	12.4 Biological Herbicides	183
	12.5 Chapter Summary	185
13	**Managing Competition Between Turf and its Pests**	**187**
	13.1 How Can We Affect Relationships Between Predator and Prey?	187
	13.2 Predator/Prey Relationships	187
	13.3 Managing Turfgrass Predators	188
	13.4 Affecting Relationships Between Predators and Turf	193
	13.5 Making a Management Plan	196
	13.6 Chapter Summary	199

14	**Making the Right Decisions**	**201**
	14.1 Learn to Use Your Knowledge to Your Advantage	201
	14.2 Planning the Project	202
	14.3 Turfgrass Establishment	205
	14.4 Making a Management Plan	207
	14.4 Chapter Summary	210
References		**211**
Index		**227**

About the Author

Dr. Greg Bell is the Wayne and Jean Huffine Endowed Professor of Turfgrass Science at Oklahoma State University, Stillwater, Oklahoma. After many years as an industrial supervisor and business owner, Dr. Bell returned to school at the Ohio State University to finish a B.S. in Turfgrass Management, followed by a Masters and Ph.D. in Agronomy with specialty in turfgrass science. Greg has published a number of scientific papers, book chapters and trade journal articles in turfgrass science, and was part of a research team that was awarded the United States Department of Agriculture Secretary's Honor Award in 2002. He was named the Outstanding Undergraduate Student Advisor in the College of Agricultural Sciences and Natural Resources at Oklahoma State University in 2001 and was awarded the Faculty Phoenix Award dedicated to the outstanding graduate student advisor of the year at Oklahoma State University in 2003. Dr. Bell teaches classes in Introductory Turfgrass Management, Turfgrass Physiology and Ecology, and Personnel and Financial Management in Horticulture.

1 Diagnosing Plant Need

Key Terms

Best management practices (BMP) for turfgrass are the most effective techniques known at any given time for managing high-quality turf with little or no off-site effects.

Capillary action is the result of adhesive and cohesive forces that cause water to rise upward against gravity in a very small tube.

Osmosis is the movement of water through a permeable membrane from high water potential to low water potential such as when roots take up soil water.

Photosynthesis is the process plants use to convert light energy into chemical energy.

Respiration is the process plants use to release the energy trapped by photosynthesis.

Transpiration is the transport and evaporation of water that is used to dissipate the heat that is produced in the plant as a result of its metabolic processes, such as photosynthesis and respiration.

Turfgrass biology is the set of growth and developmental patterns that can be expected of turfgrass plants in response to all environments conducive to life.

Turfgrass ecology is the interaction of turfgrasses with other living organisms and other non-living entities, both natural and artificial, within the turfgrass environment.

A **turfgrass environment** is the sum of all natural and artificial conditions under which a particular stand of turf is expected to live and grow.

Turfgrass physiology is the awareness of the inner physical mechanisms and chemical processes that perpetuate life in a turfgrass plant.

1.1 An Introduction to Using Scientific Concepts

We all learn from experience. In fact, we may be able to learn everything we ever need to know from experience. However, if we combine our knowledge learned from classrooms, from books, from friends and family and from other sources with our experiences, we can become truly wise.

There is no secret to managing turfgrass reasonably well most of the time. Just add fertilizer and water. However, if you intend to be a turfgrass manager, your job is to manage turf exceptionally well all of the time and do it economically with the least amount of input and with little or no impact on people, animals and off-site locations. Consequently, you need to know when to add fertilizer and water and how much to add. You also need to know how the turfgrass should react to that fertilizer and water and how to keep it healthy under a variety of both natural and artificial conditions.

Start with the power of observation, add some experience, factor in some knowledge, and assign some responsibility to yourself and you have the beginning of a formula for diagnosing plant need. First, you have to know what the plant should look like at any given time. This you learn mostly from experience and a little knowledge. Second, you need to know how to make the plant look better and grow better if it is not meeting your customer's expectations. This you do mostly with knowledge and a little from experience. Finally, you need to know how to identify the problem if the plant is not responding to your input. That is why not just anybody can do this job. That is why your customer needs you.

1.2 A Look at Terminology

As an instructor, I am constantly reminded that undergraduate students, sometimes graduate students and, often, practitioners, do not like to hear terms like "plant biology", "plant physiology" and "plant ecology". Perhaps that is because they do not fully understand what those terms mean. There are lots of popular terms that are confusing and don't seem to make sense. Let's take a look at three terms that are often used to describe conditions or situations, but are not clearly defined.

Best management practices

One of those terms that begs for a clear definition is "best management practice" (BMP). Literally, BMP identifies a supervisory or technical procedure that cannot be improved upon because it is already the best. However, that is not what the term really means. The true connotation of the term is not evident in its literary meaning. According to Balogh and Walker in *Golf Course Management and Construction: Environmental Issues* (Balogh and Walker, 1992) and McCarty *et al.* in *Best Golf Course Management Practices* (McCarty, 2001), a BMP used in golf course management, and presumably in other turf endeavors, should achieve five basic goals: (i) decrease off-site transport of pesticides and nutrients; (ii) control the application of pesticides and nutrients; (iii) decrease total chemical loads; (iv) use both biological and mechanical soil conservation plans; and (iv) educate both managers and the public about the relationship between environmental issues and golf course management. Although we may still be a little confused about BMPs, we at least know that the term does not necessarily refer to a "best" practice unless it happens to be the best practice known at that particular time for that particular location. We also now know that a BMP specifically refers to chemical application management – or does it? Actually, the term was originally used to refer to techniques used to manage waste water and storm water runoff. Its first use may have been in the 1987 amendments to the U.S. Clean Water Act (U.S. Congress, 1987). The term later evolved to include practices that reduce chemical contamination of surface and ground water following application. However, it appears that BMP has evolved further into its more literal translation to include more than just water-protection practices. Dr. McCarty and others, for instance, authored the outstanding book that I referred to earlier using the term BMP to include a wide range of environmental management techniques. Those techniques may or may not include chemical applications and water-protection practices. So a contemporary definition of BMP for agronomic purposes could be: the best-known procedure for encouraging or maintaining plant health without detrimentally affecting off-site conditions. In simple terms, contemporary turf BMPs are protocols that we use to grow good grass without hurting anything else in our environment.

Environment

"Environment" is another of those constantly evolving terms. In this case, however, the term has also been highly misused and abused. Environment has a very simple definition: the aggregate of all surrounding conditions that influence an organism (Beard, 1973). Environment can be used to describe the conditions that exist on a dish, either a petri dish or your dinner dish, or it can be used to describe conditions that exist across an entire country or around the world. The petri dish or dinner dish are better applications of the term than the more commonly used version, which refers to all natural entities that exist over a nearly unlimited location such as a country, a continent or the world. In fact, an environment does not have to consist of natural conditions. If you place the petri dish in an oven, it now exists in an artificial environment, not a natural one. If microbes happen to be living on that dish, they are natural organisms living in an artificial environment. The same could be said for certain turfgrass conditions. For instance, a golf course putting green built to United States Golf Association recommendations could be said to include natural organisms, one of which is the grass, living in an artificial environment (USGA Green Section Staff, 1993). So the question becomes: if the environment is artificial, who or what is controlling it?

On a USGA golf course putting green, the environment is primarily controlled by the grounds manager or golf course superintendent. There are some conditions, usually temperature for instance, that the manager cannot control or has minimal influence over. There are other conditions, such as shade and traffic, that the manager could control and would like to control but may be forced to accept. Because the manager can control or influence nearly all of the environmental conditions on the golf course putting green, the turfgrass growing there is at the mercy of the manager. If the manager is knowledgeable in the needs of the turf, the turf will thrive, if not, it will probably die. The same could be said for a home lawn, a park or any other location where turfgrass is cultivated for human use. The putting green is a good example because even the growing medium is highly artificial and because the conditions required to satisfy the golfer's expectations are severe. In fact, the closer the turf is to death, the better the golfers seem to like it. Especially competent turfgrass managers like to work with these artificial environments because they control, or nearly control, most, or all, of the environmental conditions. The more environmental influence the knowledgeable manager has, the better the manager likes it. Less competent managers, homeowners for instance, would rather manage plants under more natural conditions. It is easier and it requires less knowledge. The best looking home lawns are managed by knowledgeable homeowners or by professional lawn-care managers.

Best looking

"Best looking" is a relative term. It is a term relative to an individual and, therefore, to location. The best looking home lawn in Europe is not the same as the best looking home lawn in Australia, is not the same as the best looking home lawn in the USA. The lawns differ because of differing customer expectations, different rules and regulations, and different environments. Notice that the term "environments" is not used here to describe the environment in Europe or the environment in Australia, it is used to describe the environment of the best looking lawn in Europe or the best looking lawn in Australia. Consequently, if we can identify the best looking lawn, we can characterize the best environment for growing home-lawn turf on that particular medium in that particular vicinity. We can also identify the best management practices for growing home-lawn turf on that particular medium in that particular vicinity and we could possibly improve on those BMPs with a general knowledge of plant biology, plant physiology and plant ecology.

1.3 Using the Science

This discussion began with an introduction of three terms, "plant biology", "plant physiology" and "ecology". After the three scientific terms were introduced, three popular terms were introduced. It should be clear that "BMP" has a lofty connotation and represents highly desirable principles, but it tends to take on the definition of the person using the term at the time. The same is true of "environment".

"Environment" was never meant to mean all natural entities in a given area. According to Merriam-Webster (2009), environment is the circumstances, objects or conditions by which one is surrounded. For our purposes, a more scientific definition is appropriate and can also be found in Merriam-Webster: "the complex of physical, chemical, and biotic factors (as climate, soil, and living things) that act upon an organism or an ecological community and ultimately determine its form and survival". Apparently, if a BMP is going to have a positive effect on turfgrass growth or survival, it must also affect the environment in which the turf lives. Consequently, a BMP has to affect the turfgrass environment but should not affect the surrounding environment detrimentally.

Our third popular term, "best looking", is extremely important to turfgrass management but it is the hardest to define. "Best looking" is a matter of opinion or personal preference and can differ substantially by individual. However, turfgrass and other horticultural plants are used not only for their functional characteristics but also for their aesthetic value. Therefore, "best looking" is not just a result of good management, it is a condition that turfgrass managers strive to affect. "Best looking" often determines the BMPs that the turfgrass manager incorporates into the turfgrass management plan. "Best looking" determines when to irrigate, when to fertilize, and when and how to apply other management practices. Yet, "best looking" is a matter of personal preference and is seldom the same for different locations or cultures. European turfgrass managers, for instance, do not share the same preference for dark green turf that their counterparts in the USA desire.

If these last few paragraphs were somewhat confusing, don't be alarmed. They were meant to be. I introduced a term that was coined not for its literary value, but for its "feel good" effects, BMP. I also discussed a term that has been twisted by society to mean things that it was never supposed to mean, environment, and finally, a term that has a substantially different meaning depending on who is using it at the time, best looking. Certainly, scientific terms such as plant biology, plant physiology and plant ecology are not so confusing as these three popular terms and others like them that we are comfortable using daily.

In order to be the best turfgrass managers that you can be, you have to make decisions based on plant biology, plant physiology and plant ecology. We already know the basic principles that we need to know to make decisions based on these scientific factors. In the following 200 pages or so, I will help you relearn these basic principles, not in any great detail, but in enough detail that you can use them successfully. We all learn best from experience, but we can learn so much more from our experiences if we use our knowledge as well. It is good to know how to accomplish a task successfully. It is better to know why the procedures that we used to complete the task worked so well. For the time being, let's forget about how you do something and focus on why you do it instead.

Plant biology

Plant biology is the same as botany and can be defined as a science involving the study of plants (Stern, 1991). I suppose that means that if you stare at a plant for a few hours then you are performing botany. Actually, no; if you stare at a plant for a few hours, you are performing basically nothing, which is not a bad thing as long as you don't make a habit of it. As botany is a science, it involves more than staring and as it is a science, there is actually more research involved than there is learning. People like university instructors do the research so that people like university students can learn from it. Obviously, the instructors, researchers and scientists who did this research did it so that they and newer instructors could learn from it and pass it on to their students. Consequently, we need to know a little about research to understand botany. As you might have guessed, we are not interested in research done on all of the plants in the universe; we are only interested in the research done on turfgrass.

Turfgrass science

I was giving a presentation to a group of scientists several years ago when I used the term "turfgrass science". One of the crop production scientists in the front row jokingly made the comment "isn't turfgrass science something of an oxymoron?". I immediately flashed up a slide of researchers applying treatments to a golf course and before I could say anything another turfgrass scientist said "that sure beats working in a hay field doesn't it?". Yes, we do practice science and most of those techniques for managing turfgrass that you have learned on and off the job came from us. As scientists go, turfgrass scientists tend to do a lot of work for little money and we tend to do a lot more applied work, research that can be used in the field, than we do basic work, research that leads to greater understanding but is not immediately applicable in the field. Both types of research are important. As you will see, we are going to use basic science throughout this book to help us make decisions that we can apply in the field.

Turfgrass research

We use research for a variety of reasons, including testing new products, testing new grasses, testing new maintenance procedures alone or in combination with old procedures, and gaining knowledge in plant growth or metabolism that may lead to improved varieties, improved cultural practices, improved weed control and other improvements. Experiments are effective because they measure treatment differences under comparable conditions, because they answer questions posed by specific objectives and because they answer questions deemed important. An experiment is a planned inquiry to obtain new facts or to test previous results that aid in decision-making (Waddington *et al.*, 1992). Consequently, scientific experiments are governed by a fixed set of research protocols. Although a great deal of useful information has been accumulated by simple observation, an equal amount of misinformation has probably been accumulated in the same manner. Research protocols are specifically designed to minimize misinformation and are consequently more likely to overlook a positive result than to verify a result that is not true. Research can be as simple as applying a herbicide to see how many weeds are killed or it can be as complex as mapping the human genome. Although the complexity of research projects varies widely, the same scientific protocol applies in all cases. Based on scientific protocol, if a turfgrass scientist reports that a particular herbicide is effective for controlling a particular weed, the scientist should be right 95% of the time. That will always be true, regardless of the experiment, as long as the scientist works within the generally accepted 95% confidence interval.

Most of what we know about turfgrass management has been learned either by simple observation or by experimentation. What we know about turfgrass biology, however, has all been learned through experimentation. In other words, we aren't guessing. We are basing our conclusions on experiences that occur repeatedly under the same circumstances. Each biological principle has had to withstand rigorous tests. If we are aware of these principles, we can expect a plant to respond to a given stimulus in the same way each time the stimulus occurs, provided that the same environmental conditions are present each time. Using scientific information, we know when the plant will grow rapidly, when it is likely to be dormant or quiescent, when it is susceptible to damage and, most importantly, when it needs fertilizer and water. We know how low we can mow it under a given set of conditions, we know when we can perform aggressive cultural management, and we know when we should leave it alone. That's right; doing nothing can sometimes be an excellent management practice; one that is often overlooked.

Biologically speaking, we can expect a turfgrass plant to respond to a given set of environmental conditions in the same way each time that those conditions occur. Therefore, managing turfgrass is easy. All we have to do is perform the same management practices at the same time in the same amount each season and our turfgrass will always grow well and look good; or will it? Biologically speaking, that is true provided that the same set of environmental conditions occurs each year at the same time and in the same amount. That doesn't happen. Yet we are still looking for that magic set of management practices that makes our turf perform to our expectations year after year regardless of environmental differences. That magic recipe doesn't exist, but if we consider the environment independently of what we did last year or last week, we can make decisions based on simple biological principles that will help our grasses perform as best as can be expected nearly all of the time. In this text we will pursue that goal. The biological principles needed are often as simple as "turfgrass grows well when it has water but not well when there is too much water or too little water". It is not the biological principles that are difficult, it is the flexibility to make decisions and the confidence to know you are right that is difficult.

Plant physiology

Plant physiology is the science that studies plant function (Salisbury and Ross, 1992). That is a simple, yet extremely broad, definition similar to the definition that we used for plant biology. Fortunately, we do not need to know a great deal about the physiological processes of turfgrass in order to make sound management decisions. Photosynthesis, respiration and transpiration are physiological plant processes. They are activities that take place inside the plant but result in external activity or aesthetics that we can often see or measure. It is not difficult for an experienced manager to determine when transpiration has slowed and the turf needs water. However, it is difficult for a manager to determine when photosynthesis is occurring too slowly or respiration has been compromised. Poor photosynthesis or poor respiration manifest themselves in basically the same symptoms that occur when turfgrass is under nearly any type of environmental stress. Like disease symptoms, there are nuances that can help us identify the source of the stress, but the symptoms are generally similar because they all affect photosynthesis, respiration and/or transpiration.

When a turfgrass is exposed to saturated soil conditions for an extended period of time, it turns yellow, then brown, then dies in a matter of days depending on species. How does this stress from saturated soil differ from the stress experienced when turfgrass is exposed to very dry soil for an extended period of time? Actually, there is very little difference in the effects of very wet and very dry soil. In fact, in both cases, the turf dies of drought. When the soil is dry, there is no water for the plant to take up. When the soil is wet, the roots are surrounded by water but they have no air. No air means no oxygen, no oxygen means no respiration, no respiration means no energy to take up water and the plant dies of drought. How do you keep this turf alive in saturated soil until the soil has time to drain?

All that you need to know to solve this problem is a miniscule amount of plant physiology. You need to know that roots have to respire to take up water and that oxygen has to be present for the roots to respire. Once you have realized that these conditions are not present, you know that you have to get water to the plants or they will die. Consequently, your only option is to apply water in foliar sprays that the plants can take up through the leaves. The long-term fix is to provide better drainage to the area, but the short-term fix is to frequently spray water on to the leaves of plants whose roots are sitting in saturated soil. The only way that you can make that conclusion is through a simple knowledge of plant physiology. Neither experience nor common sense would lead you to the conclusion that you should spray water on to turf that is sitting in saturated soil. However, as a person who knows a little about plant physiology, a simple assessment of the problem should lead you to that conclusion even if you have never experienced such a condition. That is the power of simple scientific knowledge.

Plant ecology

Ecology is the study of the interactions of organisms with one another and with their environment (Starr, 2000). That does not sound like much, but ecology is powerful. Understanding which environmental factors favor a weed and which favor turf can save a turfgrass manager considerable time and money. The objective of the turfgrass manager is to influence the environmental conditions so that they favor the turf at the expense of competing organisms. Disease, for instance, occurs when the environmental conditions favor the growth of a pathogen more than they favor the growth of turf. If the manager can minimize or postpone the occurrence of certain environmental conditions that favor the pathogen instead of the turf, the manager can reduce the need for pesticides and minimize the amount of disease that occurs.

For purposes of this book, we will refer to plant ecology as turfgrass ecology, thereby limiting the interactions in a given environment to those most important to turf. There could be a hundred or more species living in an ecological system containing turf, but we are primarily concerned with those species that have the greatest influence on our turf. This ecological system, simply called an ecosystem, contains microbes, insects, plants and other living organisms constantly competing for nutrients and space (Danneberger, 1993). The organisms that are best adapted to the environment of the ecosystem flourish and those that are not so well adapted decline. Theoretically, given enough time, only the species that are best adapted will survive, and because intraspecific competition also occurs, only the strongest individuals of that species and their offspring will remain. Although I would not want to be the one to try to prove this theory, the concept is a good one and extremely useful for plant management.

In order to use ecology to your advantage, you have to have a sound knowledge of your target plant, your turf, and at least a basic knowledge of its competitors, as well as of the neutral, predatory and beneficial organisms with which it shares their environment. That sounds complex and it is. Ecological management is not easy but we all practice it. The more knowledge you can use and the more experiences that you have, the easier it becomes.

1.4 Observing Turf Performance

Turfgrass performance varies considerably according to species, climate, weather, time of year and several other factors. Therefore, it is difficult to know exactly what your turfgrass should look like at any given time unless you have considerable experience with a particular turf in a specific location. Site-specific factors such as low areas, shaded areas, south-facing slopes and many other conditions also affect turfgrass performance. As part of our research (Xiong *et al.*, 2007), we observed the performance of a cool-season grass, creeping bentgrass (*Agrostis stolonifera*), and a warm-season grass, bermudagrass (*Cynodon dactylon*), over two growing seasons in the south-central USA using the normalized difference vegetation index (NDVI), an objective measure of reflected irradiance, to determine the approximate color of those grasses over two growing seasons (Fig. 1.1). An NDVI of 0.6 is considered green but slightly yellow and an NDVI of 0.9 is very dark green. Based on our observations, the NDVI and the color of these grasses changed considerably during the growing season in spite of the fact that they were fertilized consistently and watered sufficiently. Consequently, unless you are very familiar with a certain grass in your location, it is unlikely that you could accurately predict what that grass should look like during a particular time of year. In addition, if weather conditions are different this year from last year, and they almost always are, it is likely that these patterns will be, at the very least, slightly different between the two seasons. Making turfgrass management decisions based on visual assessment requires a highly experienced manager, yet that is how most of us attempt to do it.

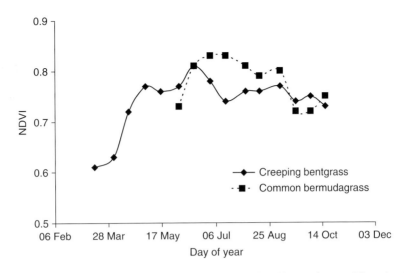

Fig. 1.1. Approximate color of creeping bentgrass (*Agrostis stolonifera*) and bermudagrass (*Cynodon dactylon*) measured using the normalized difference vegetation index (NDVI) averaged over two growing seasons in Stillwater, Oklahoma. A score of 0.6 is slightly yellow-green turf and a score of 0.9 is very dark green turf.

Earlier in this chapter, I suggested the use of biology, physiology and ecology for making management decisions pertaining to turf. According to biology, a grass should respond to a given set of environmental conditions in approximately the same way every time those conditions are present. However, plant biology is obviously a very complex method for making turfgrass management decisions, especially if we consider the huge scope of a turfgrass environment and the number of changes that could occur. We need a decision-making system that is simpler than basic plant biology and yet more attainable than relying on a multitude of experience. In order to find that simpler system, let us look a little deeper into the cause-and-effect relationship between biology and environment.

According to science, biological responses are primarily predetermined by genetic code. The genetic code encrypted on the plant's DNA regulates which proteins and enzymes, and the amount of those proteins and enzymes, that are present in the plant under a given set of conditions. The enzymes produced encourage chemical reactions that enable or disable metabolic pathways. These metabolic pathways determine the health of a plant, its appearance, how well it competes with other organisms and how quickly it grows and spreads. The study of these metabolic pathways is called plant physiology and it can help us identify plant need and manage turf accordingly.

There are three physiological processes that are of particular importance when making plant management decisions. Those are photosynthesis, respiration and transpiration. Photosynthesis is the means by which the plant makes food, respiration is the means by which the plant turns that food into energy and transpiration is the means by which the plant cools itself sufficiently to perform photosynthesis and respiration continuously. There are an enormous number of physiological processes that occur in a grass plant but, for now, those three processes will be sufficient for our purposes.

Physiological processes are far from simple. They are extremely complex. Scientists can spend their entire lives trying to unravel the biochemical reactions in a small portion of the photosynthetic pathway. Those reactions are in constant flux and the pathways that regulate those reactions can be more complex than the reactions themselves. It is understandable why many students get confused by these systems. However, the presentation of photosynthesis, respiration and transpiration need not be complex. It is far beyond the scope of this text to try to dissect these three processes. Instead, we only need to know the basics of these systems to make them useful to us.

1.5 Photosynthesis

There are two types of photosynthesis, warm-season and cool-season photosynthesis, that occur in grasses. Warm-season photosynthesis is called C_4 photosynthesis and cool-season photosynthesis is called C_3 photosynthesis. Like it or not, these distinctions are important to you and we will discuss them in a later chapter. In simple terms, photosynthesis is the means the plant uses to turn radiant energy into chemical energy. In fact, all of the biological energy on our planet comes from the sun and is channeled to us by the process of photosynthesis. Plants trap energy, then we eat the plants or we eat an animal that ate the plants and that is how we survive. Consequently, photosynthesis is not just important to you because you want to be a turfgrass manager, it is also important because it keeps you alive.

Photosynthesis requires light, water and carbon dioxide. If one of these important components is low or missing, photosynthesis will slow or stop. If the condition does not improve, the plant will become sick and eventually die. Other nutrients are required to perform photosynthesis as well but they are rarely in low supply and are not nearly as important as light, water and carbon dioxide. Consequently, it is our job, above all else, to make sure that our turf has sufficient light, water and carbon dioxide. You may believe that light and carbon dioxide are beyond your control, but that is not correct. We can have a substantial influence on those components in most situations. If, for some reason, it is impossible to deliver sufficient light, water or carbon dioxide to an area where you want to grow turf, give up and mulch it.

Photosynthesis takes a portion of the energy the plant requires in light, converts it to chemical energy, and stores it in carbon bonds. You are probably familiar with the general principle represented in the following chemical equation:

$$6CO_2 + 6H_2O + light \rightarrow C_6H_{12}O_6 + 6O_2$$

which shows that carbon dioxide (CO_2) plus water (H_2O) in the presence of light yields glucose ($C_6H_{12}O_6$) and oxygen (O_2). Glucose, of course, is a sugar and it has five carbon bonds that store energy.

The glucose equation represents the major purpose and events in photosynthesis, but it is an oversimplification. In fact, much of the energy from light is converted into glucose, which is temporarily stored as starch in the chloroplast where it was produced. However, glucose is not the only sugar produced by photosynthesis. The other sugars produced are primarily used to synthesize other compounds that the plant needs. Yes, photosynthesis produces the energy that powers all living things. However, it also produces the sugars that are used to synthesize amino acids and fatty acids for the construction of proteins, lipids and the other compounds necessary for plant survival. Glucose is also combined to make cellulose and other products used to form structural plant tissue. The glucose that is not used for energy or the synthesis of other products in the cell where it was produced is converted to sucrose and transported to areas of the plant where energy is needed. If the plant has an excess of energy, the sucrose is transported to cells, usually in the stems and roots, where it is converted back to glucose and stored as starch (C_4 plants) or is converted to fructose and glucose and stored as fructans (C_3 plants). There the stored products remain until energy is needed or it is used for conversion to other compounds.

Because a plant can photosynthesize, it does not have to forage for food and it can make everything it needs as long as sufficient nutrients are present. All the plant requires is a medium (soil) that contains the minimum nutrients necessary for survival, plus light, water and carbon dioxide.

Photosynthesis is so important that it should always be our primary focus. Nothing else occurs without it. We must always maintain sufficient light, water and carbon dioxide for our turf. If the turf declines, we should first eliminate the possibility that light, water or carbon dioxide are not sufficient before we look for deficiencies elsewhere.

1.6 Respiration

Respiration is the process by which the plant releases the energy stored in the chemical bonds produced by photosynthesis, and oxygen is required for respiration to occur. Other nutrients, especially phosphorus, must also be present but oxygen is the most important. Respiration consumes oxygen and gives off carbon dioxide, as represented in the following equation:

$$C_6H_{12}O_6 + 6O_2 + light \rightarrow 6CO_2 + 6H_2O + energy$$

which shows that glucose ($C_6H_{12}O_6$) plus oxygen (O_2) yields carbon dioxide (CO_2) plus water (H_2O) plus energy. For that reason, respiration complements photosynthesis. Photosynthesis takes in carbon dioxide and gives off oxygen, whereas respiration takes in oxygen and gives off carbon dioxide. It would seem that a turfgrass plant would be in complete balance with its environment. It takes in what it needs and gives it back in another process. However, we must remember that respiration produces nothing except carbon dioxide, water and energy for plant functions. Photosynthesis, however, produces sugars for energy and sugars used to synthesize other compounds as well as all of the structural components of the plant. Consequently, photosynthesis must take in considerably more carbon dioxide than respiration produces.

Respiration occurs in all plant parts. There is more than sufficient oxygen for respiration to occur in the plant parts located above ground. After all, the air is 21% oxygen. The area where oxygen could be deficient is below the soil surface. Compacted or saturated soils are examples of situations where sufficient oxygen is not present for adequate root respiration. That is why turf declines rapidly when those conditions occur and why sports field managers aerify at every opportunity. Aerification, of course, is a procedure where the manager opens the soil to air infiltration thereby helping to provide oxygen to the roots. Aerification is also used to break up soil layers and to encourage water infiltration, but those are subjects for another chapter. Suffice to say, that if we perceive no barriers to photosynthesis above ground then we should be making sure that there are no barriers to respiration below ground.

1.7 Transpiration

Transpiration is a simple plant system, but also a very important one. In basic terms, it is a movement of water from the soil through a plant and into the air. It occurs because of osmosis, cohesive forces, adhesive forces and evaporation, and is fueled by water potential. If you have had a soils class, you have studied water potential and you know that water moves from high water potential to low water potential. So for transpiration to work, the water potential in the plant has to be lower than the water potential in the soil, and the water potential of the air has to be lower than the water potential of the plant. During transpiration, water moves from the soil into the plant by diffusion, through the plant by cohesion, adhesion and water potential, and into the air by evaporation.

Water uptake for transpiration

If you add salt, sugar, or any other soluble compound to water, it lowers its water potential. So if you were to place a permeable barrier between pure water and water containing solutes, the pure water would move toward the water containing solutes. The water contained in a plant's soil medium is high in solutes. In fact, many of the solutes in soil water are plant nutrients. Consequently, these solutes are very important to our plant. The soil water has a water potential lower than pure water. However, the water inside the plant contains even more solutes than the water in the soil and nearly always has a lower water potential than soil water. Consequently, the soil water is attracted through the roots into the turfgrass. Once inside the turfgrass, the plant can remove whatever nutrients it needs and fill its cells with water. Most of the water, however, is not used for metabolism; it is drawn through the grass and evaporated into the air. This evaporative process is used to cool the plant. Otherwise it would die from the heat produced by its own metabolic processes.

Energy transfer is not a perfectly efficient process. Any time that energy is transferred or transformed, a certain amount of the initial energy is lost. A huge number of chemical reactions are required to perform plant photosynthesis, respiration and other important functions. Each time one of these reactions occurs, a small amount of energy is lost. It follows then that the amount of light energy accumulated by photosynthesis is not the same amount that is released through respiration. Only a portion of that energy is available for use by the plant, the rest is mostly heat. That heat has to be dissipated otherwise our plant will die of something like a very high fever. When we work hard, our metabolism increases and we produce excessive heat. We dissipate that heat through perspiration. A plant dissipates heat through transpiration. Some heat is also dissipated through conduction and convection, but transpiration is the most important system for plant cooling.

Water movement from roots to leaves

Once soil water is in the plant, it circulates among and around the cells mostly by diffusion, a relatively slow process. It can also move upward into the leaves and other plant parts as it is drawn by the relatively low water potential of the air. You may be familiar with the capillary tubes that are used for a variety of purposes and often to demonstrate adhesive and cohesive forces in chemistry class. A capillary tube is a glass tube with a very small opening. You could call it a little glass pipe. If you put this capillary tube in water, the water fills the tube even if you are holding the tube vertical and only the low end is in the water. So the tube causes the water to move upward against gravity. That is called capillary action. Plant scientists are quick to point out that capillary action is not the means by which plants move water for transpiration, but the same principles are used to help it along. The water molecules adhere or stick to the sides of the tube. They also stick to each other, a force called cohesion. The adhesive and cohesive forces hold the water to the sides of the tube and to itself so strongly that the water overcomes gravity and moves up the tube. If the tube was larger, the cohesive forces would not be strong enough to hold the water together across the width of the tube. In that case, a little bit of water would stick to the sides of the tube but the rest would fall to the bottom. The xylem that partially makes up the veins of a plant is like a capillary tube, but it is not open to the air, which is another requirement of capillary action. Consequently, capillary action does not drive transpiration but the two forces that explain capillary action must be present for transpiration to occur. It is generally believed that water potential drives transpiration and cohesive and adhesive forces help hold the water columns together in the plant and keep them from falling. There is also a root pressure effect that sometimes forces water upward, usually during the night, but this is not a major mechanism for the movement of water in transpiration.

We are all familiar with the cooling effect of evaporation, but did you also realize that the air has a water potential like the water potential of the soil and the water potential of the plant? Typically, the water potential of the air is very low and evaporation occurs readily. However, when humidity is high, the water potential increases and evaporation occurs more slowly. Hence, hot, humid conditions are difficult for plants because the humidity inhibits transpiration and the heat causes the internal plant temperature to build more rapidly.

1.8 Diagnosing Plant Need

Why is turf difficult to manage when it is surrounded by buildings even when it gets sufficient light? Why is turf difficult to manage in the shade of trees? Why does turf require more water when it is warm and windy than it does when it is cool and calm?

We have already answered those questions, not literally, but logically, in the preceding sections of this chapter. Diagnosing the needs of your turf requires some thought. The preceding three questions are relatively easy compared with those we often face in the field. However, they still require thought.

Turf surrounded by buildings is difficult to manage because the air movement at ground level is restricted. Our planet's air is only 0.035% carbon dioxide. Photosynthesis requires light, carbon dioxide and water. In this situation, we have light, we can add water, but we are limited in the amount of carbon dioxide available. As photosynthesis proceeds, our plants take in carbon dioxide and give off oxygen. If air movement is restricted, fresh air, hence fresh carbon dioxide, is not available. Consequently, the air above the turf becomes high in oxygen and low in carbon dioxide, and photosynthesis stalls.

Turfgrasses don't grow well in the shade of trees because sufficient light is not present. Again, we have a problem with photosynthesis. Our turf needs more water when it is warm and windy because transpiration is based on evaporation and the warmer it is and the windier it is, the faster evaporation occurs. There are other reasons as well, that you can probably figure out, but we will save those for later chapters.

Biological problems are nearly always complicated and more is rarely better. Plants are biological systems just like us. Food is a good thing for us but too much makes us unhealthy. If we have a headache, taking one or two aspirin might relieve the pain and make us feel better, but taking ten or 12 aspirin will almost certainly be harmful. Exercise is good for us but working too much and too hard is harmful. Our turfgrass has the same qualities that we have. Light is good but too much light can be harmful, water is good but too much water is harmful, and fertilizer is good but too much fertilizer is harmful. We not only have to know what our grass needs, we have to know how much is too much.

Single-tactic systems

If you are familiar with integrated pest management (IPM), you have probably heard of the dangers of relying on single-tactic systems. IPM refers to a system of pest control measures that combine for effective control of invading pests. For us, healthy turf is always our number one pest control measure. If a particular pest problem is severe, we may combine management practices that improve the health of our turf with mechanical practices, cultural practices, biological control techniques and pesticides that are designed to specifically manage a particular pest. A single-tactic system would refer to one of these techniques, usually pesticides, as our only control measure. Single-tactic systems, even pesticides, are ineffective in comparison with IPM. So are single-tactic diagnoses. If we are having problems managing our turf to customer expectations, there will nearly always be more than one deficiency that needs to be addressed. We might blame the problem on the soil, on the weed competition, on a disease problem or on the lack of some nutrient, but it is almost always a combination of problems that exists. Therefore it is difficult to address the needs of a particular stand of turf based entirely on things that we have done at other times at other locations. We also need to think through the situation for this particular time and this particular place. We need some independent assessment. Think photosynthesis, respiration and transpiration. Are sufficient light, water and carbon dioxide present? If so, is there sufficient oxygen in the soil? If we have sufficient water, are the environmental conditions conducive for evaporation to occur rapidly enough to cool our plants? Once we have eliminated or minimized barriers to photosynthesis, respiration and transpiration we can start to consider other deficiencies.

1.9 Chapter Summary

The objective of this text is to teach the reader how to use scientific or logical thought to make technical decisions. In other words, it is designed to demonstrate why management techniques work rather than how they work. The most important part of any decision is gathering information. The more information available, the more likely the manager will be able to make a sound decision. However, we rarely have as much information as we would like to have to make important decisions and we certainly cannot predict the future. Consequently, our knowledge and decision-making procedures must be sound enough to help overcome our lack of information. To this end, turfgrass management decisions and practices can be improved by the study of the basic scientific concepts and plant characteristics introduced in this chapter.

Suggested Reading

Introductory turfgrass textbooks

Beard, J.B. (1973) *Turfgrass Science and Culture*. Prentice-Hall, Englewood Cliffs, New Jersey.

Christians, N.E. (2007) *Fundamentals of Turfgrass Management*, 3rd edn. John Wiley and Sons, Hoboken, New Jersey.

Duble, R.L. (1989) *Southern Turfgrasses: Their Management and Use*. TexScape, College Station, Texas.

Emmons, R.D. (2008) *Turfgrass Science and Management*, 4th edn. Thompson Delmar Learning, Clifton Park, New York.

Turgeon, A.J. (2008) *Turfgrass Management*, 8th edn. Pearson Prentice Hall, Upper Saddle River, New Jersey.

2 Understanding Photosynthesis

Key Terms

nm is an abbreviation for nanometer, 1 billionth of a meter.
Light is the sun's radiant energy that we can see and that plants can use for photosynthesis. Ultraviolet and infrared radiance are not light.
PAR is an abbreviation for photosynthetically active radiation and is the energy in light or radiance.
PPF is an abbreviation for photosynthetic photon flux and refers to the particles or quantum in light or radiance.
Ultraviolet refers to the sun's radiance below the 400 nm wavelength. These wavelengths have very high energy and can be destructive to biological tissues.
Infrared refers to the sun's radiance above the 700 nm wavelength. These wavelengths are relatively low in radiant energy and very close to the energy level of heat.
Affinity is the force of attraction between objects or particles that causes them to enter into a physical or chemical combination.
NADPH (the reduced form of nicotinamide adenine dinucleotide phosphate) is a cellular energy molecule in which the light reaction of photosynthesis stores light energy. It is formed by the donation of electrons to **NADP$^+$** (nicotinamide adenine dinucleotide phosphate).
ATP (adenosine triphosphate) is a second cellular energy molecule in which light energy is stored by the light reaction of photosynthesis. It is formed from **ADP** (adenosine diphosphate) during the process of photophosphorylation.
Thylakoid is an inner compartment of a chloroplast separated from the stroma by a membrane where the light reaction takes place.
Stroma is the name given to the cytoplasm of the chloroplast.
A **sink** is a part of a plant, such as a root, that uses more energy than it produces.
A **source** is a part of a plant, such as a mature leaf, that produces more energy than it uses.
RuBP carboxylase, also called **Rubisco** is the enzyme that catalyzes the combination of ribulose bisphosphate (**RuBP**) and carbon dioxide in the dark reaction of photosynthesis to produce sugars for energy storage and for plant use in other processes.
A **boundary layer** refers to a layer of air that surrounds turfgrass leaves that, owing to the occurrence of photosynthesis and transpiration, is low in carbon dioxide and high in oxygen and water vapor.

2.1 Encourage Photosynthesis to Encourage Turfgrass Growth

Photosynthesis can be an extremely complex subject. Biologists have been studying photosynthesis for 100 years or more and have learned many of its finest details. For our purposes, there is no need to go into great detail about this important plant process, but we will study the basic concepts so that you can use them to your advantage.

As mentioned earlier, photosynthesis is simply defined as the process plants use to convert light energy into chemical energy. This chemical energy is then used to power metabolic processes. The energy may be used in the cell where it was produced, it may be transported to another portion of the plant, or it may be transported to a storage location. Regardless, if photosynthesis is proceeding normally at near maximum rate, the plant will typically be healthy and capable of resisting stress.

As you know, light, water and carbon dioxide are the most important components required for photosynthesis.

2.2 The Importance of Light

We know that the sun's radiance exists in a range from at least 50 nm to over 3000 nm in wavelength (NASA/Cool Cosmos EPO Group, 2010). The term "wavelength" refers to radiant energy's existence in waves of a specific length (Giancoli, 1998) (Fig. 2.1). The abbreviation "nm" stands for nanometer, or 1 billionth of a meter, a very short distance. Not all of the sun's radiance is light. The term "light" refers to radiance that we can see. We cannot see individual light waves. In fact, we cannot see much of the sun's radiance at all. For purposes of photosynthesis, light occurs in a spectral band from the 400 nm wavelength to the 700 nm wavelength and is also called PAR (photosynthetically active radiation) or PPF (photosynthetic photon flux). Radiance that we can see is slightly different from what plants use for photosynthesis. Visible light occurs from about the 400 nm wavelength to about the 750 nm wavelength (Starr, 2000). We cannot see wavelengths below 380 nm, called ultraviolet, and although we can see light up to around 750 nm, we do not see wavelengths above 700 nm very well (Kandel *et al.*, 1991). Radiance above 700 nm is called infrared. If we could see any part of the infrared spectrum as well as we can see green light (500–600 nm), plants would look dark red, not green. That is because plants reflect substantially more infrared radiance than they do green light (Chappelle *et al.*, 1992). Because we have such difficulty distinguishing light below 400 nm and above 700 nm, many people, including some instructors, refer to the visible spectrum as the same as PAR. Technically, that is not correct, but it is easy to remember and in a practical sense it works.

The light spectrum includes all of the colors we see and ranges from the highest energy light, violet, to lowest energy light, red (Fig. 2.2). When all of the light energy in the light spectrum is combined, we call it white light, because we cannot differentiate the colors. Consequently, a perfectly white object is reflecting all colors simultaneously. A perfectly black object absorbs all colors and reflects none. The portion of the light that we see and know as PAR (photosynthetically active radiation) not only differs by wavelength (400–700 nm), it also differs by color.

You will notice in Fig. 2.2 that there is more PAR, measured in watts per meter squared, in the blue–green range of the spectrum than any other. That might lead you to believe that more blue–green light is used for photosynthesis than any other. However, that would not be correct. In fact, less green light is used for photosynthesis than light from any other part of the PAR (Zscheile and Comar, 1941). Peak light absorption by plants is believed to occur in two wavelength bands, one in the blue range of the spectrum and one in the red range. For that reason, and for simplicity, we normally divide the plant absorption spectrum into only three colors: blue, green and red (Fig. 2.3).

The sun's radiance not only exists in wave bands of energy, it also exists as particles (Giancoli, 1998). We call those particles photons and it is the photons that plants absorb. When a plant absorbs a photon, it also absorbs the energy in that photon. Consequently, when a plant absorbs a blue photon, it absorbs more energy than it does when it absorbs a red photon. We call the particle distribution of light the photon flux, and this is usually measured as micromoles per meter squared per second. The distribution of photon flux within the spectrum is different than the light energy distribution in the spectrum (Fig. 2.4). There is more green and red photon flux available than there is blue. So, presumably, under normal conditions, more green and red photon flux would be absorbed by a plant than blue.

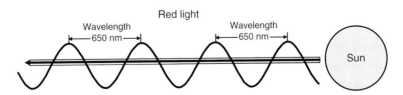

Fig. 2.1. Radiant energy exists in waves of specific length. Shorter wavelengths contain higher energy than long wavelengths.

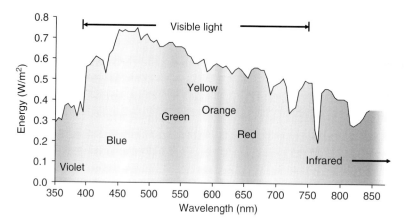

Fig. 2.2. The approximate amount of energy in visible light that strikes the earth's surface on a clear day in a temperate climate. Although radiance in the violet band has the highest energy, only a small amount of violet radiance occurs in sunlight.

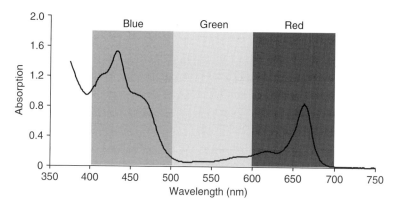

Fig. 2.3. Peak absorption of PAR (photosynthetically active radiation) wavelengths by creeping bentgrass (*Agrostis stolonifera*). All of the PAR spectrum can be used for photosynthesis, but portions of the blue band and the red band are preferred.

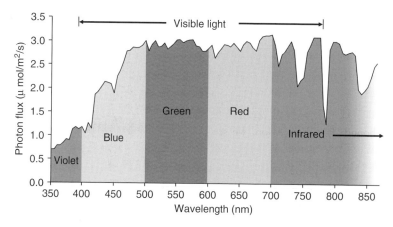

Fig. 2.4. Photon flux at solar noon on a sunny day in Columbus, Ohio. Notice that there are more green and red photons available for photosynthesis than there are blue photons.

Green light is used for photosynthesis, but chlorophyll, the plant pigment most closely associated with energy absorption in plants, accepts blue or red photons in preference to green. A biochemist would say that chlorophyll has a higher affinity for blue or red photons than it does for green photons. That does not mean that green photons are not absorbed. It means that if blue, green and red photons are equidistant from a chlorophyll molecule, the chlorophyll will bind the blue photons or the red photons in preference to the green photons. However, if the green photons are considerably closer, the chlorophyll will bind the green photons.

This same type of affinity occurs with enzymes. Many enzymes will bind more than one compound. However, the receptor site of an enzyme usually has a strong affinity for only one compound or may be specific for only one compound. If the enzyme receptor is specific for only one compound, then no other compound will bind. If the enzyme receptor is not specific for a particular compound, then it will have a strong affinity for one compound but will also bind others. If a compound with a lesser affinity is near, the enzyme receptor may bind that compound rather than the one it has the most affinity for. This concept is very important and we will visit it over and over throughout this text.

Remember that plant processes and all chemical reactions are dynamic. Molecules are constantly in motion, and the environment near a chlorophyll molecule, an enzyme or even a soil particle changes moment by moment. Consequently, given equal distributions of red, blue and green photons, red or blue photons will be bound by chlorophyll most often. However, if there are substantially more green photons present than red or blue, the likelihood of a chlorophyll molecule binding a green photon increases.

In the case of chlorophyll and PPF (photosynthetic photon flux), there are more red photons available than blue and chlorophyll has a higher affinity for red than green, so red photons are the most likely to be bound. For that reason, we see plants as green or blue–green, not red. Red photons are highly available and highly absorbed. There are two types of chlorophyll, chlorophyll *a* and chlorophyll *b*. Some plants have more chlorophyll *a* than others, and chlorophyll *a* has a relatively low affinity for blue photons compared with red (Zscheile and Comar, 1941). Consequently, plants high in chlorophyll *a* reflect a larger portion of the blue band than plants high in chlorophyll *b*. However, all plants that contain chlorophyll reflect green light in preference to red or blue and, consequently, plants look green or blue–green. Why a plant is green has minor importance, but the concept of affinity is extremely important and very useful (Fig. 2.5).

In the demonstration in Fig. 2.5, the enzyme has twice the affinity for the black molecule as it does for the white molecules, but the white molecules are half the distance from the enzyme receptor site. If the black molecule was closer, the likelihood of the enzyme binding the black molecule would increase because the black molecule would have preference. However, because the distance from the receptor cancels the affinity, all of the molecules have an equal opportunity for binding. That means that the probability of a white molecule binding is two out of three and the probability of a black molecule binding is one out of three. Remember this concept, it will be useful.

Based on the knowledge of chlorophyll absorption that was just discussed, you should be able to answer this question: if you were managing an athletic field and you wanted to paint a logo on the field that was red, blue or green, which color would be least likely to damage your turf? Take a moment to consider the question.

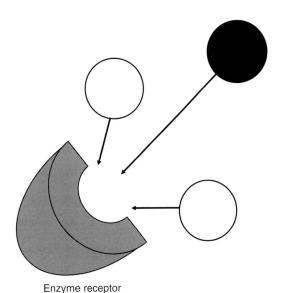

Enzyme receptor

Fig. 2.5. This enzyme has twice the affinity for the black compound as it does the white compounds, but the white compounds are half the distance from the receptor site. Which compound will the enzyme bind?

If the paint reflects green light before it can be absorbed by the plants, that would be the least damaging to the turf because green is the least likely to be needed for photosynthesis. Red paint would be most damaging because the plant depends on the presence of a substantial number of red photons. Blue would be intermediate. You could answer that question and be reasonably certain that you were correct simply by applying the knowledge you just learned. That has value.

The light reaction

The process of photosynthesis requires two distinctly different pathways. They are called the light reaction or z-scheme and the dark reaction or Calvin cycle. The light reaction converts light energy into chemical energy and the dark reaction stores the energy in sugar molecules for later use according to plant need. You have probably seen a diagram of the z-scheme before. I include it here to refresh your memory (Fig. 2.6).

The purpose of the light reaction is to capture energy for use in the dark reaction. The initial receptor of light energy is chlorophyll. Chlorophyll is one of many plant pigments. Green plants also contain significant amounts of carotenoids, and smaller amounts of phycoerythrin, phycocyanin and others. All of these pigments are in substantially lower concentrations than chlorophyll. As already explained, the reflection of green light from chlorophyll determines the green color of green plants. When chlorophyll degrades as a result of stress or seasonal change, turfgrass turns yellow. The yellow color is reflected by xanthophylls, a specific form of carotenoids. Under normal circumstances, chlorophyll is in a high enough concentration to mask the color of the xanthophylls. Other plants, such as trees, reflect yellow, orange or red light when chlorophyll degrades, depending on which normally masked pigments are present in greatest concentration. We can encourage the production of chlorophyll by applying iron (Fe), a precursor for the chlorophyll synthesis pathway, and nitrogen (N), a necessary component of the chlorophyll molecule and of components of the chlorophyll synthesis pathway, and sometimes by applying magnesium (Mg), another important component of the chlorophyll molecule (Fig. 2.7). A simple knowledge of biochemistry is sometimes useful for the turfgrass manager.

When a light photon combines with a chlorophyll molecule, the molecule becomes excited as it absorbs the light energy of the photon. The amount of energy absorbed depends on the wavelength of the photon. As you know, the light reaction depends on electrons being passed among the components of the z-scheme. However, chlorophyll is

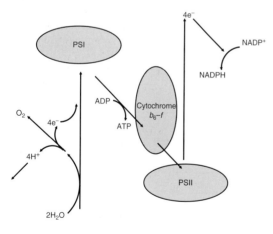

Fig. 2.6. A basic diagram of the light reaction or z-scheme. The purpose of the light reaction is to capture energy for use in the dark reaction. ADP = adenosine diphosphate; ATP = adenosine triphosphate; NADP$^+$ = nicotinamide adenine dinucleotide phosphate$^+$; NADPH = reduced form of nicotinamide adenine dinucleotide phosphate; PSI = photosystem I; PSII = photosystem II.

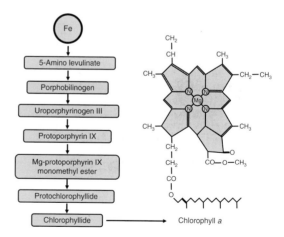

Fig. 2.7. Iron (Fe) is an important precursor for the chlorophyll synthesis pathway shown here, and nitrogen (N) and magnesium (Mg) are important components.

not the component that donates electrons, it is simply an energy carrier (Voet and Voet, 1990). The energy of the photon causes the chlorophyll to enter a highly excited state. In this state, the chlorophyll becomes unstable and will pass the energy on to another chlorophyll molecule or to other compounds that it touches. Hence, the light energy is passed among chlorophyll molecules until it is accepted by one of the reaction centers, PSI (photosystem I) or PSII (photosystem II), in the z-scheme (Fig. 2.8).

A photon of blue light may have energy equivalent to that contained in a 450 nm wavelength. A photon of red light will have considerably less energy such as the energy equivalent to that of a 650 nm wavelength (Fig. 2.2). The reaction centers, PSI and PSII, cannot accept energy of that magnitude. PSI accepts energy equivalent to approximately 700 nm and PSII accepts energy equivalent to 680 nm. Consequently, the blue light and red light in this example are too powerful for the reaction centers to accept. The light energy absorbed by the chlorophyll cannot be absorbed by one of the reaction centers until the energy degrades to an acceptable magnitude. When a chlorophyll molecule absorbs a photon the energy degradation process begins immediately (Salisbury and Ross, 1992). Every time the energy passes from one chlorophyll molecule to another, a portion of the energy is lost. This energy can be lost as phosphorescence or fluorescence, both forms of radiance, but it is most commonly lost as heat (Salisbury and Ross, 1992). Hence, transpiration is required to cool the plant.

Photosynthesis can only proceed temporarily unless transpiration occurs simultaneously as the process of photosynthesis produces too much heat. The plant has to be cooled. Otherwise, it will, in simple terms, burn itself up. Turfgrasses, even warm-season species, yellow during the summer as a matter of seasonal acclimation (Xiong et al., 2007). This yellowing is caused by high photosynthesis and high energy levels among the chlorophyll molecules resulting in chlorophyll degradation (Demmig-Adams, 1990). The long day lengths and intense light common to summer reduce the need for chlorophyll and increase the need for carotenoid pigments that help to quench excess energy (Adams and Demmig-Adams, 1992). During summer, grasses need less chlorophyll to absorb light energy because light is available at high intensity and for long periods. However, the plants need more carotenoid pigments to quench excess light energy when the photosynthesis pathway becomes overloaded. Consequently, your turf turns yellow–green instead of the dark-to-medium green color that occurs in the spring and usually in the fall. It also becomes resistant to fertilization. During summer, less fertilizer is needed to encourage plant metabolism. Because of the high heat of summer that dramatically increases the rate of chemical reactions, and because light and day length are driving photosynthesis to optimum and beyond, little fertilizer is needed (Xiong et al., 2007).

In summer, light is optimal, and except in shade, light is present in greater amounts than turfgrass requires. However, water for transpiration becomes increasingly important because of the high heat and the high rate of activity in the plant. High transpiration rates are common and required for plant cooling. Carbon dioxide is also needed in larger quantities during the summer because of the high rate of activity, especially the increases in photosynthesis that accompany increases in light intensity and day length (Adams and Demmig-Adams, 1992).

2.3 The Importance of Carbon Dioxide

As mentioned earlier, the light reaction provides the energy that fuels the dark reaction or Calvin cycle. As electrons pass through the z-scheme, energy rich molecules called NADPH (the reduced form of

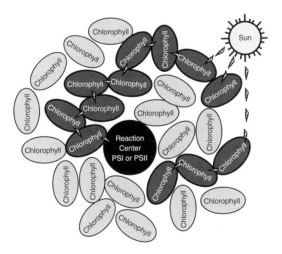

Fig. 2.8. Light energy from the sun is absorbed by chlorophyll and passed among chlorophyll molecules until it is absorbed by a reaction center, either PSI (photosystem I) or PSII (photosystem II).

nicotinamide adenine dinucleotide phosphate) and ATP (adenosine triphosphate) are produced (Fig. 2.6). The NADPH is produced when $NADP^+$ (nicotinamide adenine dinucleotide phosphate) is reduced by an electron from the z-scheme. A chemical reduction is the name given to a chemical reaction when an electron is accepted by a molecule. A chemical oxidation occurs when an electron is released from a molecule. Molecules of ATP are formed from ADP (adenosine diphosphate) by a proton pump that is activated when electrons flow through the z-scheme. The energy in these molecules is used to convert carbon dioxide to sugar in the dark reaction.

The actions of proton pumps and similar mechanisms that work across cellular membranes, organelles and internal membranes are very important concepts in biology and deserve consideration. The light reaction of photosynthesis occurs in the wall of a membrane inside the chloroplast. This membrane is called the thylakoid membrane and it separates the thylakoid, an inner compartment of the chloroplast, from the stroma. The stroma is the name given to the cytoplasm of the chloroplast. The electrons that pass through the z-scheme inside the thylakoid membrane are obtained from water. Water inside the thylakoid is split by a complex called the oxygen-evolving complex (Fig. 2.9). This splitting of water is what causes oxygen to be released during photosynthesis. When two water molecules are split, four electrons (e^-) are released along with four protons (H^+) and one oxygen (O_2) molecule. This activity helps to make the inside of the thylakoid positively charged compared with the outside. Reactions that occur in the z-scheme also drive a proton-producing process by chemical reaction. Each time four electrons pass through the z-scheme eight protons are transferred from the stroma to the thylakoid, thus further increasing the difference in polarity between the inside and outside. Because of this electrical gradient, protons are attracted to the stroma and negative solutes are attracted to the thylakoid.

Because the inside of the thylakoid is relatively positive and the outside is relatively negative, a hole in the membrane would be all that is needed to provide the passage of positive solutes out and negative solutes in. However, as a hole in the membrane would destroy the cell, special carrier proteins are employed to move protons from the inside to the outside. This movement provides the energy necessary to synthesize ATP from ADP. All proton pumps work on this principle and are important for various functions in plant metabolism. As you

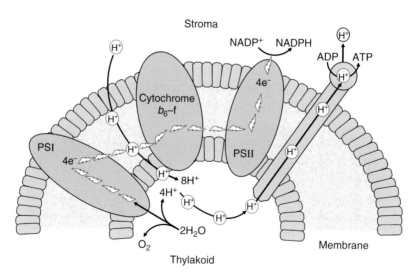

Fig. 2.9. The oxygen-evolving complex. The z-scheme in the light reaction of photosynthesis produces ATP (adenosine triphosphate) by the action of a proton pump. The splitting of water and the passage of electrons along the z-scheme cause an electrical difference across the thylakoid membrane. The electrical difference, which is positive inside, negative outside, provides the energy that fuels the proton pump. ADP = adenosine diphosphate; ATP = adenosine triphosphate; $NADP^+$ = nicotinamide adenine dinucleotide phosphate; NADPH = reduced form of nicotinamide adenine dinucleotide phosphate; PSI = photosystem I; PSII = photosystem II.

will learn, turf maintenance practices or environmental conditions that affect the passage of solutes through membranes or through the proton pumps in membranes can drastically affect plant health. If the proton pumps are damaged or holes occur in the thylakoid membrane, photosynthesis no longer works. When all is working properly, each time three protons pass through the pump their passage provides sufficient energy to convert one molecule of ADP to one of ATP.

The dark reaction

The dark reaction of photosynthesis is a complex process that provides the plant with a number of three-, five-, six- and seven-carbon sugars that are used for various functions (Fig. 2.10). The dark reaction provides the chemical bonding of carbon molecules that store or use the energy sequestered in the light reaction. Carbon dioxide provides the carbon that the dark reaction needs to produce sugars. Many of these sugars will be converted to sucrose for transport to plant parts that need energy. As you probably know, we call those plant parts that require energy "sinks" and those plant parts that produce energy "sources". Other sucrose molecules will be used to form starch or fructans for long-term energy storage. This energy storage occurs primarily in chloroplasts, in roots, or in plant stems. Turfgrasses have three types of stems. The stem that holds the reproductive parts – the flowers and seeds, stems called rhizomes and stems called stolons. We rarely see the seed stalks because most turfgrasses will not produce flowers or seeds at the height at which we mow. In addition, some grasses do not produce rhizomes or stolons, and store their energy mostly in roots. All of these stems are produced by the turfgrass crown, and you should recall from introductory turfgrass classes that the crown is the source of all turfgrass growth. We will study these processes further in Chapter 4.

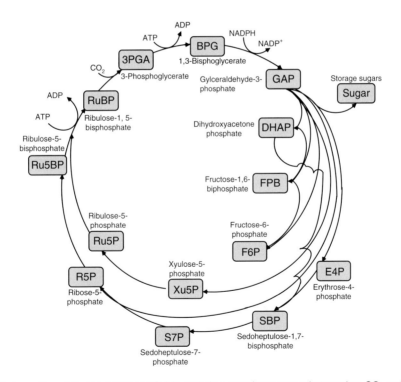

Fig. 2.10. The dark reaction of photosynthesis or Calvin cycle is a complex process that requires CO_2 and both ATP (adenosine triphosphate) and NADPH (the reduced form of nicotinamide adenine dinucleotide phosphate) from the light reaction. The ATP and NADPH are converted back to ADP (adenosine diphosphate) and $NADP^+$ (nicotinamide adenine dinucleotide phosphate), and the energy so released is used to drive the Calvin cycle.

In this chapter, we are primarily concerned with the processes that sequester energy for future plant use. The production of energy storage compounds and sugars for use in other processes requires the combination of carbon dioxide (CO_2) with ribulose bisphosphate (RuBP), a 5-carbon sugar. This combination is catalyzed by an enzyme called RuBP carboxylase, commonly called Rubisco, and believed to be the most pervasive enzyme in existence. Needless to say, Rubisco is important.

The first product of the dark reaction of photosynthesis is an unstable 6-carbon sugar that immediately breaks down into two 3-carbon sugar molecules called PGA (phosphoglycerate). Although PGA is produced each time carbon dioxide combines with RuBP, not all of the PGA produced is used to sequester energy. As RuBP has to be preserved so that the cycle can continue and because the cycle also produces sugars that are needed to synthesize tissue and other compounds, it actually takes three cycles to produce one PGA or six cycles to produce one 6-carbon sugar such as glucose. The production of two PGA molecules requires the energy from 18 ATP molecules and 12 NADPH molecules that came from the light reaction (Voet and Voet, 1990). The light reaction may be producing these energy molecules at a very high rate and all of the constituents of the pathway may be available, but the dark reaction does not occur unless carbon dioxide is present.

Maintaining the presence of carbon dioxide

The presence of carbon dioxide should be paramount in any high-quality turfgrass maintenance plan. Light may be available, water may be available, but if carbon dioxide is not available, photosynthesis does not occur. Carbon dioxide is often the limiting factor for photosynthesis.

Boundary layers

Carbon dioxide is most often deficient in areas where air movement across the turf is restricted. The air is only 0.035% carbon dioxide. Therefore, during daylight when sufficient water is present, a dense turf can use up the carbon dioxide in stagnant air rather quickly. As photosynthesis proceeds, the carbon dioxide in this stagnant air is replaced with oxygen. As the plant transpires to cool itself from the heat of photosynthesis, the stagnant air also becomes high in water vapor. The air surrounding leaves that is high in oxygen and water vapor and low in carbon dioxide is called a boundary layer (Turgeon, 2008). There is always a boundary layer surrounding plant leaves, and because this layer is high in oxygen and low in carbon dioxide, it deters photosynthesis. Because it is also high in water vapor, it slows transpiration. Consequently, poor air movement reduces two of the three most important plant functions, photosynthesis and transpiration.

A thin boundary layer is normal and not particularly detrimental. A thick boundary layer, however, can slow or even halt photosynthesis entirely. When this occurs, the lack of carbon dioxide has become so limiting that photosynthesis will no longer occur in spite of otherwise conducive environmental conditions. In this case, you have failed as a turfgrass manager. It is imperative that you provide enough air movement across your turf to allow adequate photosynthesis to occur.

Scientists have sophisticated instruments that can be used to estimate photosynthesis in the field. However, it is difficult for us to determine how much photosynthesis is enough under a given situation and changing environmental conditions. Unless technological advances occur, turfgrass managers will not have the equipment to measure photosynthesis anytime soon. If you suspect that air movement across your turfgrass is not sufficient to perform adequate photosynthesis, it probably isn't. Your turfgrass will tell you. If carbon dioxide is deficient the turf will yellow more quickly than normal when day lengths extend beyond optimum. The turf will be weak. It will be easily damaged by traffic, disease, insects or any other stress. Once damaged, it will be very slow to recover. Root mass and root depth will be poor. Density will suffer as there is not enough plant energy available to produce new tillers and/or daughter plants. These plant responses are typical of turfgrass stress caused by a number of different conditions. However, in this case, the symptoms will be pervasive, extensive and contiguous. Growing conditions normally conducive to plant health will have little effect. Other areas of turfgrass managed similarly will respond well to management, but the same management will have no effect on the air-restricted turfgrass. Planting a different species, applying more or less water, more or less fertilizer or more pesticides is not the answer. In this case, as in most, you have to fix the problem, not just treat the symptoms.

Improving air movement

Restricted air movement is usually caused by surrounding vegetation, buildings or problems with the contour of the land in the immediate vicinity. Surrounding vegetation is the easiest to deal with. All it takes is removal. You don't necessarily have to remove trees, only low-growing vegetation. Removing low vegetation and debris does not usually meet with a great deal of customer resistance. In fact, in many cases it makes the site look better. All low-growing vegetation and debris should be removed. Fences and hedges are particularly restrictive. It is most important to remove vegetation and debris facing the predominant winds and on the leeward side of the site. It usually does not do any good to remove restrictions on only one side. Restrictions on both sides should be removed to allow air to flow across the site. Clearing only one side of the site is similar to opening a door on one side of a house. Opening a single door has a minor effect on ventilation but opening doors or windows on opposite sides of the house allows air to circulate freely into one side of the house and out of the other. If possible, remove low-growing vegetation and debris on all sides of the site.

Although trees do not necessarily need to be removed to improve air circulation, low-hanging branches are restrictive. Remove all branches within at least 10 feet (3.2 m) of the ground. Higher removal is better but more difficult and not often necessary. Coniferous trees are usually more restrictive than deciduous trees and more likely to require additional trimming. A small, inexpensive anemometer (wind gauge) is useful during and after this operation. Use the anemometer to measure wind speed just above the turf at the wind-restricted site and compare that measurement with wind speed at an unrestricted site immediately before or after measuring the restricted site. Complete this operation a few different times each day for several days. Record the measurements each time they are made and compare the average wind speed at the restricted site with the average wind speed at the unrestricted site. The difference may be substantial. A 50% decline is not unlikely. Continue to make these measurements as you clear the restricted site and after you think that you have improved the site sufficiently. There is no hard, fast number short of 0.0% decline in air movement that is sure to be effective. As in most situations involving turfgrass management, it depends on the climate and the situation. However, a major improvement in air movement at the site is likely to result in substantially healthier looking turf. If not, there may be additional negative factors to consider.

Buildings can be a real problem to deal with. Courtyards and stadiums present unique problems to air movement over turfgrass. In this case, man-made structures surround the turf. Sometimes the solution is as simple as opening or removing doors on opposite ends of the courtyard or stadium. Opening doors or windows to the outside may not solve an air restriction problem completely, but it will certainly help. In some instances open passageways to the outside can be constructed fairly easily, but in most cases a major reconstruction is required. In that case it depends on the value of the turf, and sometimes on your salesmanship, to convince the owners of the importance of new passages. Fans are a last resort. Fans are expensive and require electrical installation. They are noisy and need constant upkeep and repair. However, if the turf is of significant value, a fan or multiple fans will keep it alive and make it easier to manage and maintain at an acceptable level.

There are times when the contour of the land does not allow air movement across your turf. I have, in fact, seen golf courses with putting greens designed for placement in such a situation. It simply does not work to place a high-value stand of turf that is required to be maintained at near-perfect condition in the bottom of a hole. If it can be maintained at an acceptable level, the manpower and materials required are not worth the effort. Fortunately there are few situations where land contour restricts air movement severely enough to influence turfgrass growth. In those few instances where improving air movement is required, the only good option is to regrade the site. Fans can also be useful. In most situations, the turfgrass can be managed, but it will always be weak and require more than normal care. If you learn your lessons well and take advantage of as much practical experience as possible, you will be able to handle it.

2.4 The Importance of Water

Water is important for all plant functions. Turfgrass leaves, where most of the photosynthesis of the plant occurs, are mostly water. Dry clippings of turfgrass leaves, for instance, weigh considerably less than fresh clippings. Not only is water necessary to provide the medium for all chemical reactions in photosynthesis and to cool the plant because of the heat produced in photosynthesis, it also has to provide the electrons for the light reaction of photosynthesis. Too little water can also cause indirect consequences for photosynthesis. As you will learn later in the text, dry conditions cause plant stomates to close, thus limiting the amount of carbon dioxide that enters the leaf and severely reducing photosynthesis. The importance of water will be a recurring theme throughout this text.

2.5 Chapter Summary

Photosynthesis is necessary for plant survival and for the survival of all living things on earth. Like all biological physiology, it is an extremely complex process of molecular activity. The concept, however, is relatively simple. The process of photosynthesis is the physiological pathway that plants use to transform light energy into chemical energy. This energy may be used immediately, it may be transported to other plant parts to be used as energy or converted to important plant products, or it may be transferred to a storage area and sequestered for later use.

There are two major pathways that constitute photosynthesis, the light reaction or z-scheme and the dark reaction or Calvin cycle. Light and water are the major components required to fuel the light reaction. Light provides the energy that is sequestered by the plant during the process and water provides the electrons that carry the energy through the z-scheme. Water also provides a medium for chemical reactions to occur and is consumed by the transpiration process that cools the plant of the excessive heat produced by the light reaction.

Light called PAR (photosynthetically active radiation) is the source of energy for photosynthesis. PAR occurs in a spectral band from 400 nm to 700 nm wavelengths and is divided into three major colors, blue, 400 to 500 nm, green 500 to 600 nm, and red 600 to 700 nm. All three colors are active for photosynthesis, but red and blue are preferred.

PAR not only contains energy but also occurs in particles called photons. These photons are the particles actually absorbed by chlorophyll. So chlorophyll does not actually absorb PAR, it absorbs PPF (photosynthetic photon flux). When chlorophyll absorbs a photon, it also absorbs its energy. Blue photons have high energy, red photons have low energy and green photons have intermediate energy. Nearly all of the photons absorbed by chlorophyll have energy that is too high for use in the light reaction. Therefore, this energy has to degrade to a level equivalent to either 680 nm for use in the PSII (photosystem II) pathway or 700 nm for use in the PSI (photosystem I) pathway. As this energy degrades it gives off heat. That is why transpiration is so important for photosynthesis. The plant must be protected from the heat produced during this release of energy.

Once energy is absorbed by the reaction centers, PSI and PSII, the z-scheme is activated and the light reaction begins. Water is split by the oxygen-evolving complex, producing four electrons, four protons, and one oxygen molecule for every two water molecules. The electrons enter the z-scheme and carry the energy through the pathway. The protons enter the thylakoid where they and additional protons produced by the passage of electrons through the z-scheme cause a difference in polarity across the thylakoid membrane. Because of the difference in polarity between the inside of the thylakoid and the outside protons move across the thylakoid membrane through a proton pump and provide the energy necessary to produce ATP (adenosine triphosphate) from ADP (adenosine diphosphate). The ATP and NADPH (the reduced form of nicotinamide adenine nucleotide phosphate) are the two energy molecules formed during the light reaction. NADPH is produced from the reduction of NADP+ (nicotinamide adenine nucleotide phosphate) by an electron from the z-scheme.

The ATP and NADPH are used to provide the energy to fuel the dark reaction or Calvin cycle. Light is not required in the dark reaction but carbon dioxide must be present. During the dark reaction, RuBP carboxylase, an enzyme more commonly called Rubisco, catalyzes the combination of carbon dioxide with RuBP (ribulose bisphosphate), a 5-carbon sugar. Each time one carbon dioxide is bound, one more carbon enters the cycle. This carbon accumulation results in the synthesis of many different sugars for a variety of plant uses.

The combination of RuBP and carbon dioxide results in 3-carbon sugar molecules called PGA (phosphoglycerate). As the Calvin cycle proceeds, some PGA is used to synthesize glucose and fructose, which can be combined to make sucrose. The sucrose is transported throughout the plant as an energy source or combined into long chains of starch or fructans for energy storage and later use.

Carbon dioxide is imperative for the performance of the Calvin cycle. It is up to the turfgrass manager to make certain that a lack of carbon dioxide does not interfere with photosynthesis. Carbon dioxide becomes limiting when a boundary layer high in oxygen and water vapor and low in carbon dioxide builds up around the turf. Managing for good air movement across the turf to refresh the air and bring in new carbon dioxide is a necessary practice in some situations. Low-growing tree limbs and other vegetation and debris may need to be removed from a site to enhance air movement. Sometimes regrading of the site is necessary. In courtyards or stadiums air channels from the inside to the outside may need to be constructed. There is always a thin boundary layer around plant leaves as they perform photosynthesis, but when the boundary layer gets too thick, photosynthesis is affected and steps must be taken to create fresh air circulation. Otherwise, the turfgrass may become so weak that it can not be managed.

As turfgrass managers, one of our most important functions is to make sure that the light, water and carbon dioxide necessary for healthy photosynthesis are present in our turfgrass environment. If this essential process cannot be performed at a high level on a near constant basis, our turf will always be weak or nonexistent.

Suggested Reading

Kelly, G.J. and Latzko, E. (2006) *Thirty Years of Photosynthesis*. Springer, New York.

Starr, C., Evers, C.A. and Starr, L. (2006) *Biology: Concepts and Applications*, 6th edn. Thomas, Brooks/Cole, Belmont, California.

Stern, K.R., Bidlack, J. and Jansky, S. (2008) *Introductory Plant Biology*, 11th edn. McGraw-Hill Higher Education, New York.

Voet, D.V. and Voet, J.G. (2004) *Biochemistry*, 3rd edn. John Wiley and Sons, New York.

Suggested Websites

NASA/Cool Cosmos EPO (Education and Public Outreach) (2010) Infrared Astronomy: Near, Mid and Far Infrared. Available at: http://coolcosmos.ipac.caltech.edu/cosmic_classroom/ir_tutorial/irregions.html (accessed 27 July 2010).

3 Why C₃ and C₄ Grasses Require Different Management

Key Terms

A **warm-season grass** is a grass that uses the C_4 photosynthesis pathway.
A **cool-season grass** is a grass that uses the C_3 photosynthesis pathway.
Photorespiration occurs when Rubisco, the enzyme that binds carbon dioxide in the Calvin cycle, binds oxygen instead.
Mitochondria are organelles where oxidative respiration takes place in both plants and animals. Mitochondria also function in photorespiration in C_3 plants.
A **peroxisome** is a plant organelle that, among other things, is part of the photorespiration pathway.
The **light saturation point** is the amount of light present when photosynthesis is occurring as rapidly as possible.
Vapor pressure is the pressure exerted by a vapor in a closed container that is in dynamic equilibrium with its liquid form.
Dynamic equilibrium is a state of balance in which opposing processes occur at the same rate at the same time so that it appears as if nothing is happening. In fact, a great deal is happening but there is no net effect.
Kranz anatomy is the name given to the special anatomy of a C_4 plant that allows the light reaction of photosynthesis and the dark reaction of photosynthesis to take place in two separate cells.
A **bundle sheath cell** is a small cell located near the vascular system of grasses where the Calvin cycle takes place in C_4 plants. The bundle sheath cell is the basic component of Kranz anatomy.

3.1 Use Species Adaptation to Your Advantage

The C_4 (warm-season) pathway of photosynthesis is rare in plant species in general but quite common in turfgrasses (Matsuoka *et al.*, 2001). The grasses that we call warm-season, the C_4 plants, and the grasses that we call cool-season, the C_3 plants, perform photosynthesis differently. They also differ in their adaptation to climatic conditions. Whether warm-season grasses adapted to the climate that C_4 photosynthesis was best suited for, or the grasses adopted C_4 photosynthesis because it was more efficient for their climate is unknown. However, we know for certain that C_3 and C_4 grasses require different management procedures because of the differences in photosynthesis and climatic adaptation between them.

I have been told by at least one biochemist and at least one plant physiologist that C_4 photosynthesis is more efficient than C_3 photosynthesis. In each case, I was living in a temperate climate at the time. Consequently, I asked them both to tell me why there weren't any perennial C_4 grasses growing outside and they both explained that most C_4 grasses could not survive cold temperatures. Then I asked why, if the C_4 pathway is more efficient, don't cold-adapted plants use it as well as warm-adapted plants? Neither scientist had an explanation for that. Both of those scientists know a lot more about photosynthesis and other plant functions than I do, but until they can answer that question for me, I will continue to believe that C_3 photosynthesis is more efficient in cool temperatures and C_4 photosynthesis is more efficient in warm temperatures. If you are a homeowner in a temperate climate, you will probably want to know how to manage cool-season grasses. If you are a homeowner in a tropical climate you will want to

Table 3.1. Some of the most common cool- and warm-season grasses used for turf throughout the world. Common names are those used in the USA. This list does not include all grasses used for turfgrass.

Common name	Scientific name
Cool-season (C_3) grasses	
Velvet bentgrass	*Agrostis canina* L.
Colonial bentgrass	*Agrostis capillaris* L.
Creeping bentgrass	*Agrostis stolonifera* L.
Tall fescue	*Festuca arundinacea* Schreb.; synonym, *Schedonorus phoenix* (Scop.) Holub.
Fine fescues	
Sheep fescue	*Festuca ovina* L.
Creeping red fescue	*Festuca rubra* L.
Chewing's fescue	*Festuca rubra* L. ssp. *fallax* (Thuill.) Nyman; synonym *Festuca rubra* L. ssp. *commutata* Gaudin
Hard fescue	*Festuca trachyphylla* (Hackel) Krajina; synonym, *Festuca brevipila* R. Tracey
Annual ryegrass	*Lolium multiflorum* Lam.
Perennial ryegrass	*Lolium perenne* L.
Alpine bluegrass	*Poa alpina* L.
Annual bluegrass	*Poa annua* L.
Texas bluegrass	*Poa arachnifera* Torr.
Bulbous bluegrass	*Poa bulbosa* L.
Canada bluegrass	*Poa compressa* L.
Kentucky bluegrass	*Poa pratensis* L.
Supina bluegrass	*Poa supina* L.
Roughstalk bluegrass	*Poa trivialis* L.
Warm-season (C_4) grasses	
Common carpetgrass	*Axonopus affinis* Chase
Tropical carpetgrass	*Axonopus compressus* (Swartz) Beauv.
Buffalograss	*Buchloe dactyloides* (Nutt.) Engelm.; synonym, *Bouteloua dactyloides* (Nutt.) Columbus
Bermudagrass	*Cynodon dactylon* (L.) Pers.
African bermudagrass	*Cynodon transvaalensis* Burtt-Davy
Centipedegrass	*Eremochloa ophiuroides* (Munro) Hack.
Bahiagrass	*Paspalum notatum* Flugge
Seashore paspalum	*Paspalum vaginatum* Swartz
Kikuyagrass	*Pennisetum clandestinum* Hochst ex Chiov.
St. Augustinegrass	*Stenotaphrum secundatum* S. (Walt.) Kuntze
Zoysiagrasses	
Japanese lawngrass	*Zoysia japonica* Steud.
Manilagrass	*Zoysia matrella* (L.) Merr.
Mascarenegrass	*Zoysia tenuifolia* Willd. ex Thiele; synonym, *Zoysia pacifica* Willd.

know how to manage warm-season grasses. If you are a turfgrass manager, you need to know how to manage both types of grasses and how to manage them both well (Table 3.1).

3.2 Photorespiration

In Chapter 2, you learned about Rubisco, technically called ribulose bisphosphate carboxylase, the enzyme that binds carbon dioxide in the dark reactions of photosynthesis, also called the Calvin Cycle. Rubisco is more correctly called ribulose bisphosphate carboxylase oxygenase because it not only binds carbon dioxide, it binds oxygen. When Rubisco binds oxygen instead of carbon dioxide, the resulting process is called photorespiration – an inefficient, energy-consuming process. Considering that our air is 21% oxygen but only 0.035% carbon dioxide, Rubisco would be considerably more likely to bind oxygen than carbon dioxide, except that it has a much higher affinity for carbon dioxide than it does for oxygen (Fig. 3.1). Enzyme affinity is a concept that you also learned about in Chapter 2 (Fig. 2.5). Rubisco's high affinity for carbon dioxide overcomes the amount and proximity of oxygen and increases the likelihood that carbon dioxide will be bound in preference to oxygen. However, sometimes, environmental conditions – mainly intense light and high temperature – increase the likelihood for oxygen to be bound by Rubisco and for photorespiration to occur.

Only C_3 plants photorespire, and the likelihood for photorespiration to occur increases with increasing temperature and usually with increasing light. Although C_4 plants have the pathways and enzymes to accommodate photorespiration, it does not occur in a measurable amount, and we assume that it is nonexistent in C_4 plants. In order to understand why photorespiration occurs and how C_4 plants avoid it, you need to learn a little about the process.

Photorespiration is an inefficient form of plant respiration that occurs when Rubisco binds oxygen instead of carbon dioxide. As you know, carbon assimilation has to occur during photosynthesis or there is no conversion of light energy to chemical energy. When the Calvin cycle is working properly, Rubisco binds carbon dioxide and RuBP (ribulose bisphosphate), a combination resulting in an unstable 6-carbon molecule that immediately splits into two 3-carbon molecules of PGA (phosphoglycerate). When Rubisco binds oxygen instead of carbon dioxide, a different result occurs. The combination of RuBP and oxygen results in one molecule of PGA and one molecule of PGL (phosphoglycolate), a 2-carbon molecule:

$$\text{Carbon assimilation} = \text{RuBP} + \text{CO}_2 \xrightarrow{\text{Rubisco}} 2\text{PGA}$$

$$\text{Oxygen assimilation} = \text{RuBP} + \text{O}_2 \xrightarrow{\text{Rubisco}} \text{PGA} + \text{PGL}$$

The phosphoglycolate that results from the combination of RuBP and oxygen is not useless but it must be processed through a complicated pathway, after which it is converted to PGA and enters the Calvin cycle. This is actually a respiratory process, hence the name photorespiration, and except for the end product, PGA, it has little to do with photosynthesis. The process requires energy and results in the loss of one carbon dioxide molecule. To salvage photosynthetic carbon from glycolate, photorespiration combines two glycolates to produce one PGA and one carbon dioxide. The PGA enters the Calvin cycle and the carbon dioxide is lost to the atmosphere. Consequently, two combinations of RuBP and oxygen result in three PGAs and the loss of one carbon. If Rubisco had bound carbon dioxide twice, the result would be four PGAs for the Calvin cycle, with no loss of carbon and no need for additional energy from ATP (adenosine triphosphate) to recover carbon otherwise lost as PGL. Therefore, you could call photorespiration a destructive process.

The photorespiratory process occurs in three different organelles, the chloroplast, the mitochondrion and the peroxisome (Fig. 3.2). Phosphoglycolate is transported from a chloroplast to a peroxisome, then to a mitochondrion (via glycine), back to a peroxisome (via serine) and, finally, back to a chloroplast (via glycerate), where it enters the Calvin cycle as PGA. During photorespiration then, two PGLs are converted to one PGA, with an expenditure of energy, and one carbon dioxide is lost. Because less carbon is accumulated and energy is actually used when photorespiration occurs, the plant may have to use stored energy reserves for metabolic processes, even though conditions are suitable for photosynthesis.

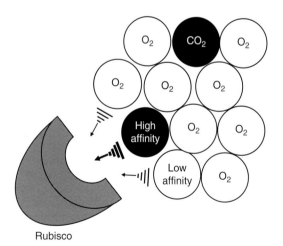

Fig. 3.1. Under normal environmental conditions, considerably more oxygen (O_2) than carbon dioxide (CO_2) is in close proximity to the enzyme Rubisco (ribulose bisphosphate carboxylase), but Rubisco has such a strong affinity for CO_2 that it is more likely to bind CO_2 than to bind O_2.

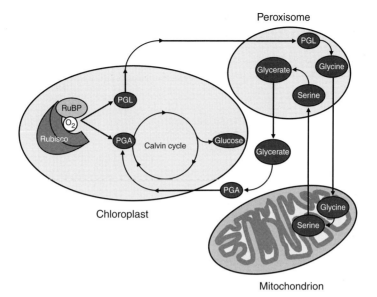

Fig. 3.2. Photorespiration is a complicated process that requires phosphoglycolate (PGL) to be transferred from a chloroplast to a peroxisome, then to a mitochondrion (via glycine), back to a peroxisome (via serine) and finally, back to a chloroplast (via glycerate), where it enters the Calvin cycle as phosphoglycerate (PGA). RuBP = ribulose bisphosphate; Rubisco = ribulose bisphosphate carboxylase.

Light affects photorespiration

Photorespiration is most common under conditions of high light and high heat. When light is intense, the light reactions of photosynthesis occur rapidly and oxygen is released in high concentrations. Because carbon dioxide is being assimilated rapidly, the air in and around our plants becomes high in oxygen and low in carbon dioxide. These are perfect conditions for photorespiration. Warm-season grasses have a definite advantage over cool-season grasses under such conditions because they do not photorespire and because they can assimilate carbon dioxide even when its concentration in the surrounding air is very low (Krenzer and Moss, 1969).

In Chapter 2 we called this high oxygen low carbon dioxide condition a boundary layer and discussed it as a barrier to photosynthesis where air movement was restricted. In this case, the boundary layer is not a result of restricted air movement but of rapid photosynthesis. It follows that when the air holds higher than normal oxygen and lower than normal carbon dioxide Rubisco is more likely than normal to bind oxygen and less likely to bind carbon dioxide. As photorespiration occurs when Rubisco binds oxygen instead of carbon dioxide, boundary layers inhibit photosynthesis and enhance photorespiration.

These conditions do not occur during the winter or during the spring or fall. Under natural conditions, the formation of boundary layers due to rapid photosynthesis can only occur during the summer. Summer is when sunlight is most intense and day length is the longest. Consequently, if you are managing cool-season grasses, you must prepare them by doing the right things in the fall, winter and spring so that they can survive the next summer. Summer is a very stressful period for cool-season grasses (Liu and Huang, 2001). If your plants are not healthy in late spring, they will not survive the summer.

Light saturation

You may be familiar with the "light saturation point". The light saturation point of a particular plant is the amount of light present when photosynthesis is occurring as rapidly as possible. An increase in light beyond the light saturation point will not result in an increase in photosynthesis. Light saturation points are affected by a number of factors, including temperature and carbon dioxide availability. Consequently, a light saturation point is not constant for a particular species and may not even be constant for two different leaves on the same plant (McLendon and McMillen, 1982). Therefore, the

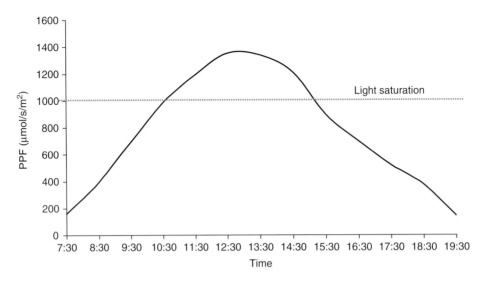

Fig. 3.3. Photosynthetic photon flux (PPF) during a summer day compared with the approximate light saturation point of creeping bentgrass (*Agrostis stolonifera*). Notice that PPF exceeds the light saturation point for about 5 hours from 10:30 to 15:30. This situation encourages photorespiration in cool-season grasses. Source: Gaussoin *et al.* (2005).

light saturation point is a value that can be measured on a given leaf, plant or group of plants at a single point in time that will change when internal or external changes occur. Nonetheless, light saturation is a useful concept. Increases in light beyond the light saturation point do not result in greater photosynthesis (Fig. 3.3). In fact, increases in light beyond the light saturation point are detrimental. Just like all biological conditions, too much light can be as detrimental as too little light. Photorespiration is one of those detrimental conditions encouraged by too much light. You will learn of others in succeeding chapters.

Temperature affects photorespiration

High temperature encourages photorespiration. Remember back in chemistry class when you learned that molecular activity increases with increasing temperature. You learned a term called "vapor pressure" and you learned that vapor pressure increases with increasing temperature (Brown *et al.*, 1991). Consequently, you can heat a liquid such as water until it boils and turns to steam and the water will quickly evaporate away. However, water also evaporates without being heated. Unless it is in a sealed container under pressure, water evaporates as long as it is not frozen. It evaporates because there are always a certain number of molecules escaping the liquid. If you put a liquid in a sealed container, the pressure of the gaseous form of that liquid trying to escape the container is called vapor pressure. Liquids like alcohol and gasoline have higher vapor pressures than water. That means that they evaporate more easily and in a sealed container there will be more pressure above alcohol or gasoline than there will be above water at the same temperature (Fig. 3.4).

Dynamic equilibrium

When a liquid in a closed container reaches a steady vapor pressure it has achieved a dynamic

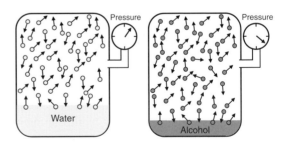

Fig. 3.4. In a sealed container partially filled with alcohol there will be more vapor pressure above the alcohol than there will be above the same amount of water under the same environmental conditions. Heating the water or the alcohol causes more molecules to escape the liquid, and vapor pressure increases.

equilibrium. Dynamic equilibrium is an important concept. Remember from Chapter 2 that during the discussion about affinity you learned that chemical reactions and activities are dynamic, meaning that they are constantly occurring and often changing with changes in the environment. When a steady vapor pressure is achieved, that does not mean that the molecules of liquid stop evaporating and that the molecules of vapor stop condensing. It means that the number of liquid molecules that are becoming vapor is the same as the number of vapor molecules that are becoming liquid in a particular period of time. Consequently, there is a lot of activity occurring but there is no net change in the amount of liquid or the amount of vapor present. Because there is no change in the amount of liquid or vapor, it appears that nothing is happening. However, the vapor and water molecules are in constant motion, making the equilibrium dynamic, not static.

Because of this vapor pressure property of liquids, water in a sealed container does not evaporate beyond equilibrium. However, water exposed to the atmosphere evaporates readily. In a sealed container, the water vapor is in equilibrium with the water and the air above the water is at 100% humidity. In the open, the air above the water is continually refreshed. Therefore, water and water vapor cannot reach dynamic equilibrium unless all of the air in the surrounding area is at 100% humidity. So, except under rare circumstances, water exposed to the atmosphere is evaporating continuously. Warm air can hold more water than cold air so water evaporates faster when the air above it is warm. Warm water also evaporates faster than cool water.

As we know, water molecules are in constant motion. As water warms, movement increases. It is this increase in movement that allows a water molecule to attain a high enough velocity to literally fly out of the water and into the air. Water molecules in the air are also in constant motion, and if they strike a body of water, they are likely to stick to it and become part of it. Hence, there is a constant exchange of water molecules in both directions between the liquid water and the air. That exchange becomes faster as the temperature of the water increases. Heat is a form of energy, so when water absorbs heat and becomes warm it has more energy than it does when it is cold. Consequently, water molecules become more active and move greater distances with greater velocity as they get warmer. Because of this increase in activity, it is more likely for warm water molecules to escape from the liquid into the air than it is for cold water molecules.

Oxygen and carbon dioxide in water

The recent discussion concerning temperature, evaporation, vapor pressure and dynamic equilibrium was leading up to this point: oxygen and carbon dioxide exist in water. They dissolve in water and they also exist as whole molecules in water. When water warms, every molecule in the water, including oxygen and carbon dioxide, becomes more active and more likely to leave the liquid. The reason that water temperatures in your plant affect photorespiration so greatly is because carbon dioxide is not bound as tightly by water as is oxygen. As water gets warm the concentration of carbon dioxide in the water decreases more rapidly than the concentration of oxygen (Hull, 1992). The ratio of carbon dioxide to water becomes smaller and the difference between the concentrations of carbon dioxide and oxygen in the water becomes greater. Therefore, there is greater likelihood for oxygen to be bound by Rubisco when the temperature is high than there is when the temperature is low. As temperature increases from 50°F to 95°F (10°C to 35°C), the rate of photorespiration in C_3 plants more than doubles (Hall and Keys, 1983). Because all of the chemical reactions in the plant take place in water, the properties of water have a direct influence on what occurs in the plant. As water heats up, both oxygen and carbon dioxide escape the liquid, but carbon dioxide escapes more rapidly and the remaining concentrations are relatively higher in oxygen and lower in carbon dioxide than they were when the water was cool. A scientist would say that the ratio of carbon dioxide to oxygen in water decreases as the temperature of the water increases.

High light and high heat encourage photorespiration in C_3 plants. Therefore summer is a stressful period for C_3 grasses and is the season when they are most difficult to manage. Photosynthesis in C_4 grasses is more efficient than in C_3 grasses when daytime temperature is high, but that does not mean that C_4 grasses are unaffected by high temperature. It means that they can perform photosynthesis efficiently when C_3 grasses cannot. This also means that C_4 grasses can metabolize normally at higher temperatures than C_3 grasses. It does not mean that C_4 grasses are completely unaffected by high temperature. It is widely believed that C_3 grasses perform best at temperatures up to about 80 °F (27 °C) and that C_4 grasses perform best at temperatures up to about 95 °F (35 °C) (Beard, 1973). Beyond 95 °F, healthy growth of C_4 grasses begins to decline, but the decline is not a result of photorespiration.

3.3 Kranz Anatomy

Warm-season C_4 grasses not only perform photosynthesis differently, they have different anatomy from cool-season C_3 grasses. Photosynthesis is compartmentalized in C_4 grasses. In C_3 grasses, both the light reaction of photosynthesis and the dark reaction of photosynthesis occur in the same cell. In C_4 grasses the light reaction and the dark reaction occur in different cells. C_4 grasses also bind carbon differently. In C_3 grasses, carbon is bound only in a mesophyll cell, usually near the surface of the leaf, whereas in C_4 grasses it is bound first in a mesophyll cell and is then transferred to a bundle sheath cell where it is released as carbon dioxide. Although this process requires extra steps to bind the carbon and transfer it from one cell to another, it is more efficient than C_3 photosynthesis when plants are exposed to high light and high heat.

Bundle sheath cells

The cells that make the C_4 pathway work so well are called bundle sheath cells (Rogers et al., 1976). Cool-season plants may have bundle sheath cells, but they do not perform photosynthesis in them. Bundle sheath cells are small cells located deep in the leaf near the vascular system (Fig. 3.5). You may recall that the xylem and phloem make up the vascular system of a plant. The xylem primarily transfers water through the plant for transpiration. The phloem carries carbohydrates and other essential compounds through the plant from source to sink. Mature leaves are sources because that is where most photosynthesis occurs. In grasses, photosynthesis may also occur in stolons, in aboveground stems and in sheaths, but production of carbohydrates at these sites is slight compared with that in mature leaves. Most stems, as well as roots, flowers and seeds are sinks. They don't produce carbohydrates, they only use them. Because the bundle sheath cells are located right next to the phloem, carbohydrates produced in the bundle sheath cells are easily transferred to the phloem for distribution.

In C_4 plants, the Calvin cycle takes place in the bundle sheath cells. It is the same Calvin cycle that occurs in C_3 plants but it is located in these

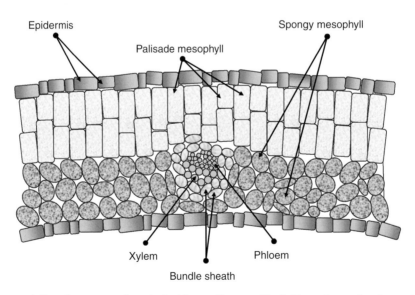

Fig. 3.5. A representation of a cross section of a leaf from a C_4 grass. The light reaction and carbon uptake occur primarily in the palisade mesophyll cells near the upper surface of the leaf. The dark reaction takes place in the bundle sheath cells near the vascular system, which is composed of the xylem and phloem.

specialized cells. The only reason for the existence of a bundle sheath cell appears to be to perform the dark reaction of photosynthesis in C_4 plants. Because bundle sheath cells are located deep in the leaf they are relatively unaffected by light and by atmospheric oxygen compared with mesophyll cells. In addition, the main source of oxygen, the light reaction of photosynthesis, does not occur in the bundle sheath cells. Consequently, although the same Rubisco enzyme is used to bind carbon dioxide for the Calvin cycle in the bundle sheath cells as the Rubisco that is prone to photorespiration in C_3 plants, the Rubisco in the bundle sheath cell does not, or at least only rarely, binds oxygen. The explanation for that is quite simple. There is very little oxygen present to be bound.

Bundle sheath cells are small. They do not rely on atmospheric carbon dioxide for use in the Calvin cycle. Instead, all of the carbon dioxide used in the cycle is transferred to the bundle sheath cells after being assimilated in the mesophyll cells. As bundle sheath cells are not directly exposed to atmospheric oxygen and because there is no oxygen production from the light reaction in the bundle sheath cells, the cytoplasm of a bundle sheath cell is highly concentrated in carbon dioxide. High carbon dioxide concentration and low oxygen is a perfect situation for Rubisco. Its high affinity for carbon dioxide in an environment rich in carbon dioxide makes it highly unlikely that Rubisco will ever bind oxygen. Hence, C_4 plants do not photorespire.

Carbon binding in C_4 plants

Now that you know what a bundle sheath cell is and why the Calvin cycle is so efficient in C_4 plants, we need to discuss how the carbon dioxide for the Calvin cycle gets to the bundle sheath cells. First let's talk about the light reaction. The light reaction in C_4 plants does not differ from the light reaction in C_3 plants. It takes place in the mesophyll cells near the surface of the leaves because that is where the light is most intense. Just as in C_3 plants, chlorophyll in the mesophyll cells absorbs light energy and transfers that energy to reaction centers where it enters the z-scheme. Water is split producing oxygen that vents to the air through stomates and electrons that carry the light energy through the z-scheme. The z-scheme produces ATP and NADPH (the reduced form of nicotinamide adenine dinucleotide) just as it does in C_3 plants, and the ATP and NADPH are used to fuel the Calvin cycle and other plant processes. There is no Calvin cycle in the mesophyll cells of a C_4 plant. The Calvin cycle, the dark reaction, only occurs in the bundle sheath cells.

Carbon is bound in the mesophyll cells of C_4 plants just as it is in C_3 plants but a different process is used to accomplish this binding. As you recall, in C_3 plants the Calvin cycle exists in the mesophyll cells so carbon dioxide is bound by Rubisco as it enters the cell. In C_4 plants, the mesophyll cells do not bind carbon dioxide, they bind bicarbonate. Bicarbonate (HCO_3^-) is carbon dioxide dissolved in water. Consequently, unless the pH of the water is extremely low there is always a substantial amount of bicarbonate in water exposed to air. A pH low enough to restrict the formation of bicarbonate would also be low enough to kill the plant. As bicarbonate enters the mesophyll cells, it is bound by an enzyme called PEP (phosphoenolpyruvate) carboxylase. The PEP carboxylase enzyme catalyzes the combination of bicarbonate and PEP, resulting in a 4-carbon molecule; hence, the name C_4 pathway. The C_3 pathway is so named because the first stable molecule produced by fixing carbon in cool-season photosynthesis is a 3-carbon compound, PGA. The C_4 pathway is named C_4 because the first product of fixing carbon in warm-season photosynthesis is a 4-carbon compound, which is called oxaloacetate. Oxaloacetate cannot move through cell membranes so it is converted to malate, another 4-carbon compound, and shipped off to a bundle sheath cell (Fig. 3.6). In the bundle sheath cell, the malate is split into a molecule of pyruvate, which returns to a mesophyll cell, and a molecule of carbon dioxide, which enters the Calvin cycle. When the pyruvate returns to a mesophyll cell, it is phosphorylated to resynthesize PEP and the process starts again.

We have already discussed the advantages of Kranz anatomy. There are also advantages to binding carbon by combining bicarbonate and PEP. In water, carbon dioxide exists in three forms, carbonic acid, bicarbonate and carbon dioxide (Box 3.1). Unless the pH of the water is extremely high or low, bicarbonate is the predominate form. Carbon dioxide and bicarbonate exist in equilibrium, but the equilibrium strongly favors bicarbonate. Consequently, we believe that the concentration of bicarbonate in the cytoplasm of the mesophyll cells is always sufficient to perform sustained photosynthesis regardless of temperature.

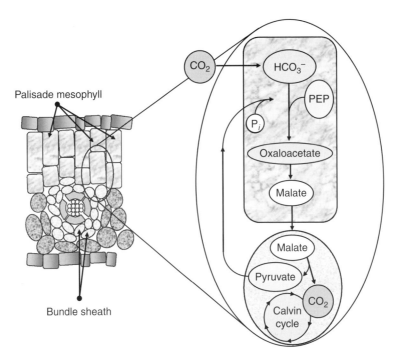

Fig. 3.6. In C_4 plants, carbon is fixed in a mesophyll cell by combining bicarbonate (HCO_3^-) and phosphoenolpyruvate (PEP). This combination results in oxaloacetate, a 4-carbon compound that is converted to malate and shipped to a bundle sheath cell. In the bundle sheath cell, the malate splits into pyruvate and carbon dioxide. The carbon dioxide enters the Calvin cycle and the pyruvate returns to a mesophyll cell where it is again converted to PEP by combination with phosphate (P_i), and the process starts again.

Box 3.1. Is the pH of pure water really 7.0?

Presumably the pH of pure water is 7.0. You know that because you have been told that many times by chemists, physiologists and botanists. However, you may not know that the difference between a pH of 7.0 for pure water and a pH of 5.7 is simply the presence of air. Carbon dioxide (CO_2) dissolves in water. When carbon dioxide dissolves in water, it forms carbonic acid ($H_2O + CO_2 \leftrightarrow H_2CO_3$). The carbonic acid and carbon dioxide reach equilibrium with some of the carbon dioxide remaining in the water as a gas, but most becomes carbonic acid. Most of the carbonic acid dissociates to bicarbonate ($H_2CO_3 \leftrightarrow H^+ + HCO_3^-$), and so the pH of otherwise pure water in the presence of air becomes about 5.7. If we were to add more acid to that water the equilibrium would move in reverse and carbon dioxide gas would evolve. If the concentration of carbon dioxide in the air was to increase, more carbon dioxide would enter the water and form more carbonic acid and more bicarbonate, decreasing the pH further. In chemistry, assuming pure water has a pH of 7.0 works very well, and that assumption is the right one to make. In nature, water should have a measured pH of about 5.7, but because water in nature is not pure, it rarely does.

Therefore, carbon is rarely a limiting factor for C_4 photosynthesis. Although air movement over C_4 grasses needs to be maintained, restricted air movement over these grasses is not as great a concern as restricted air movement over C_3 grasses.

Warm-season grasses easily outperform cool-season grasses when atmospheric carbon dioxide concentrations are low (Moss and Smith, 1972). In fact, C_4 photosynthesis is maintained at carbon dioxide concentrations as low as 2% of normal

(Hesketh, 1963). Although high temperature can cause a high enough loss of carbon dioxide from water to restrict C_3 photosynthesis, bicarbonate in the water will still be plentiful and C_4 photosynthesis will not be restricted.

The enzyme PEP carboxylase does not bind oxygen. What does that mean? It means that C_4 grasses can perform photosynthesis in intense light for long periods. It means that no matter how long the day or how intense the sun, C_4 photosynthesis will not be affected as long as carbon dioxide is present to dissolve and maintain sufficient bicarbonate levels in water. No other molecules compete for the binding site of the PEP carboxylase enzyme. Therefore, oxygen and other compounds do not affect the rate of carbon binding in the mesophyll cells. Even when plant photosynthesis is evolving oxygen rapidly, the air will not get rich enough in oxygen for PEP carboxylase to bind it. Therefore, C_4 photosynthesis eliminates the perceived deficiencies of Rubisco. The pathway sequesters Rubisco in small cells deep in the interior of the leaf and surrounds the enzyme with highly concentrated carbon dioxide. Rubisco is unlikely to bind oxygen in a bundle sheath cell. Instead of using Rubisco to bind carbon initially, C_4 plants use PEP carboxylase, which binds bicarbonate exclusively, so it is not affected by high oxygen concentrations. The use of bicarbonate as a carbon source ensures a sufficient supply of carbon even when temperatures get exceptionally warm. The net result is that C_4 photosynthesis is considerably more effective in the intense light and high heat of summer in temperate climates and in tropical climates.

3.4 Characteristics of C_3 and C_4 Plants

After reading to this point, you probably have the impression that C_4 photosynthesis is a more efficient process than C_3 photosynthesis. That would seem to be true and many experts believe it to be true. However, the fact remains that the number of C_4 plants known to exist is very small compared with the number of C_3 plants in existence. All of these plants have had considerable time to adapt to their environments, and the C_3 pathway is still the major system of photosynthesis. Nature has determined that C_4 photosynthesis is best for plants in tropical regions. The performance of grasses during midsummer periods in temperate regions certainly demonstrates that C_4 is the most effective pathway for conditions of intense light, long day length and temperatures exceeding roughly 80 °F (27 °C). However, nature appears to have chosen the C_3 pathway for plants in temperate climates. Buffalograss (*Buchloe dactyloides*), for instance, is the most cold-tolerant warm-season (C_4) species that we use for turfgrass in the USA (Turgeon, 2008). Although it survives extremely cold temperatures, such as −20 °F (−29 °C) or lower, it has a very short growing season in temperate climates compared with C_3 grasses. The same can be said of Japanese lawngrass (*Zoysia japonica*), another relatively cold-tolerant C_4 grass. Its growing season is considerably shorter in temperate climates than the growing season of C_3 grasses in the same location. Perhaps C_3 grasses perform better than C_4 grasses in cool weather because the C_4 pathway includes extra binding steps that require more energy (Black *et al.*, 1973). A C_3 plant requires 3 ATP molecules to bind a carbon dioxide molecule. After first binding carbon in the mesophyll cell, and its then transfer, release and rebinding by Rubisco in the bundle sheath cell, the C_4 pathway requires 5 ATP molecules. It is possible that a C_3 plant must be photorespiring at a particularly high rate before the extra energy required to perform C_4 photosynthesis equals the energy lost to photorespiration in a C_3 plant. Although we know a great deal about photosynthesis, there are still questions to be answered.

Both C_3 and C_4 plants have unique characteristics that influence management practices. In the following section, we will discuss the general differences between cool- and warm-season grasses. Specific management techniques will be discussed later in individual sections of the text.

Temperature tolerance

It is clear that warm-season grasses are more heat tolerant than cool-season grasses and that cool-season grasses are more cold-tolerant than warm-season grasses. Cool-season grasses differ in their tolerance for cold and warm-season grasses differ in their tolerance for heat (Turgeon, 2008). If you live near the equator or near the Arctic Circle that would be very important to you as there are few grasses that can tolerate those extremes. However, most of us are more concerned about the heat tolerance of our cool-season grasses and the cold tolerance of our warm-season grasses. Warm-season grasses differ in their tolerance to cold and cool season grasses differ in their tolerance to heat (Turgeon, 2008). However, how you manage the C_3 pathway when temperatures are high for your area, and how you manage the C_4 pathway when temperatures are low for your area differs little by species.

Adaptation to light

Generally speaking, cool-season grasses are better adapted to shade or cloudy conditions and tend to grow better in those situations than do warm-season grasses. You learned earlier that warm-season grasses are much better adapted to intense light and long days than cool-season grasses. Also, generally speaking, cool-season grasses grow better in low light and short days than warm-season grasses. There is no such thing as a shade-loving grass. All grasses like full sun but some grasses like more sun than others. Most cool-season species are somewhat shade tolerant, and because of photorespiration prefer clouds rather than intense sunlight. Warm-season plants tend to decline when cloudy conditions prevail for more than a few days. St. Augustinegrass (*Stenotaphrum secundatum*) is an exception to these statements. This grass is a warm-season grass that, in certain cultivars, exhibits shade tolerance that rivals some of the cool-season grasses (Beard, 1973). Select species and cultivars of zoysiagrass (*Zoysia* spp.) also exhibit some shade tolerance, but they do not appear to be as shade tolerant as most cool-season grasses (Duble, 1989).

Drought tolerance and water use rates

If a plant is drought tolerant that does not necessarily mean that it has a low water use rate. Some of the most drought-tolerant grasses escape drought by entering a state of semi-dormancy. Warm-season grasses like bermudagrass (*Cynodon dactylon*) and buffalograss and cool-season grasses like Kentucky bluegrass (*Poa pratensis*) are good examples of plants that can remain dormant for long periods of time during dry conditions. When these plants are active, they do not necessarily have low water use rates but they are relatively good at surviving a drought. A cool-season plant like tall fescue (*Festuca arundinacea*) has very good drought tolerance for a cool-season grass, but a high water use rate when water is plentiful (Qian *et al.*, 1997). Low water use rates and drought tolerance are important characteristics of modern turfgrasses. Irrigation water is in high demand in most parts of the world, and water use rates and drought tolerance will become increasingly important in the future, although it is important to note that they are two different characteristics. The perfect turfgrass for dry conditions would have good aesthetic value, a low water use rate, a high drought tolerance and remain green during dry periods.

Of the grasses that we use most commonly for turfgrass, the warm-season grasses tend to have lower water use rates and higher drought tolerances. Our warm-season grasses are genetically adapted for producing aggressive root systems that provide drought tolerance. It is not uncommon for warm-season grasses to develop root systems that extend over 3 feet (91 cm) deep (Christians, 2007). Consequently, these grasses can reach water when the upper soil layers have dried out. Their low water use rates can be explained by the efficiency of the C_4 pathway (Black *et al.*, 1969). Warm-season grasses transpire less water (Kneebone *et al.*, 1992). Because C_4 grasses can bind bicarbonate, which is usually plentiful in water, stomates can be partially closed when needed to conserve water without affecting the supply of carbon dioxide.

Nitrogen use efficiency

Warm season grasses use nitrogen more efficiently than cool-season grasses (Brown, 1978). The leaves of C_4 grasses contain less nitrogen per unit area or mass than C_3 grasses, suggesting that less nitrogen is required for each carbon assimilated during photosynthesis (Hull, 1992). The reason for the C_4 increase in nitrogen efficiency could be because a considerably smaller amount of Rubisco is needed in C_4 leaves than in C_3 leaves (Brown, 1978). A second possible explanation could be the allocation of nitrate reductase and assimilation enzymes between mesophyll and bundle sheath cells (Moore and Black, 1979). Regardless, we don't know the whole story.

Although research demonstrates that warm-season grasses are more efficient users of nitrogen compared with cool-season grasses, such is not the case in practice. For instance, bermudagrass is a C_4 grass that presumably is an efficient user of nitrogen but, in practice, it requires 0.8 to 1.5 pounds N/1000 square feet each growing month (39–73 kg/ha a month) to perform up to most customer expectations (Beard, 1973; Duble 1989). A quick review of introductory turfgrass textbooks (see Suggested Reading, Chapter 1) reveals that the nitrogen recommendations for C_3 grasses do not exceed the nitrogen recommendations for C_4 grasses. In this case, we are missing something. We have failed to consider other factors. Bermudagrass is extremely aggressive. It has excellent recuperative ability, spreads rapidly, forms a deep, extensive root system, and a vast network of stolons and rhizomes. Perhaps it needs more nitrogen because of these qualities, yet uses it efficiently. I honestly don't know. Again, we don't know everything that we would like to know.

In this case, knowing the physiology, the fact that C_4 grasses are more efficient users of nitrogen, does not help us in the field. The physiology leads us to believe that C_4 grasses require less nitrogen fertilizer for optimum performance. Such is not the case. We need more information. Usually the physiology will lead you in the right direction. In this case it does not; there are always exceptions.

3.5 Chapter Summary

Photorespiration occurs when Rubisco binds oxygen. This process only occurs in C_3 (cool-season) plants. Because of photorespiration, cool-season grasses do not perform well when temperatures are high and light is intense. Rapid photosynthesis in intense light leads to high oxygen and low carbon dioxide concentrations in and around the plant, making the binding of oxygen by Rubisco more likely. Also, high temperatures reduce the concentration of gases, including oxygen and carbon dioxide, in water. Carbon dioxide is less tightly bound by water than oxygen. Consequently, as water gets warm the ratio of carbon dioxide to oxygen in the water decreases, further increasing the likelihood that Rubisco (ribulose bisphosphate carboxylase) will bind oxygen.

Warm-season (C_4) grasses avoid photorespiration by adding extra steps in carbon assimilation and by performing the light reaction and the dark reaction of photosynthesis in separate cells. Warm-season grasses have Kranz anatomy, meaning that they have bundle sheath cells located near their leaf vascular system where the Calvin cycle occurs. The light reaction takes place in the mesophyll cells near the surface of the leaf and the dark reaction takes place in the bundle sheath cells. Both reactions are the same as the reactions performed in C_3 plants, but they occur in different cells. The bundle sheath cell where Rubisco resides is highly concentrated in carbon dioxide so Rubisco does not bind oxygen. Warm-season plants concentrate the carbon dioxide by assimilating carbon from bicarbonate in the mesophyll cells and transferring it to the bundle sheath cells where they release it as carbon dioxide. This C_4 process eliminates photorespiration and is considerably more efficient than C_3 photosynthesis when temperatures are high and light is intense. Consequently, C_4 photosynthesis is better adapted to tropical and subtropical climates than C_3 photosynthesis. Because of the C_4 pathway, warm-season grasses use water and nitrogen more efficiently and they can perform photosynthesis when the carbon dioxide concentrations in the air are very low. However, they do not adapt well to low temperatures and low light, suggesting that the C_4 pathway is not as efficient under those conditions as the C_3 pathway.

The warm-season grasses have either adopted the C_4 pathway of photosynthesis because of their environment or they have adapted to their environment because of the C_4 pathway. In either case, the warm-season grasses have characteristics not directly related to photosynthesis that differ from cool-season grasses. The C_4 grasses tend to have deeper root systems than C_3 grasses and tend to be more drought tolerant. In general, they do not perform well in low light and cannot survive extremely low temperatures. Consequently, warm-season grasses require different management strategies than cool-season grasses. The turfgrasses selected for a particular site and the management practices required to meet customer expectations at that site are largely determined by the pathway that they use for photosynthesis and the adaptation that accompanies the pathway.

Suggested Reading

Hull, R.J. (1992) Energy relations and carbohydrate partitioning in turfgrass. In: Waddington, D.V., Carrow, R.N. and Shearman, R.C. (eds) *Turfgrass*. ASA-CSSA-ASSA (American Society of Agronomy-Crop Science Society of America-Soil Science Society of America), Madison, Wisconsin, pp. 175–206.

Salisbury, F.B. and Ross, C.W. (1992) *Plant Physiology*, 4th edn. Wadsworth Publishing, Belmont, California.

Suggested Websites

NTEP (National Turfgrass Evaluation Program) (2009) Available at: http://www.ntep.org (accessed 17 December 2009).

4 Respiration and Transpiration

Key Terms

A **source** is a part of a plant such as a mature leaf that produces more energy than it uses.

A **sink** is a part of a plant such as a root that uses more energy than it produces.

Carbohydrate partitioning refers to the allocation of carbohydrates from source leaves to other organs such as roots and stems in a priority determined by season and environment.

Respiration is the breakdown of carbohydrates and other compounds into smaller units of cellular energy such as ATP (adenosine triphosphate).

Stomates are small holes in grass leaves controlled by guard cells that open and close as necessary to permit or restrict transpiration.

Transpiration is the transport and evaporation of water that is used to dissipate the heat that is produced in the plant as a result of its metabolic processes, such as photosynthesis and respiration.

Photosynthates are the carbohydrate products of photosynthesis.

Starch is a long chain of glucose molecules and is the storage carbohydrate used by C_4 grasses.

Sucrose is table sugar and results from combinations of a glucose molecule and a fructose molecule. Sucrose is the predominate form of carbohydrate movement through a plant.

Amyloplasts are organelles where starch is stored.

Fructans are long chains of fructose molecules and are the storage carbohydrate used by C_3 grasses. Fructans are soluble, starch is not.

Vacuoles are organelles responsible for maintaining cell turgor pressure and are the site where most fructans are stored.

Carbon dioxide compensation point is the amount of carbon dioxide present when the binding of carbon dioxide for photosynthesis equals the evolution of carbon dioxide during respiration.

Phloem consists of sieve tube elements and companion cells and is the part of the vascular bundle, also called a vein, through which plants move sucrose to where it is needed.

A **pathogen** is a disease-causing organism.

Passive transport means that plant energy is not required to move compounds from one location to another.

Active transport means that plant energy is required to move compounds from one place to another.

The **Apoplast** is the area between cells also called intercellular space. It is, in fact, everything outside the cell membranes including cell walls and xylem. It is the dead part of the plant.

The **Symplast** is the interior of a cell. It is the living part of the plant or the protoplasm inside the cell membranes.

Aerobic means that air, in the case of respiration, specifically oxygen, is necessary.

Anaerobic means that air, in the case of respiration, specifically oxygen, is not necessary.

NADH (the reduced form of nicotinamide adenine dinucleotide) is an energy-rich electron donor molecule used in respiration much like **NADPH** (the reduced form of nicotinamide adenine dinucleotide phosphate) is used in photosynthesis. It is formed by the donation of electrons to its oxidized form, **NAD$^+$**.

Capillary action is the result of adhesive and cohesive forces that cause water to rise upward against gravity in a very small tube.

Xylem is the tubular system of tracheids and vessels that carries water from the roots to other parts of the plant.

Symplastic movement is movement from cell to cell as occurs through phloem, or plasmodesmata.

Apoplastic movement is movement through intercellular spaces, through cell walls, or through xylem.

4.1 Encouraging Efficient Use of Carbohydrates and Water

Plants have only five organs; leaves, stems, roots, flowers and seeds. Of the five plant organs only leaves, except for a minor amount in stolons, perform photosynthesis. Leaves are a source; the other organs are sinks. The source provides chemical energy in the form of carbohydrates. The sinks use it or store it. The translocation and fate of plant carbohydrates is called carbohydrate partitioning. The sink that is growing the fastest during periods of growth or storing the most during periods of storage has priority. Grasses, both C_3 (cool season) and C_4 (warm season), for instance, slow shoot growth and produce carbohydrates for storage in their roots and stems as winter approaches (McKell et al., 1969).

Release of chemical energy stored in carbohydrates for use by the plant is called respiration. Respiration is not a particularly efficient process and much of the chemical energy in the respired carbohydrates is lost as heat. When you work hard you perspire because respiration in your cells is occurring rapidly and heat is building up inside your body. Perspiration is one of your body's methods for cooling itself. Photosynthesis and respiration in plants create heat so plants cool themselves by transpiration. Plants move water from their roots to their leaves where it exits the plant through stomates and evaporates. This translocation and evaporation for cooling purposes is called transpiration and occurs consistently during daylight hours and sometimes but rarely during warm nights. If you have done everything that you can do to encourage healthy photosynthesis in your turfgrass, it is time to look at respiration and transpiration. Aside from photosynthesis, respiration and transpiration are the two physiological processes that can have the most influence on the health of your turf.

4.2 Plant Respiration

In order for respiration to occur there must be carbohydrate or another form of chemical energy present. Plants can also respire lipids and proteins, but carbohydrates are preferred (Plaxton, 1996). If a plant is respiring more than a small amount of (usually damaged) lipids and proteins, it is probably under stress. In order for plants to use respiration effectively, the respiration must take place in plant locations where the energy is most useful. Consequently, respiration is of little use unless the carbohydrates of photosynthesis are translocated to where energy is needed.

Components of plant energy

Carbohydrates formed by active photosynthesis can either be respired by the photosynthesizing cell for energy in that cell or be translocated immediately to other plant parts. However, most of the photosynthates produced are stored as starch in the chloroplast where they were formed, and then translocated during the night (Geiger and Servaites, 1994). Both warm- and cool-season grasses use starch as their temporary storage mechanism in chloroplasts, and translocate carbohydrates as sucrose (table sugar). If carbohydrate supply exceeds demand, sugar is translocated mostly to stems and to roots for long-term storage and future use. In warm-season grasses, the sugar is combined into units of starch (long chains of glucose) and stored in organelles called amyloplasts (Chatterton et al., 1989). In cool-season grasses, the sugar is combined into fructans (long chains of fructose) and stored in vacuoles (Pollock and Cairns, 1991). Vacuoles are organelles responsible for maintaining cell turgor and are unique to plants. Fructans are soluble, but starch is not. Consequently, the storage of fructans in vacuoles aids in maintaining cell turgor pressure in cool-season plants. Warm-season plants use other solutes such as potassium to maintain turgor pressure in their vacuoles and cells.

As mesophyll cells in C_3 plants or bundle sheath cells in C_4 plants fix carbon dioxide, the resulting carbohydrates are stored in the chloroplasts as starch. Plants respire continuously during both day and night, and a cell performing photosynthesis is also performing respiration at the same time (Krömer, 1995). Consequently, some photosynthates are used or translocated immediately to accommodate normal respiration. However, if photosynthesis is proceeding at a rate that exceeds the carbon dioxide compensation point, which is the point at which carbon dioxide fixation equals carbon dioxide evolution, excess photosynthates are produced. These photosynthates are stored as starch in the chloroplast where they were produced and translocated later as they are needed.

Starch is a huge molecule and it cannot be transported through cell walls and membranes. So in order for the plant to translocate starch, it

> **Box 4.1. How do vacuoles help to maintain cell turgor pressure?**
>
> Pure water in the presence of air contains carbon dioxide in the form of bicarbonate as a solute. No other solutes are present in pure water. What happens to osmotic water potential when we add a solute such as sugar or salt to otherwise pure water? Any general chemistry book will tell you that at least four things happen: (i) the freezing point of the solution decreases; (ii) the boiling point increases; (iii) the vapor pressure decreases; and (iv) the osmotic pressure increases. Chemists call these colligative properties. All of those colligative properties are important to us in the study of biological systems and in understanding how best to manage them. Everything that we have discussed so far in this text depends on osmosis. Osmosis may seem confusing, but it is actually quite simple. Colligative properties occur when a solute is mixed with a solvent. In our case, the solvent is always water. Anything that dissolves in water is a solute. The solutes most important to us in terms of osmosis are sugars, salts and nutrients. Potassium, for instance, is a solute. You will learn shortly that potassium is instrumental in the opening and closing of stomates by managing osmosis. In biology, water is never pure. It always contains solutes, so it is always a solution. If a solution has a lot of potassium in it, we say it has a high concentration of potassium. If a solution with a high concentration of potassium is separated by a cell membrane permeable to water but not to potassium, and there is a solution on the other side of the cell membrane with a low concentration of potassium, water will move through the membrane toward the higher potassium concentration until both concentrations are equal. In other words, the water will move until the ratio of solute to water is the same on both sides of the membrane. Consequently, there will be less water on the (originally) low potassium concentration side of the membrane and more water on the (originally) high potassium concentration side so that the potassium to water ratio is the same on both sides. In the case of guard cells (cells that surround the pore of a stomate), the movement of water from outside a guard cell to inside that guard cell causes the cell to swell which, in turn, causes the stomate to open. Remember: water always moves from a low concentration of a particular solute to a high concentration of that solute. Solutes attract water. In a tug of war for water, the high concentration of solute always wins. Vacuoles maintain high concentrations of solutes, mostly sugar and/or potassium. The sugar and potassium draw water through the vacuole membranes so that the vacuole can maintain turgor pressure. That is how vacuoles help to maintain cell turgor pressure in a plant.

must first be converted to molecules that can be moved. The transport carbohydrate of choice in most plants is sucrose, a sugar made of one molecule of glucose combined with one molecule of fructose. Sucrose is what we know as table sugar. It dissolves readily in water and moves easily through cell walls and membranes. Some plants also transport raffinose, stachyose or verbascose, but turfgrasses transport mostly sucrose (Lalonde et al., 2004).

Starch is broken down (hydrolyzed) one glucose unit at a time in a chloroplast through chemical processes catalyzed by enzymes called amylase and starch phosphorylase. The resulting glucose is then transported through the chloroplast membrane into the cytoplasm. In the cytoplasm some of the glucose is converted to fructose and the two simple sugars are combined to make sucrose. The sucrose is loaded into the phloem and transported to a sink where it is unloaded and either converted to another carbohydrate, an amino acid or a lipid, used for cellular energy or combined into starch or fructans for storage and later use (Plaxton, 1996) (Fig. 4.1).

Sources and sinks

Mature grass leaves are sources, meaning that they produce more photosynthates than they use (Fig. 4.2). Very young leaves are sinks because they have yet to develop their full photosynthetic potential. Very old leaves have outlived their usefulness and are also sinks for a short time until they senesce, a plant science term meaning that they die and fall away. The sheaths of turfgrass plants are normally green and can be a source of photosynthesis, but they are not a major source of photosynthates. The same can be said for stolons. Most stolons are green and perform photosynthesis, but they are not major sources of photosynthates for plant metabolism. Plant roots are sinks. They require energy for growth and for water

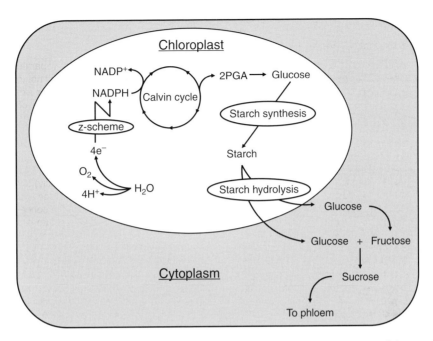

Fig. 4.1. Starch in a chloroplast is hydrolyzed to glucose and enters the cytosol (the fluid part of the cytoplasm) where some of it is converted to fructose and combined with glucose to form sucrose. The sucrose is loaded into phloem cells and transported to areas of need called sinks. $NADP^+$ = nicotinamide adenine dinucleotide phosphate; NADPH = reduced form of $NADP^+$; PGA = phosphoglycerate.

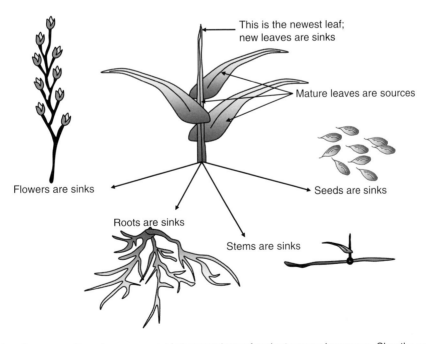

Fig. 4.2. Mature leaves are the primary source of photosynthates for plant energy in grasses. Sheaths and stolons may provide some photosynthates but are not major sources. All other plant parts are sinks.

Respiration and Transpiration

uptake. The root tip where meristematic growth occurs is an active sink. Energy is also necessary for root water uptake so if photosynthates or oxygen are not available for respiration, water uptake does not occur.

Reproductive parts are major sinks. Turfgrass growth slows, sometimes stops, as plants develop flowers and seeds. Many crop plants are annual grasses. They produce considerable vegetation in the weeks after establishment, but growth slows and stops as flowers and seeds are produced (Barden *et al.*, 1987). With few exceptions, turfgrasses do not flower and seed at the mowing heights that we commonly use. Those few that do waste energy. We would prefer that those plants produce vegetation and spread rather than produce flowers and seeds as flowers and seeds are strong sinks for energy that could be used for other purposes. A good example of a plant that wastes energy is annual bluegrass (*Poa annua*).

Annual bluegrass is ubiquitous throughout the world and is often considered a weed. However, under some conditions it makes an exceptionally nice turf (Huff, 1998). Annual bluegrass is best suited for golf course putting greens in areas where summers are cool, and it is competitive with perennial grasses such as creeping bentgrass (*Agrostis stolonifera*) and others. Under the right situation, annual bluegrass can exist as an annual or a perennial (Turgeon, 2008). There is also a perennial species of the plant that we call annual bluegrass (*Poa annua* f. *reptans*). A perennial plant called "annual" bluegrass is a misnomer but, nonetheless, the name exists. One reason that annual bluegrass persists as a weed and as a turf in putting greens is because it can seed at extremely low mowing heights. As a volunteer plant, annual bluegrass germinates in the fall and continues rapid vegetative growth until winter temperatures become too low for healthy metabolism. During the fall and early spring annual bluegrass appears green and healthy. As spring progresses, however, the species begins to flower and seed prolifically. As the reproductive phase continues, the plants turn yellow and look unhealthy because most of the available photosynthate energy goes toward reproduction rather than tissue growth and maintenance (Fig. 4.3). This rapid vegetative growth stage followed by prolific seeding is common in annual plants. In annual bluegrass, however, it is common in both annual and perennial biotypes, and is a good example of how much energy is required to produce flowers and seeds.

Technically, sinks are organs that use more energy than they produce. Sinks do not have to be organs that are rapidly using energy, they can be organs that are storing energy. Grasses primarily store carbohydrate energy in stems and secondarily in roots (Hull, 1992). During seasonal periods or periods of low carbohydrate accumulation, such as summer for C_3 plants, these stems and roots can become sources rather than sinks (Pollock and Cairns, 1991). Spring, for instance, is a period of rapid shoot growth and root growth in grasses. More vegetative growth occurs in our turfgrasses during the spring than at any other time of year. Photosynthesis is occurring rapidly during periods of sunlight, but spring days are relatively short and often cloudy in many locations. So where does all this energy for spring growth come from? It comes from storage.

Phloem transport is not always downward from shoots to roots. In the spring, it is often upward from roots to shoots. Every fall, turfgrasses slow in growth and move carbohydrates into storage to maintain winter metabolism (Hull, 1976). During winter, photosynthesis slows and often stops depending on the species and climate. Plants use stored carbohydrates to maintain life over the winter and use what's left in the spring to spread and grow. If we understand this movement and when it occurs, we can use it to our advantage. For instance, spring is not the best time to apply a foliar-absorbed herbicide, but it is the best time to apply a root-absorbed herbicide. Most herbicides are meant to be applied to actively growing weeds. In the fall, turfgrass weeds move photosynthates downward and foliar-absorbed pesticides move with them. In the spring, phloem movement is upward and foliar-applied pesticides do not work quite as well. Although today's herbicides are effective in both spring and fall, you can increase their effectiveness by applying them at the right time of year. Unfortunately, you don't usually have a choice.

Phloem transport

Phloem transport occurs from source to sink. Phloem is the portion of the plant's vascular system responsible for the translocation of sucrose and other sugars to areas of need. The phloem also transports larger molecules such as proteins, RNA

Fig. 4.3. Annual bluegrass (*Poa annua*) surrounded by a Kentucky bluegrass (*Poa pratensis*) lawn in early spring (a) and late spring (b). Notice the difference in the healthy appearance compared with the lawn in early spring versus a relatively unhealthy appearance in the late spring.

(ribonucleic acid) and even viruses (Oparka and Cruz, 2000). The phloem is under tremendous pressure because of the difference in solute concentration (remember osmosis) at the source compared with its concentration at the sink. This pressure creates problems for us when we mow. When we cut the phloem the nutritious contents leak out and become food for any number of microorganisms, including pathogens (disease-causing agents).

However, this pressure also encourages rapid passive movement of phloem sap from source to sink; the word "passive" means that no plant energy is needed for transport. "Active" transport means that plant energy is required for transport. Phloem loading is an active process resulting in high pressure at the source but phloem transport is mostly passive.

The phloem consists of sieve cells and companion cells (Fig. 4.4). Macromolecules such as proteins

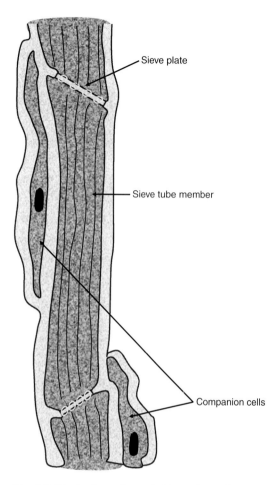

Fig. 4.4. The basic anatomy of phloem tissue. In the plant's vascular system, the phloem transports photosynthates, protein, RNA and, sometimes, by accident, microorganisms.

and RNA that are transported in the phloem are synthesized in the companion cells. This is necessary because those molecules are too large to pass through typical plant cell walls and membranes. The membrane between the sieve cell and companion cell is not a normal cell membrane. It is believed to consist of a thin line of organelles (Oparka and Turgeon, 1999). The gaps between these organelles permit the passage of large molecules.

Phloem loading is both a passive and active process in grasses. Sucrose passes relatively easily through plasmodesmata, the intersecting cell-to-cell passive transport system. As long as the sucrose concentration in the cytoplasm of source cells is greater than the concentration of sucrose in the phloem near those cells, sucrose will move passively toward the phloem and enter the phloem. A second and more common passage is through the apoplast, or intercellular space, and through cell walls and membranes by active transport. In this case, the sucrose is part of a co-transport system with protons in a process similar but backward to the ATP (adenosine triphosphate) pumps in the thylakoid membrane that you learned about in the z-scheme of photosynthesis (Lalonde *et al.*, 2004). In this case, ATP is used rather than made. The energy in ATP is used to fuel a proton pump. The ATP pump, activated by an ATPase enzyme, is in the membrane of the phloem companion cells. The ATPase enables the pumping of protons from inside the cell to outside the cell. As a result, the apoplast becomes rich in protons that diffuse back into the symplast (the interior of the cell) as a result of the acid gradient formed across the membrane. Sucrose enters the cell with the protons in a co-transport relationship. Consequently, the sucrose concentration in the phloem can be substantially higher than it is in surrounding cells. That means that the source cells do not have to maintain the high concentration of sucrose necessary to perpetuate a passive transport gradient into the phloem. Owing to the systems of active and passive transport into the phloem, plants are able to perpetuate high concentrations of sucrose in the phloem for transport from sources to sinks during both day and night.

As sucrose concentrations are high at the sources and low at the sinks, sucrose can move passively through the phloem traveling with the concentration gradient. The faster the sink is using or storing sucrose, the greater the gradient becomes and the more pressure that is placed on the phloem. In fact, there is some evidence that sink demand can actually influence the rate of photosynthesis in leaves (Gifford and Evans, 1981).

The basic idea behind phloem transport is fairly simple if you remember your chemistry. Remember that water moves toward the lowest water potential and that water potential decreases with increasing solute concentration. So if you add table sugar to a glass of water, its water potential becomes lower. If you have one glass of water straight from the tap and a second glass of tap water in which you added two tablespoons of sugar, the water potential in the sugar water is lower than the water potential of the pure tap water. What would happen if you were to put tap water into a larger container, then take a membrane permeable to water but not to dissolved sugar, partially submerge the membrane in the new container, and pour the sugar water into it? The tap water would move into the membrane until the water potential in the membrane was equal to the water potential outside of the membrane. In this case, nearly all of the tap water would probably move into the membrane because the tap water is not likely to have enough solutes in it to equilibrate with two tablespoons of sugar. So, except for some water clinging to the sides of the container, nearly all of it would move into the membrane to dilute the sugar (Fig. 4.5). If the permeable membrane was rigid and could not expand, the water moving into it would create a great deal of pressure against the side of the membrane and sugar water would push out any available opening except back into the tap water.

Because plant cells have walls and those walls are relatively rigid, osmotic pressure can build within them as it did inside the membrane in the tap water example above. The osmotic pressure that builds near sources as sucrose fills companion cells is the primary force behind phloem transport (Box 4.2). Although the pressure is obviously the result of an energy source, plant energy is not used to move sap through the phloem and the transport is considered passive. This pressure buildup can also be passive if the only sucrose entering companion cells is the result of the concentration gradient from source cell to companion cell. More likely, however, the high concentration of sucrose in the companion cell is the result of active phloem loading by ATPase pumps, in which case phloem loading is considered active but transport is still considered passive.

Carbohydrate partitioning

Carbohydrate partitioning is the term used to describe the priority distribution of carbohydrates throughout a plant. We actually know little about carbohydrate partitioning. Many scientific projects have attempted to use carbohydrate content as a measure of turfgrass health. However, after multiple attempts over many years it appears that the amount of carbohydrate in a plant or plant part does not always correlate with plant health or visual turf quality (Xu and Huang, 2003). In some cases, for the measurement of certain parameters under specific conditions, there may be a relationship, but more often, there is not. We have known for many years that the amount of photosynthate translocated to roots is negatively correlated with shoot growth (Mehall *et al.*, 1984). Consequently, as we reduce mowing height, we encourage shoot

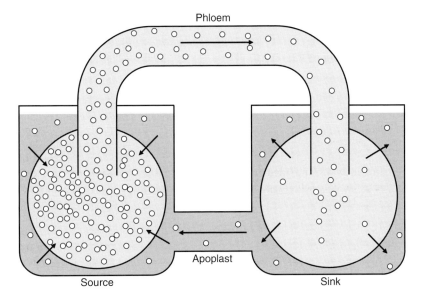

Fig. 4.5. As sucrose moves into the companion cells of the phloem either by active or passive transport the sucrose solute lowers the water potential of the companion cell and attracts water from outside the cell. The water moving into the cell causes pressure that forces water and solutes into the sieve tubes and instigates rapid flow toward the sinks.

> **Box 4.2. The pressure of the phloem.**
>
> How much pressure is actually present in phloem tissue? Considerably more pressure is exerted on the sides of plant phloem tissue than is exerted within the arteries and veins of the human circulatory system. It is a good thing that plant cells have walls. Otherwise this pressure could not be contained. The velocity of phloem transport is in the neighborhood of 20–40 inches/h (500–1000 mm/h) (Salisbury and Ross, 1992). The pressure on the walls of the sieve tubes in weeping willow (*Salix babylonica*) was estimated to be about 145 pounds per square inch (1.0 MPa) (Wright and Fisher, 1980). However, that pressure varies by height and varies considerably by species according to the size of the sieve tubes and other variables.
>
> It is also interesting to note that the phloem pressure concept was first introduced by a German scientist back in 1927 and has changed little since then (Münch, 1927, 1930).

growth to replenish the portion lost to mowing and root growth suffers. The amount of carbohydrate relative to root mass has not changed. However, the roots are fewer and shorter and the plant is less tolerant of dry conditions and less capable of nutrient uptake. We have also known for many years that total nonstructural carbohydrate (TNC) relative to tissue mass declines with increasing shoot growth, suggesting that healthy plants can have low TNC (Youngner and Nudge, 1976).

In 1996 and 1997, we attempted to use total TNC content of entire grass plants as an objective measure of shade stress (Bell and Danneberger, 1999). There were four treatments, morning shade, afternoon shade, all day shade and no shade. The no shade treatment was dense and healthy and the all day shade treatment had extremely poor cover and appeared unhealthy, yet the TNC content did not differ between the shade and no shade treatments. We reasoned that grass plants probably do not store carbohydrates during the growing season beyond a certain predetermined level. Instead, they use whatever excess carbohydrates that are available to form new tillers and daughter plants. In fact, as stated earlier, research indicates that during periods of rapid growth TNC levels in grass plants are lower (Hull, 1992). Our conclusion may or may not be true, but based on past research findings, it is a reasonable explanation and a good example of the thought process and information-gathering effort required to make decisions when there is no prior experience with a given situation. Had we measured TNC content in the shoots and TNC content in the roots, we might have found more TNC in the shoots of the shade plants and more TNC in the roots of the sun plants, as shade tends to encourage shoot growth at the expense of root growth (Dudeck and Peacock, 1992). It is probably also reasonable to assume that the plant parts that are growing fastest are the plant parts that are receiving the most carbohydrates. However, those plant parts will not have the highest TNC unless they are receiving carbohydrates faster than they are using them for growth.

Glycolysis

Once an energy source such as sucrose reaches its destination in a plant, it can be used for energy, used for the synthesis of another compound or stored for later use. When a compound is broken down for its energy, we call that respiration. Respiration releases the energy stored in chemical bonds and converts it to cellular energy, usually ATP that is used to fuel cellular metabolism. There are four kinds of plant respiration: the photorespiration that you already learned about, glycolysis, pentose phosphate respiration and aerobic respiration. As you know, photorespiration is an effort to salvage as much carbon and energy as possible when Rubisco (ribulose bisphosphate carboxylase) binds oxygen. The other three types of respiration provide energy and the synthesis of plant compounds for plant growth and maintenance.

To extract all of the energy possible from a molecule of carbohydrate, protein or lipid, the compound or its components must pass through the citric acid cycle and through an electron transport system in a process of aerobic respiration. In order for aerobic respiration to occur, air, specifically oxygen, must be present. Glycolysis is an anaerobic process, meaning that oxygen is not required. Glycolysis provides cellular energy in small doses very rapidly. Its strength is that it can provide energy under the anaerobic conditions that can occur when respiration is moving so rapidly that oxygen is used faster than it can be refreshed.

Plant and animal respiration are not quite the same, but they are very similar. If you were to lift a heavy weight repeatedly with the same arm in the same way, your muscles would soon tire. You would begin to feel pain as your muscles exhausted themselves and, sooner or later, you would no longer be able to lift the weight. What you would have experienced is a balancing act between aerobic and anaerobic respiration. Eventually, if you repeated the exercise long enough, your arm would have tried but failed to lift the weight and you would have had to give up. If you had waited a few minutes, no matter how tired your arm became, you would have been able to lift the weight again. What happened?

Your muscles lifted the weight easily the first time you tried because they were well supplied with oxygen and aerobic respiration was not inhibited. Plenty of energy was available to lift the weight. As you lifted repeatedly, respiration went into overdrive in an effort to supply the energy needed to complete the weight-lifting act. Soon, your muscles were using more oxygen than could be refreshed and aerobic respiration began to slow. Glycolysis was also providing energy during the activity, and as respiration slowed, glycolysis was able to provide enough energy to complete the activity. However, as the oxygen content became lower and lower in the muscles, aerobic respiration became completely ineffective and glycolysis was required to support the energy needed for the entire task. As glycolysis can only provide a limited supply of energy, it was quickly over-tasked, the ATP in the muscle cells was exhausted, and you could no longer lift the weight. The pain in your muscles was caused by fermentation.

As far as metabolic pathways are concerned, glycolysis is quite simple (Fig. 4.6). The components of sucrose, glucose and fructose, can enter glycolysis as glucose 6-phosphate or fructose 1,6-bisphosphate, both of which have to use a molecule of ATP in their preparation. Another molecule of ATP is required to convert fructose 1,6-bisphosphate to glyceraldehyde 3-phosphate (GAP) or to dihydroxyacetone phosphate (DHAP). You may recognize GAP and DHAP as important components of the Calvin cycle. These compounds can enter glycolysis directly from the Calvin cycle if needed. At this point, the six-carbon sugars glucose and fructose have been split into two three-carbon compounds so there are two GAPs passing through glycolysis for every glucose or fructose that entered the pathway. Near the end of the pathway, 2-phosphoglycerate is converted to phosphoenolpyruvate (PEP), the molecule that is so important for C_4 photosynthesis. A PEP molecule can enter late in the pathway and generate one molecule of ATP when it is converted to pyruvate. Pyruvate is the end product of glycolysis. Two ATP molecules are required to convert glucose or fructose to two GAP molecules. The two GAP molecules move through glycolysis in a series of three-carbon compounds, with ultimate conversion to two pyruvates. In this process, the two GAPs generate four ATPs, so the net product of glycolysis is two ATPs.

$$\text{Glucose} + 2ATP \rightarrow 2 \text{ pyruvate} + 4ATP + 2NADH + 2H_2O$$

The NADH (reduced form of nicotinamide adenine dinucleotide, an electron donor like NADPH, nicotinamide adenine dinucleotide phosphate) produced in glycolysis can enter the electron transport system of aerobic respiration where each will yield two ATPs.

Glycolysis is capable of producing pyruvate faster than the citric acid cycle can accept it. That is what happened when your arm started to hurt. The buildup of pyruvate reached a key level and because it could not enter the citric acid cycle, it entered the fermentation process and was converted to lactic acid (lactate). In animals, fermentation usually results in lactic acid, in plants it usually results in ethanol. Both products are toxic at high levels and should be metabolized quickly. The buildup of toxin is what caused your arm to hurt. Once the toxin had been partially metabolized and more oxygen was supplied to your muscles, you could lift the weight again. Glycolysis produces energy rapidly but it only produces two ATPs per six-carbon sugar. The glycolysis process can provide energy when oxygen is not present, such as in waterlogged soil, but it cannot sustain a plant for very long. If fermentation does not occur, the NADH produced during glycolysis is not consumed and will produce two ATPs for each NADH in aerobic respiration. If aerobic respiration cannot occur, the NADH is consumed by fermentation and glycolysis produces only two ATPs. The process of aerobic respiration, including the products of glycolysis, can convert a six-carbon sugar to 36 ATPs, which is 18 times more energy than glycolysis alone.

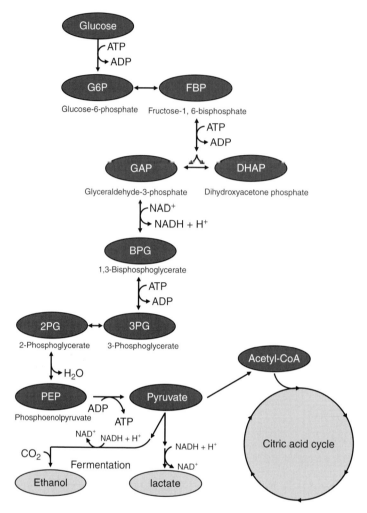

Fig. 4.6. Glycolysis is an anaerobic process that results in pyruvate, which can either be combined with carbon dioxide or converted to acetyl-CoA and enter the citric acid cycle for aerobic respiration. However, if the citric acid cycle is overloaded and cannot accept acetyl-CoA, the pyruvate degrades by fermentation to lactic acid or ethanol. Both products are toxic at high concentrations and must be metabolized quickly. ADP = adenosine diphosphate; ATP = adenosine triphosphate; NAD$^+$ = nicotinamide adenine dinucleotide; NADH = reduced form of NAD$^+$.

Many of the components of glycolysis are also precursors for other plant compounds (Table 4.1).

The pentose phosphate pathway

A second form of respiration is the pentose phosphate pathway (Fig. 4.7). As you examine Fig. 4.7, you probably recognize many of those compounds from earlier discussions. The pentose phosphate pathway is much like the Calvin cycle. The Calvin cycle, however, is in the business of sugar synthesis, whereas the pentose phosphate pathway is in the business of degrading sugars. A quick study of this pathway and you probably realize that these compounds could be shared with the Calvin cycle if the two processes occurred in the same place. Apparently, the Calvin cycle provides these important compounds for the chloroplast or bundle sheath cell. The pentose phosphate pathway provides them in the cytosol. The Calvin cycle uses these processes to synthesize sugars and the pentose phosphate pathway uses the same reactions in reverse to break sugars

Table 4.1. The products of glycolysis are used to synthesize other plant compounds.

Product	Fate
Triose phosphates (phosphoglycerate (PGA), etc.)	Glycerol (fats, oils, phospholipids), serine, cysteine
Phosphoenolpyruvate (PEP)	Phenolic amino acids, auxin
Pyruvate	Ethanol, lactic acid, alanine
Acetyl-CoA	Fatty acids, carotenoids, gibberellins, terpenes

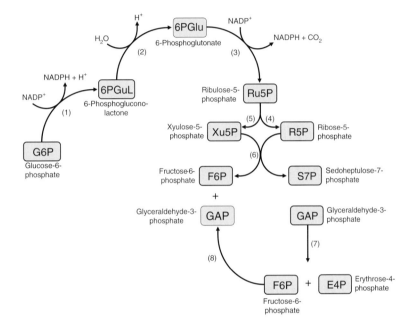

Fig. 4.7. The pentose phosphate pathway contains many of the same compounds that appear in the Calvin cycle. $NADP^+$ = nicotinamide adenine dinucleotide phosphate; NADPH = reduced form of $NADP^+$.

down. Many of these compounds are important for the synthesis of other products and for the functioning of other pathways (Table 4.2).

The GAP produced by the pentose phosphate pathway can enter glycolysis directly and potentially yield 19 ATPs after full aerobic respiration. Fructose-6-phosphate (F6P) can also enter glycolysis, potentially yielding 36 ATPs. NADPH can be oxidized in the mitochondria to produce ATP but it is also necessary as an electron donor in pathways that synthesize fatty acids and in other important processes. Erythrose-4-phosphate (E4P) is a necessary precursor for many phenolic compounds, and ribose-5-phosphate is a required component for the synthesis of RNA and DNA.

Table 4.2. The products of the pentose phosphate pathway are important components of other plant functions and precursors for other important plant products such as ATP (adenosine triphosphate), DNA and RNA.

Product	Fate
GAP (glyceraldehyde 3-phosphate)	Glycolysis to ATP
F6P (fructose-6-phosphate)	Glycolysis to ATP
NADPH (reduced form of nicotinamide adenine dinucleotide phosphate)	Electron donor reactions Mitochondria to ATP
E4P (eryththrose-4-phosphate)	Phenolic compounds
R5P (ribose-5-phosphate)	DNA, RNA

Aerobic respiration

Although plants have at least one other method for producing compounds that fuel aerobic respiration, glycolysis is the dominant pathway for that purpose (Plaxton, 1996). Aerobic respiration is a two-part process much like photosynthesis in reverse. The first process of aerobic respiration is the citric acid cycle, also called the Krebs cycle, and it is similar to the Calvin cycle in reverse. The citric acid cycle yields the electron donors NADH and ubiquinol (UQH_2). These donors fuel an electron transport system like the z scheme. As in the z-scheme the electrons traverse a series of reactions that moves hydrogen protons from one side of a membrane to the other, setting up an electrical gradient. The same type of ATP synthase spans the membrane and with the passage of protons combines ADP (adenine diphosphate) and P_i (phosphate) to form ATP. Curiously, the final electron acceptor is oxygen, and this is reduced to form water. Recall that in photosynthesis water provides electrons and oxygen is evolved. In respiration, oxygen is reduced to form water and carbon dioxide is evolved.

The citric acid cycle

The citric acid cycle or Krebs cycle initiates aerobic respiration. Although the citric acid cycle is considered part of aerobic respiration, oxygen is not required for this portion of the aerobic respiration process. Remember that in photosynthesis the electron transport system, the z-scheme, was used first to capture light energy. The captured energy was then used to power the Calvin cycle. The Calvin cycle stored whatever energy that it could recover in carbon bonds for use when it was needed to fuel metabolic processes. As the function of respiration is to release that energy into usable forms of cellular energy, primarily ATP, it would stand to reason that respiration would work backwards from photosynthesis. In respiration, the citric acid cycle occurs before electron transport, forming electron-rich NADH and UQH_2 for use in an electron transport system. When NADH and UQH_2 are oxidized they donate electrons to the respiratory electron pathway that occurs in the inner membrane of the mitochondria. The citric acid cycle also results in many components that are used in the synthesis of other compounds (Fig. 4.8).

Uses for components of aerobic respiration

Like the Calvin cycle, glycolysis and the pentose phosphate pathway, the citric acid cycle provides precursors for many important plant compounds (Fig. 4.9). Citric acid (citrate) for instance, is a precursor for cholesterol and fatty acids. Sometimes electron donor molecules like NADH from the citric acid cycle or NADPH from the pentose phosphate pathway are required for the synthesis of amino acids, fatty acids and other compounds necessary for plant life that originate from the citric acid cycle. Cellular energy, usually ATP from aerobic respiration or glycolysis is almost always required. Succinyl-CoA is a precursor for amino acids such as isoleucine, methionine and valine. It is also a precursor for odd-chain fatty acids and for porphyrins, structural components of chlorophyll and other pigments. Fumarate is a precursor for aspartic acid, phenylalanine and tyrosine, which you probably recognize as amino acids. Malate is a precursor for glucose. Oxaloacetate and α-ketoglutarate are precursors for more amino acids. A portion of the carbon fixed by photosynthesis provides the plant with energy through respiration. A large portion is also removed at various stages of the energy pathways to provide all of the structural and nonstructural components of the plant (Plaxton, 1996). Therefore, your grasses must take in considerably more carbon dioxide during photosynthesis than they give off in respiration.

The respiratory electron transport system

The final step in aerobic respiration is the electron transport system. The respiratory electron transport system is located within an inner membrane in the mitochondria. The process is very much like the z-scheme of photosynthesis (Fig. 4.10). Electrons are donated to the pathway by NADH from the citric acid cycle. UQH_2 from the citric acid cycle and NADH from glycolysis can also donate electrons. In the inner membrane of the mitochondria, the electrons pass through a series of oxidation–reduction reactions that transfer protons from the space between the inner and outer membranes (called the intermembrane space) to the space inside the inner membrane (called the matrix). An ATP synthase mechanism just like the one across the thylakoid membrane in the chloroplast converts ADP to ATP as protons move from the matrix to the intermembrane space as a result of the electrical gradient across the inner membrane.

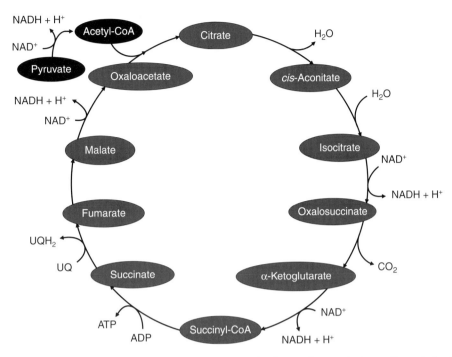

Fig. 4.8. The citric acid cycle provides electron donor molecules for the respiratory electron pathway in the inner membrane of the mitochondria. The cycle also produces components that are used in the synthesis of many other plant compounds. ADP = adenosine diphosphate; ATP = adenosine triphosphate; NAD$^+$ = nicotinamide adenine dinucleotide; NADH = reduced form of NAD$^+$; UQ = ubiquinone; UQH$_2$ = ubiquinol.

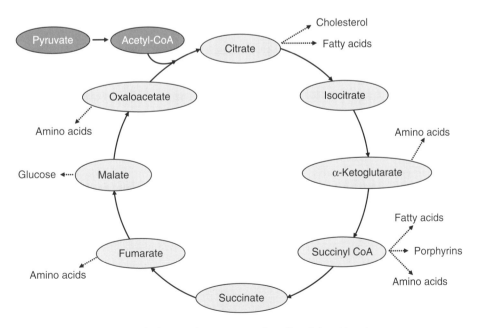

Fig. 4.9. Some of the plant compounds that require precursors from the citric acid cycle.

Respiration and Transpiration

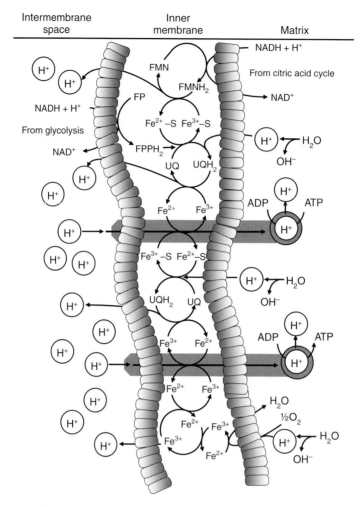

Fig. 4.10. The respiratory electron transport system in the mitochondria is much like the z-scheme of photosynthesis and synthesizes ATP in a similar manner. The oxidation of water to oxygen begins the z-scheme and the reduction of oxygen to water is the final step in respiration. ADP = adenosine diphosphate; ATP = adenosine triphosphate; FMN = flavin mononucleotide; $FMNH_2$ = reduced form of FMN (an electron donor); NAD^+ = nicotinamide adenine dinucleotide; NADH = reduced form of NAD^+; UQ = ubiquinone; UQH_2 = ubiquinol.

Oxygen must be present for the electron transport system to occur. If there is no receptor at the end of the electron transport pathway, electrons do not flow, and oxygen must be present to accept the electrons as they exit the pathway. Oxygen will accept two electrons and two hydrogen protons to form water. Remember that in photosynthesis, water was split to provide the electrons for the z-scheme. The split also resulted in two protons that increased the acidity of the space inside the thylakoid, helping to establish an electrical gradient across the thylakoid membrane. During the process, oxygen was released. You learned that this oxygen evolution can cause problems when the air around your C_3 grasses becomes high in oxygen and low in carbon dioxide, resulting in photorespiration. It is too bad that your plants cannot inject the oxygen from photosynthesis into the soil where they really need it.

Boundary layers high in oxygen and low in carbon dioxide form over turfgrass during periods of high photosynthesis and poor air movement. Air movement in soil is poor by definition, but it is adequate as long as the soil does not become compacted. We will talk more about compaction in later chapters. In short, however, compaction is the process of turning large soil pores into small soil pores through traffic or some other situation that causes pressure on the soil. Large pores drain water and are normally filled with air. Small pores hold water by capillary action and are usually filled with water. As the large pores are compacted to small pores, they stop draining and hold water. Where there is water there is no air. Consequently, there is no oxygen for root respiration. If a root cannot respire, it cannot provide energy for water uptake, which is an active process. So if the soil becomes highly compacted your turf has difficulty taking up water in spite of being surrounded by soil with plenty of water in it. If your turf has everything that it needs to perform photosynthesis, the next thing that you need to evaluate is respiration.

4.3 Transpiration

It has been estimated that 60% of the energy sequestered during photosynthesis is lost as heat during aerobic respiration (Salisbury and Ross, 1992). We can probably assume that at least 60%, probably more, of the light energy absorbed by chlorophyll is lost as heat before or during photosynthesis. Accordingly, plants get hot; plants get very hot. The temperature of a turfgrass canopy on a warm day with less than sufficient soil water uptake reaches well over 104°F (40°C), the temperature at which enzymes and proteins begin to denature (Steinke et al., 2009). Extended exposure at this temperature each day for several days will probably kill your grass. An extended temperature of 122°F (50°C) will kill cool-season grasses in 2–3 hours and warm season grasses in 10–12 hours (DiPaola and Beard, 1992). A temperature of 140°F (60°C) will kill grass within a few minutes. If transpiration stalls because soil water is not present, a grass plant must enter a state of semi-dormancy, reach deep into the soil for more water, or die.

Many botanists seem to feel that transpiration is of limited importance. In fact, it has been stated that transpirational cooling is not necessary and is a waste of water (Salisbury and Ross, 1992). If transpiration did not occur, plants presumably have other means of cooling themselves. If you have been tasked to keep C_3 turfgrasses green in temperatures greater than 90°F (32°C) or C_4 grasses green in temperatures over 105°F (41°C), you know how important it is to have transpirational cooling. Research demonstrates that grasses with high transpiration rates tolerate summer heat better than those with lower transpiration rates (Perdomo et al., 1996; Abraham et al., 2008).

The components of transpiration

To understand how transpiration works, it is necessary to revisit the concepts of the apoplast and symplast. Plant fluids exist in three different systems: the apoplast, the symplast and vacuoles (Canny, 1977). Vacuole fluids function to keep individual cells turgid and to help prevent collapse. They also have other important functions such as regulating ion concentrations and storing carbohydrates that we will discuss in later chapters. Vacuoles are separated from the cell protoplasm by a special membrane called the tonoplast. Each vacuole exists in only one cell and is not connected to any other cell. The symplast, living protoplasm, and apoplast, non-living fluid, are continuous systems. The protoplasm of a root cell is connected to the protoplasm of a leaf cell and to all other cells through the plasmodesmata (interconnecting cell ports) of all the cells in between them (Fig. 4.11). The phloem is part of the symplast, but the xylem, the system through which a plant transpires, is part of the apoplast. The symplast is separated from the apoplast by interconnecting cell membranes.

Water passes through the cell membrane readily, but most solutes are restricted. As we discussed earlier, some solutes can pass through cell membranes passively, without requiring plant energy, but most have to pass through active pumping mechanisms. These active pumping mechanisms require cellular energy, usually ATP. These ATPase pumps remove nutrients from the apoplast for biochemical activity in the symplast. Consequently, the symplast tends to be highly concentrated in solutes whereas the apoplast is relatively dilute. Water in the apoplast circulates throughout the plant crossing cell membranes passively to rehydrate the symplast when necessary. The water not used for metabolism is transpired.

The apoplast is a lot more complicated than just intercellular spaces. It includes cell walls, spaces between walls and spaces between membranes and walls. Cell walls are not impervious to water. In fact, the cell walls of a plant are always wet and their pores are filled with water. Consequently, the walls themselves are part of the apoplast. This system of cell walls and spaces is hydrated by the xylem.

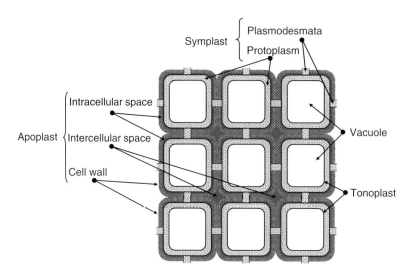

Fig. 4.11. Fluid in plants exists in three systems: a system of vacuoles, one per cell, that do not connect with each other; the symplast, living material; and the apoplast, non-living material. The three systems are interconnected. The symplast fluid inside the cell membranes interconnects from cell to cell through plasmodesmata. The fluid outside of the cell membranes includes the cell walls, the space between the membranes and walls, and the space between the walls, and is called the apoplast.

The xylem is the tubular structure, similar to the phloem that carries water quickly from the roots to the shoots. The xylem and phloem exist in the veins of the plant and are easily seen in turfgrass leaves. The veins extend throughout the plant, connecting all of the organs and tissues. Technically, a vein is called a vascular bundle and includes xylem, phloem, and bundle sheath cells and/or fibers (Fig. 4.12). In biology or plant biology it is common to study the vascular systems of trees, and considerable scientific study has also been done with vascular systems of trees. If you are familiar with the vascular system of a tree, you will notice from Fig. 4.12 that the vascular system of a grass plant is not quite the same. A tree has one large vascular bundle that forms its trunk. The shoot of a grass plant has many small bundles spaced in its sheath, similar to the spacing in the leaf. The sheath, in fact, is composed of the visible leaves wrapped around each other. The newest leaf is on the inside and the oldest leaf makes the outside layer. Consequently, the grass plant is not supported by a single stalk or stem but by multiple leaves wrapped around each other. Water does not pass up through the center of the shoot but all around it. Except for late summer (C_4) or fall (C_3), symplastic movement is downward and apoplastic movement is upward.

In grasses, as in deciduous trees, the xylem comprises systems of cells called the tracheids and vessels (Fig. 4.12). These cells have both primary (normal) and secondary walls for added strength. The secondary walls are thick and strong, but have small openings like valves to the primary wall so that passage of water and certain solutes can proceed normally to surrounding cells. Tracheids are long and narrow; vessels are short and fat. Both cell types are hollow. Once they are formed, the protoplasm is translocated to other cells, and the tracheid and vessel cells die, leaving only the walls intact. Like irrigation pipes, the small tracheids are more resistant to water flow than the larger vessels, so vessels move water much faster. Water moves up against gravity through the tracheids and vessels on principles similar to capillary action combined with the attractive forces of water potential.

The mechanisms of transpiration

Trying to understand how water moves through xylem can be confusing. Primarily because botanists do not fully agree on how water gets from the soil to the top of a 300-foot (96-m) tree. We really don't need to know how this occurs in order to manage turfgrass, but the same principles, osmosis, adhesion, cohesion and water potential apply to

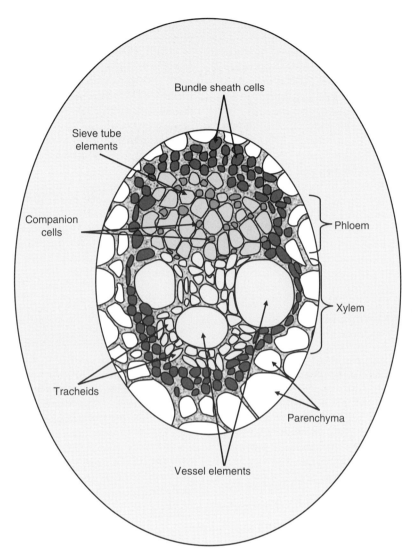

Fig. 4.12. A representation of a vascular bundle in a warm-season (C_4) grass plant. The sieve tube elements and companion cells of the phloem share the vascular bundle, also called a vein, with the tracheids and vessels of the xylem. The bundle is surrounded by bundle sheath cells where the Calvin cycle takes place in C_4 plants.

the theory of xylem transport as apply to many other processes that we must understand to become the best plant managers that we can be. Consequently, the principles of water uptake and xylem transport deserve attention.

The first thing that we need to do is realize that we are not dealing with a tall tree. We are managing a grass no more than a few inches tall. Scientists have spent considerable time trying to work out how water can get from tree roots 20 (6 m) feet below the soil surface to the leaves 300 feet above the surface. Although that is interesting, it does not concern us. There are principles that apply to grasses that cannot, according to physics, apply to tall trees. One such principle is a very important one called root pressure.

Root pressure is very similar to the osmotic potential of the phloem. Before water can move through the apoplast, it must first be absorbed from the soil. Root absorption is based on osmotic movement and water potential. If the water potential of the root is the same as the water potential of the soil surrounding it, root water uptake does not occur. The root must have a lower water potential than the soil around it. So the root lowers its water potential by pumping solutes into its cell membranes, thus encouraging water uptake by diffusion. Pumping these solutes across membranes is an active process and the reason that roots cannot take up water unless they can respire aerobically to produce ATP: no oxygen, no uptake. Once in the symplast, the water moves through the cells and passively diffuses into the apoplast, including the xylem. When the grass is transpiring, the water moves into the xylem because the xylem is relatively dry. Hence, its water potential is much lower than that of adjacent cells. At night, when the plant is not transpiring because the cells are full, water drawn into the roots has nowhere to go; not a very scientific statement but it is true, nonetheless. In this case, water is moving into the apoplast because of pressure from the excess water in the cells and we call this root pressure.

Root pressure can be substantial, but it is not great enough to move water more than several inches off the ground, it does not occur unless the soil is well watered and is rarely present during the day when stomates are open (Salisbury and Ross, 1992). Accordingly, root pressure is of no consequence in most plants. It can have very dire consequences for turfgrass, however. The dew that forms on grasses overnight does not all come from the atmosphere, some of it comes from the soil. We call it guttation fluid and it is a result of root pressure (Box 4.3) (Duell and Markus, 1977). The root pressure forces fluids from the leaves of grasses.

These fluids contain sugar, proteins and other solutes that microorganisms love to feed on. Most of these microorganisms are either neutral or beneficial, but some are pathogens. It is clear from multiple research studies that this guttation fluid, if not removed from the grass, increases the likelihood of disease occurrence (Couch, 1995; Ellram et al., 2007).

Root pressure is one principle that explains water movement in grasses, but it is not the most important. Capillary action is another possibility that presumably cannot occur in trees. Based on capillary theory and the size of tracheal tubes, capillary action can only account for water movement to approximately 1.5 feet (0.5 m) off the ground (Salisbury and Ross, 1992). That would be sufficient for grass, but capillary movement also requires that the upper end of the capillary tube be exposed to the atmosphere. Based on plant architecture and leaf hydraulics it does not seem likely that the apoplastic path is sufficiently clear to be considered open to the air in a capillary sense (Sack and Holbrook, 2006). However, the principles of adhesion and cohesion that are so important for capillary action to occur also occur in the xylem. In fact, as already mentioned, plant cell walls are porous and absorb water. So unless the plant is dehydrated and about to die, the cell walls are filled with water, thus improving adhesion. The water in the xylem is also dilute compared with the fluid in the cells, so the osmotic potential of the symplast also works to draw xylem water upward. Finally, the cohesion of water molecules is very strong and unless an air bubble occurs in a xylem vessel or tracheid, the

---

Box 4.3. What is a hydathode?

Grasses don't have to be mowed to exude guttation fluid. The guttation fluid is forced out through small pores at the tips of leaf veins called hydathodes (Stern, 1991). Although it may seem that stomates should occur at the tips of the veins, such is not the case. Stomates evaporate water from adjacent cell walls and other apoplast components. The xylem is not exposed directly to the air through stomates, which is why most botanists feel that xylem does not have sufficient exposure to the atmosphere for capillary action to be a driving force of transpiration. Hydathodes are very much like stomates, and may have evolved from stomates. However, their guard cells no longer open and close. Consequently, they appear to act as a pressure relief device. When pressure builds up in a vein, the sap forces its way through the hydathodes and onto the leaf surface.

We continue to teach students that guttation fluid from turfgrasses is exuded through hydathodes when, in fact, we mow most of them off. Each time we mow, we open the xylem and phloem to the atmosphere. These wounds close quickly just as they would on our skin, but they are weak and root pressure can force them open to exude guttation fluid. The leaves may regrow and form new hydathodes, but if we are mowing regularly, we continually cut them off.

xylem is always filled with a long column of water, each molecule bound to the other about as tightly as intermolecular bonding can occur.

The most common hypothesis of xylem water movement is called the cohesion–tension theory (Steudle, 2001). Based on hydraulic and physical principles, the cohesion–tension theory demonstrates an accumulation of forces sufficient to move water from the roots to the leaves of the tallest trees. Although it is widely accepted by botanists, critics of the theory are not uncommon (Zimmerman et al., 2000). The cohesion–tension theory includes the principles of adhesion and cohesion but is based in water potential. In short, water is pulled by the molecular forces of adhesion and cohesion but the most important molecular force of the two is the cohesive force of water. As long as water exists in a column of molecules bound to one another, the difference in water potential between the air and the roots is nearly always great enough to cause water movement through the xylem from the roots to the leaves. In the case of a grass, especially a mown grass, that is not very far. You can think of the column of water in the xylem as a rope held together by cohesive forces. The air applies tension to the rope pulling it upward through the xylem. If the rope does not break, the process continues as long as the air keeps pulling.

The regulation of transpiration

Now that you know what the apoplast is, what the symplast is, what the components of the apoplast and xylem are, and how water is transported through the xylem, it is time for the important part. How is water transport regulated? If we understand how transpiration is regulated and how environmental factors influence it, we can devise means to affect it.

The primary components of transpiration regulation are miraculous little holes in grass leaves called stomates. Stomates open to expose the apoplast to the atmosphere. Unless the air around our grass is at almost 100% relative humidity, the water potential of the air will always be less than the water potential of our plant's apoplast. Obviously, relative humidity is a very important environmental factor that affects transpiration. Many grasses cannot tolerate high humidity and most grasses are adversely affected by it. We can influence humidity to some extent by encouraging air movement, adjusting irrigation practices, reducing shade periods and through other minor management practices or management practices that have a small effect. We will discuss those practices more thoroughly in later chapters. Air movement and temperature have a definite influence on evaporation and anything that affects evaporation also affects transpiration. The faster the atmosphere evaporates water, the lower its water potential. The lower its water potential the faster the flow of water through the xylem provided that the stomates are open and the roots are providing sufficient water. With that statement, it should become clear that many factors, plant, atmospheric and soil factors included, affect transpiration. The only means that a plant uses to control its rate of transpiration is by opening and closing its stomates. Stomate control, of course, is not a conscious act, it is a response to external and internal plant conditions.

Stomates respond to light, specifically blue light (Zeiger and Hepler, 1977), to plant carbon dioxide concentration and to increases in abscisic acid (ABA). The opening and closing of stomates is controlled by guard cells on either side of the stomate (Fig. 4.13). When guard cells swell as a result of increased turgidity, the stomates close. When guard cells lose water and shrink, their stomates open. That is easy to remember. If the plant cannot get enough water and its cells begin to shrink, the guard cells get smaller and the stomates close so that more water is not lost to the atmosphere. So dry means less turgor and the stomates close; wet means more turgor and the stomates open. If you understand hydraulics,

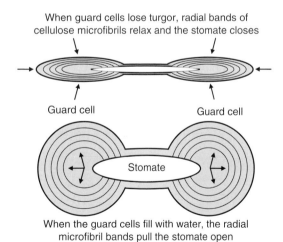

Fig. 4.13. From above, the stomates in grass plants appear like sets of barbells suspended on two bars. When the guard cells expand, bands of microtubules attached to the bars pull the bars apart, opening the stomate.

however, that procedure does not make sense. You would think that the guard cells would swell together closing the stomate, but that is not how it happens.

In most plants, guard cells are shaped like lips with the mouth serving as the stomate. The guard cells are wrapped with cellulose microfibrils that restrict swelling. Think of a long balloon wrapped with tape everywhere except on the ends. As you add air to the balloon, the tape restricts expansion and expansion is channeled to the ends. If two of these balloons are placed together longitudinally inside a cardboard box where their ends cannot expand sideways, the middle will separate as the ends expand pushing the centers of the balloons into half moon shapes in both directions leaving a hole in the middle. That is how the stomates of trees and many other plants work, but it is not how the guard cells of grasses work. In grasses, cytokinesis (cell division) stops before completion so the guard cells never fully separate. The incomplete division results in a bridge between the cells that surround the stomate (Fig. 4.13). The guard cells resemble a set of barbells with two bars instead of one. The stomate is in between the two bars. In this case, the microfibrils are wrapped around the barbells with ends attached to the bars. When the guard cells swell, the microfibrils pull the bars apart. When the guard cells shrink, the bars fall back together closing the stomate. The opening and closing of stomates is mediated by potassium ions (Zeiger and Hepler, 1977). When conditions are right for stomates to open, potassium pumps in the guard cell membranes pump potassium into the cell. The increase in solute lowers the osmotic potential of the cell and water moves through the membranes causing the cells to expand and the stomates to open.

Light causes stomates to open (Fig. 4.14). Stomates are normally closed during the night. Of course, they are hardly ever closed completely and there is almost always a small exchange between the apoplast and the air. Stomates open when guard cells sense light and not only cool the plant during photosynthesis but admit the carbon dioxide that is necessary for photosynthesis to occur. Consequently, during daylight there is a constant exchange of carbon dioxide and water between the plant and the atmosphere. Carbon dioxide concentrations in the plant also affect stomatal opening. When light levels are low, carbon dioxide concentration appears to dominate stomate response. In high light, however, the light level appears to dominate and carbon dioxide concentration can be high without affecting stomate status (Sharkey and Raschke, 1981). Stomates close when guard cell

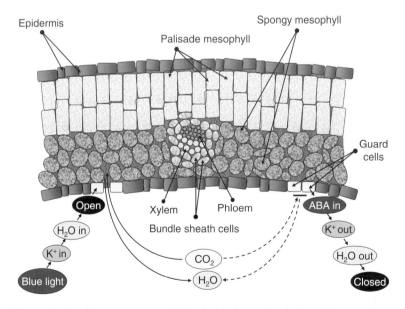

Fig. 4.14. When the guard cells of a stomate sense light, chemical reactions in the cell turn on the potassium pumps in the membranes and the cell water potential declines. As water diffuses into the guard cells, they swell and cause the stomates to open. A drought situation triggers a buildup of abscisic acid (ABA) in plant leaves that turns off the potassium pumps, allowing the guard cells to lose turgor and the stomates close to conserve water.

components sense an increase in ABA. This is a protective measure during drought stress, and it occurs regardless of light and carbon dioxide status. The ABA increases in response to drought stress in the leaf and triggers the closing of stomates (Wang *et al.*, 2004). As you will learn later in the text, heat stress is more severe when drought stress is also present (Jiang and Huang, 2000b).

4.4 Chapter Summary

Photosynthesis fixes carbon into chemical bonds providing a source of energy for use in building and maintaining a plant's physiological systems. However, photosynthesis, for the most part, only occurs in leaves. There has to be a system available in plants to transfer the photosynthetic energy to other plant parts in need. The energy is transported in units of sucrose through veins containing phloem. Phloem is a tube-like system comprising sieve tube elements and companion cells that carries sucrose to the plant parts needing it most. Plant parts that need more energy than they produce are called sinks. Plant parts like mature leaves that produce more energy than they use are called sources. So the sources produce the energy and then the energy is shipped to the sinks via the phloem.

The energy in the carbon bonds of sucrose and other plant compounds is released for use through a process called respiration. There are four types of respiration: (i) photorespiration; (ii) glycolysis; (iii) the pentose phosphate pathway; and (iv) aerobic respiration. Respiration not only produces energy in the form of ATP (adenosine triphosphate), or other energy molecules, it produces compounds that are used to synthesize other important plant components such as amino acids, lipids, RNA and DNA. Carbohydrate energy is used for all plant maintenance and growth. One of the most important processes requiring energy is root uptake of water and nutrients. Our grass leaves take in carbon dioxide and give off oxygen. Our grass roots take in oxygen and give off carbon dioxide as they take up water. If we expect our leaves to perform photosynthesis we must make carbon dioxide available. If we expect our roots to take up water, we must make oxygen available.

Our third important process is transpiration. Grasses have to have a means of cooling themselves. Photosynthesis traps the energy of the sun. Much of this energy is lost as heat before it is fixed in carbon bonds. As this energy is released for plant use, more than half is again lost, mostly as heat, during respiration. If our grasses are to survive this heat when temperatures are high, they must transpire. Not only must we provide water when necessary if we want our plants to stay green, they must also have a strong root system to take up the water they need. If we do not intend to keep our plants green during drought conditions, they still need a good root system to maintain root and crown hydration and remain alive until the rains occur.

Although photosynthesis is a very important process, it is only part of the very complicated chemical systems required to maintain plant life. In this text, we are also concerned with two other very important processes, respiration and transpiration. By learning how to encourage photosynthesis, respiration and transpiration, we also learn to maintain the other complicated systems that support these three. In the remaining chapters you will be asked to use the knowledge gained from these first four chapters to understand why management practices work and how to adjust them to your advantage. By learning the underlying factors rather than just the cosmetic symptoms that make your grasses respond the way they do, management decisions become relatively easy.

Suggested Reading

DiPaola, J.M. and Beard, J.B. (1992) Physiological effects of temperature stress. In: Waddington, D.V., Carrow, R.N. and Shearman, R.C. (eds) *Turfgrass*. ASA-CSSA-ASSA (American Society of Agronomy-Crop Science Society of America-Soil Science Society of America), Madison, Wisconsin, pp. 231–268.

Dudeck, A.E. and Peacock, C.H. (1992) Shade and turfgrass culture. In: Waddington, D.V., Carrow, R.N. and Shearman, R.C. (eds) *Turfgrass*. ASA-CSSA-ASSA (American Society of Agronomy-Crop Science Society of America-Soil Science Society of America), Madison, Wisconsin, pp. 269–284.

Hull, R.J. (1992) Energy relations and carbohydrate partitioning in turfgrass. In: Waddington, D.V., Carrow, R.N. and Shearman, R.C. (eds) *Turfgrass*. ASA-CSSA-ASSA (American Society of Agronomy-Crop Science Society of America-Soil Science Society of America), Madison, Wisconsin, pp. 175–206.

Suggested Websites

University of Hamburg (2009) Botany *Online* – The Internet Hypertextbook. Available at: http://www.biologie.uni-hamburg.de/b-online/e00/default.htm (accessed 8 May 2009).

5 Why Our Management Practices Affect Our Turf

> **Key Terms**
>
> **Xeriscapes** are landscapes designed to minimize or eliminate the need for irrigation.
> **Soil compaction** is the compression of soil pores, usually caused by traffic, resulting in less soil air and greater soil water-holding capacity.
> **Topdressing** is the act of applying a thin layer of soil material, usually sand, directly over the turf and encouraging it to penetrate the thatch (see below) by dragging a mat, heavy irrigation or some other process. Topdressing can be used to fill soil pores following core aerification, or to smooth the surface, but it is mainly used to encourage the microbial degradation of thatch.
> **Thatch** is a layer of dead and living turfgrass roots, stems and crowns that exists between the grass shoots and the soil.
> **Creeping grasses** are those that spread not only by tillers but also by stems called stolons and/or rhizomes.
> **Shoot priority** is a term used in this book to refer to the natural tendency of turfgrasses to favor shoot growth over root growth and stem growth.
> **Humus** is organic material that has been degraded over a long period to a semi-stable colloidal form.
> **Colloids** are very fine particles of one substance somewhat evenly distributed throughout another. As used in this book, clay particles and humus are colloids that help bind soil together to form structure.
> **Soil structure** is the aggregation of soil particles to form larger units with small pores that hold water. These larger units, called aggregates, have larger pores between each other that hold air.
> **Mat** is thatch mixed with soil and is located between the thatch and the soil.
> **Aerification** is a process used to help increase the air porosity of the soil. Core aerification, where circular probes penetrate the soil to remove soil cores, is the most common aerification practice in turfgrass management. However, there are several other protocols used as well. I prefer the term aerification rather than aeration so that the cultivation of soil by aerification is not confused with the process of aeration of water and other substances.

5.1 Understanding What We Know From a Turfgrass Perspective

You have spent considerable time during the first four chapters of this book learning about cellular mechanics and important plant physiological processes. You may or may not have understood the importance of these concepts but I hope that you learned them well because now you are going to begin applying them. There are several common turf management practices that are designed to either make turfgrass look better, make turfgrass more functional or keep turfgrass healthy. Many practices that make turf look good are detrimental to turfgrass health. We use the other practices to help relieve the stress of the detrimental practices and the stresses caused by the natural and artificial environments in which turfgrass is expected to persist. Highly managed grasses are not natural. They are biological entities growing in artificially managed environments. We like our turf to be weed free, dense and uniform in color and height. There are no monocultures and very little uniformity in nature. When we manage our grasses for human aesthetic value and functional sports characteristics, we have to fight nature to do it. If we are not persistent and precise, we lose.

Turgeon (2008) defines turfgrass as plants that form a more or less contiguous cover that persists under regular mowing and traffic. Notice that

Dr. Turgeon used two detrimental human practices, mowing and traffic, to describe turfgrass. Apparently turfgrass is a system of plants under continuous stress because of human intervention, and that is exactly right. According to Beard (1973) turfs have a number of functional characteristics. They reduce wind and water erosion, mud and dust. Turfs reduce glare, noise, air pollution, heat buildup and visual pollution problems (Beard, 1973). These functional characteristics, however, are not unique to the grasses that we mow. These are natural characteristics and they occur with any mature grass sward. Dr. Beard also describes turf as having recreational and ornamental value. The recreational and ornamental characteristics of turfgrass make it desirable to us for more than just functional characteristics and encourage its widespread use. Consequently, turfgrasses are used all over the world for ornamental and recreational purposes, while also providing functional advantages. In fact, the functional characteristics of turf are often forgotten in favor of other types of supposedly environmentally sound landscapes. Turfgrass managers have become so incredibly good at what they do that people have forgotten that grass does not really need our help to survive. In many locations, turfgrass is believed to require artificial irrigation and is, therefore, undesirable. That is only true in particularly arid regions for which only desert plants are adapted. In most situations, the plants that are recommended for landscapes because they require little irrigation require more water than turfgrass (Park *et al.*, 2005).

Grass needs our help to look and play as we want it to. It does not need our help to survive in areas for which it is adapted. We use certain grasses as turfgrasses for a reason. They are highly adapted to most of the regions in which they are grown and they are highly resistant to stress. If the proper turfgrass is chosen for a site, based on its adaptation, it will survive easily without our help, but there are some regions where turfgrasses are not adapted and will not grow without help.

Turfgrasses will not survive desert conditions unless we provide water. In that situation we should consider whether or not the functional characteristics of turf are sufficiently important to justify the use of water for irrigation. In most regions of the world, turfgrasses can exist in low maintenance, mow-only situations. Turfgrasses adapted to the region in which they are used will not need irrigation, fertilization or cultural management practices and will provide functional characteristics that other plants lack. Functional characteristics are further discussed in Box 5.1. The turf will not, however, look as good or play as good as we would like. As turfgrass managers, it is our responsibility to make turfgrasses look good and to provide adequate playing surfaces. They do not need our help to survive. They only need our help to survive the stress that we apply to them or to survive an environment for which they are not adapted (Fig. 5.1).

According to our adopted definition, turfgrasses have to be mown and have to survive traffic. Therefore, mowing is a management practice that has to be performed or we are not working with a turfgrass. Mowing by cattle, sheep, goats, rabbits or other animals doesn't count. Although animals were the original mowers, grasses mowed by domestic animals are now called forages and grasses mowed by wild animals are just called grasses. Were you aware that at one time, the Royal and Ancient Golf Club of St. Andrews in Scotland, believed to be the oldest surviving golf course, once calculated the number of cunnigers or rabbits (*Oryctolagus cuniculus*) required to keep the course at a particular mowing height (Beard and Beard, 2005)? It was estimated that 150 to 200 cunnigers were required to graze 2.5 acres (1 ha) of bentgrass (*Agrostis* spp.)/fescue (*Festuca* spp.) turf to 0.4 inches (10 mm). Thanks to Dr. and Mrs. Beard for that interesting tidbit and an interesting paper. The study of turf does not always have to be boring. Hopefully you can find the book that includes that paper at your library. See the list of references at the back of this book for more information.

By definition, we are also faced with traffic problems on turfgrass. Whether it is a lawn, park, golf course or athletic field, turf is meant to be walked on, sometimes driven on. If you live in an agricultural area, you know that it is considerably easier and safer to walk or drive on a grass field when it is wet than it is to walk or drive on a dirt road turned to mud. Turfgrass roots and sometimes rhizomes are extensive and extremely effective for holding soil together and providing traffic areas resistant to separation and to compaction. However, repeated traffic can overcome the root/rhizome system of turf, turning the grass field into the dirt road. The soil will separate, allowing a vehicle to sink when the soil is wet enough or the vehicle is heavy enough. Turf areas can only stand a finite amount of repeated vehicle or foot traffic until the

Box 5.1. Have people forgotten about the functional characteristics of turfgrasses and turf?

Although it may appear that I use turfgrasses and turf interchangeably they are not actually the same thing. Turfgrasses are plants. A turf is an ecological system that includes the plants, the dead material accumulated because of the plants and the soil that surrounds the turfgrass roots. Consequently, functional characteristics as well as recreational and ornamental characteristics are qualities of the turf, not the turfgrass. So why do people seem to have forgotten the functional characteristics of turf?

I have very definite opinions on this matter, but I would rather you form your own opinions and be able to support them with logical thought. Everyone knows how nice a manicured golf course, home lawn or commercial area looks under intensive turfgrass management. These areas are especially attractive if they include other ornamental plants designed to attract attention, with the manicured turf used as a background. Aesthetically it would be very difficult to improve on that system. People like it; you and I like it. In this scenario we not only are managing a good-looking area, we are perpetuating the functional characteristics of turf. However, there is a price for aesthetic value. The beauty is maintained through a system of labor, water, fertilizer and perhaps even pesticides. So I ask you: is this environmentally sound management?

Let's consider another potential system, a xeriscape: a system of drought-resistant, pest-resistant plants that requires no fertilizer growing naturally on sand or native soil. The only maintenance performed on this site is trimming and mechanical weed removal. Compare this with a low-maintenance turfgrass species that is mowed as needed to maintain a 3-inch (8-cm) mowing height and weeded on a similar interval to the xeriscape. Which is most aesthetically pleasing – the turf or the xeriscape (Fig. 5.1)? Which is most environmentally desirable? Which system has the greatest functional value? Is one system always the best alternative compared with the other? Is there some reason other than aesthetic value that a drought-tolerant turfgrass has to be irrigated? Which system would you recommend to a neighbor who does not want to irrigate, fertilize or apply pesticides? Can you support your conclusion with a logical argument? As a plant manager, these are questions that you should be comfortable with and able to discuss intelligently.

---

soil becomes compacted and the grasses thin or die. Consequently, we use cultivation techniques to help relieve soil compaction and we employ aggressive sod-forming grasses in areas where traffic is expected to be severe. We use other techniques, such as specialized mowing and topdressing to smooth playing surfaces and reduce thatch, a buildup of dead plant material unique to turf. All of our management practices affect turf either positively or negatively. The negative practices make the turf look better or make it available to us for recreation. The positive practices help us counteract the negative practices.

5.2 Mowing Causes Chronic Damage to Turf

Christians (2007) made a profound comment when he wrote "turfgrasses do not thrive on mowing – they tolerate it". Mowing is an injury. We mow our grasses according to the $\frac{1}{3}$ rule. We never like to remove more than $\frac{1}{3}$ of the canopy of our turfgrasses at any one time. For instance if our target mowing height is 2 inches we mow when the grass reaches a height of 3 inches thereby removing one third its height. The grass shoot is not all photosynthetic leaf area. It is a combination of sheath and leaves. The sheath is not a major source of photosynthates. So when we remove $\frac{1}{3}$ of the canopy measured by height from the soil surface, we are removing around $\frac{1}{2}$ of the primary photosynthetic leaf area. As photosynthesis is one of the three important plant processes that we try to encourage as managers, removing $\frac{1}{2}$ of the plant's photosynthetic capacity is highly detrimental and counterproductive to our goals. However, it is not a turfgrass if it is not mowed. Consequently, we have to mow and we have to help our turf overcome the detrimental affects of mowing. Let us consider just how detrimental mowing really is.

The negative aspects of mowing

When we mow, we not only remove a large portion of photosynthetic material, we also create an injury (Madison, 1962; Krans and Beard, 1985). This

Fig. 5.1. (a) A planned low-maintenance xeriscape, and (b) a low maintenance turf. If the turfgrass is naturally adapted to the region, it will require little if any maintenance and provide functional value that the xeriscape cannot provide. However, a beautiful xeriscape like this one also has value.

injury has to be healed. Luckily, plants, especially grasses, are very good at growing new parts. If a grass is healthy it can grow new leaves or fix damaged leaves. Even if the grass is scalped so low that all of the leaves are removed, it can grow new ones. The effort required of a previously healthy plant to grow a new leaf system may weaken it so much that it dies as a result of other damaging factors. However, it will most likely survive if growing conditions are conducive. Turfgrasses are survivors. They are not easy to kill and will normally withstand poor environmental conditions better than most ornamental plants. Imagine what would happen to most ornamental plants if you removed half their leaf area weekly. This could be a good student project; do you own any bushes that you would like to challenge?

Although grasses are quite capable of healing themselves, the frequency at which we mow exposes them to continuous, or at least recurring, chronic injury. A golf course putting green, for instance, that is double cut for a tournament has

Management Practices

lower photosynthetic efficiency than a green cut only once (Howieson and Christians, 2008). A portion of a turfgrass plant's energy is continuously required to heal its injuries from mowing. Mowing requires the removal of photosynthetic area. Mowing seriously affects photosynthesis, causing, among many other things, a serious decline in root length (Youngner and Nudge, 1976). The shorter the mowing height, the shorter the root system becomes. Imagine how much more energetic these grasses could be if we were not removing half their leaf area on a regular basis. In addition to affecting water uptake by reducing root length, the injury also creates further problems with another of the three primary physiological functions, transpiration.

Grasses control transpiration by adjusting their stomates to allow more or less transpiration to occur and more or less carbon dioxide to enter the leaf. When the leaf cells begin to dry out severely, a buildup of abscisic acid (ABA) occurs, closing stomates and conserving water. When we mow, we open the xylem and other apoplastic areas to evaporation. Substantial soil water is lost for a short time after mowing, and the closing of stomates cannot prevent it. If the soil is dry to begin with and the plant is near drought status, dehydration occurs rapidly and leaf tips dry beyond recovery (Fig. 5.2). Under those circumstances you will notice the leaf tips turn white then brown as their cells die over a period of hours or days. Shredding of the leaf tip can cause an identical condition. Dull mower blades tear rather than cut, shredding leaf blades and exposing more than twice the leaf area to desiccation as would occur with a clean sharp cut. These leaf blades desiccate and partially die even when sufficient water is available. Regardless, water use rate increases for a short time immediately following mowing, wasting soil water that could be used for normal cooling processes.

Mowing wounds allow microorganisms to enter leaf blades (Vargas, 1994). Nearly all of the economically important diseases of turfgrass are caused by fungi. Most fungi though are not pathogens. Most are either beneficial or neutral to turf, but some fungi cause nasty diseases in our grasses (Fig. 5.3). Pathogenic fungi are capable of degrading cellulose to enter grasses, but their entry is easier and faster through wounds. Consequently, turfgrasses with wounds from mowing, traffic and other injuries are more susceptible to disease. Turfgrasses whose

Fig. 5.2. Leaf tips lose water uncontrollably for a short time following mowing. If sufficient water is not available to replenish this loss quickly the cells near the injury dehydrate and die.

photosynthetic potential has been recently reduced by half and whose leaf tips are desiccated and in need of repair are increasingly susceptible to disease.

The positive aspects of mowing

Mowing disrupts two of our primary plant functions, photosynthesis and transpiration, but the mowing process is not entirely negative. Mowing has some beneficial aspects as well. These beneficial aspects are ornamental and recreational, not agronomic, and only occur because mowing is not as disruptive to photosynthesis as you might suspect. One of the principles of ecology that we will discuss later in the text has to do with maintaining

Fig. 5.3. This tall fescue (*Festuca arundinacea*) leaf is infected with brown patch disease caused by *Rhizoctonia solani*. Based on the spread of the symptoms, the pathogen probably entered through the leaf tip.

consistent plant biomass. Mature turfgrass plants strive to maintain their biomass at levels within a certain range. It helps to think that their physiology is designed to operate best when their shoot structure constitutes a particular mass. When we mow, the plant not only tries to heal the wound and grow new leaves, the mowing encourages new growth. It is believed that the mowing of the canopy allows more light to strike the lower leaves of grass plants, thus encouraging the formation of new tillers and, in the case of creeping grasses, new daughter plants (Duble, 1989). It is well known and accepted that mowing turfgrass lower within its range of adaptability results in greater shoot density (Juska and Hanson, 1961). If the grass plant cannot increase its biomass vertically, the biomass increases horizontally. So a very good way to increase the aesthetic value of your turf is to mow it near the low end of its range of adaptability, thereby increasing its density and ornamental characteristics. Increases in aesthetic value or playability, however, always seem to exact a price.

In this case, the price is root growth. The lower turf is mowed, the shorter the root system becomes (Liu and Huang, 2002b). By spreading, a grass can increase its photosynthetic potential. However, the photosynthetic potential is less than the potential that existed at its former higher height of cut. Although the photosynthetic potential increases as more shoot area is added, the transpiration potential is reduced by shorter root growth. Consequently, the plants are now more susceptible to drought stress and heat stress, and more precise irrigation management is required. Turfgrasses tend to exhibit shoot priority; this means that in most situations, shoot growth is favored at the expense of root growth. When excessive nitrogen is applied, for instance, an increase in shoot growth normally occurs, sometimes at the expense of root growth (Dunn *et al.*, 1995). When photosynthetic area is reduced, shoot area is partially maintained laterally, but root growth suffers (Tucker *et al.*, 2006). Shoot priority is a concept that you need to remember. It influences how grasses respond to stress and how they respond to your management practices.

Mowing has one more important advantage, this time an agronomic one. Mowing helps to control weed competition. If you are thinking ahead, you might want to argue that issue, as some of the weeds that compete with turfgrasses need to perceive light in order to germinate. The notorious summer annual, crabgrass (*Digitaria* spp.) and winter annual, annual bluegrass (*Poa annua*) are two of those weeds (although you may remember from Chapter 4 that annual bluegrass is a suitable turfgrass under some conditions). Opening the canopy allows more light to reach the soil surface and encourage weed seed germination. Some of that germination can be discouraged by proper irrigation practices, but nature has a way of providing those species with the conditions that they need at the time that they need them. Consequently, those weeds are likely to germinate.

Although weed pressure is severe and generally constant, we should not underestimate the aggressive nature of most turfgrasses. An immature weed is not much competition for a mature turfgrass in the act of spreading. Naturally some weed encroachment will occur when mowing height is lowered but, eventually, turfgrass density will increase and the weed population is likely to decline rather than increase. Unless herbicide application or time-consuming, back-breaking mechanical removal are practiced, 100% or even 80% weed control is unlikely. Nature does not like a monoculture. Nature likes diversity (Fig. 5.4). If we mow our grasses lower than our grass species can tolerate, weeds are likely to out-compete the turf (Voigt *et al.*, 2001). We are fortunate that most weeds cannot compete with turf at common mowing heights. So mowing alone provides a substantial amount of weed control.

Mowing management

Mowing affects photosynthesis by removing portions of leaves, the most important plant organ for photosynthesis. It also results in temporarily uncontrolled transpiration. By minimizing the detrimental effects of mowing on these two important plant functions you are doing most of what you can to help your turf survive and prosper under chronic mowing. If each time that you injure the leaves by mowing uncontrolled transpiration occurs and photosynthesis is reduced, how often should you mow?

(a)

(b)

Fig. 5.4. (a) A common bermudagrass (*Cynodon dactylon*) managed using irrigation, fertilization and herbicide weed control as well as regular mowing. (b) A common bermudagrass where the only management practice used is regular mowing. People like uniformity, nature likes diversity.

This is a good time to start thinking about such things and preparing yourself to start making decisions that solve problems. Decisions are made by considering alternatives. Management practices are often developed through trial and error, but not without forethought. You should begin to practice your skills in creating solutions to problems and in organizing plans now while you have an instructor or author to help you. All too soon, you will have to make those decisions and plans without help and with considerably greater risk. Practice your skills now while you can. Mistakes are part of learning. We learn more from our mistakes than we do from our successes.

As far as mowing frequency is concerned, you mow as infrequently as possible while maintaining the turf to your customer's expectations and abiding by the ⅓ rule (Johnson *et al.*, 1987; Soper *et al.*, 1988). That minimizes injury and water loss by prolonging the period between losses of photosynthetic material. If you thought about that, you probably came to a similar conclusion. If not, no problem, keep trying. What else can we do to reduce the negative effects of mowing? Think about it.

What else can we do to reduce the negative agronomic effects of mowing? We need to use the right equipment and the equipment needs to be in good working order and operated correctly (Box 5.2). Both operational training and safety training will be an important part of your job as a turfgrass manager. Most of you are going to have to stop being the worker and start being the manager, and that is a difficult transition. It is far easier to do the work yourself than it is to successfully encourage and train others to do it for you. You will not only be responsible for the work you do, you will be

---

Box 5.2. Know your mower.

We usually use one of three types of machines for mowing turfgrass, a reel mower, a rotary mower or a flail. Each has strengths and weaknesses that are clearly explained in introductory turfgrass texts.

Rotary mowers are the most common. They have single blades mounted horizontally on a spindle underneath the mower deck that cut as the blade rotates. All mowers have to be treated with respect to avoid accidents and potentially serious injuries, but the rotary mower is the most dangerous of the three. It is reasonably easy to slip or otherwise accidentally extend a foot under a rotary mower deck and it is common for debris to be thrown from under the deck as the blade hits it. Large rotary mowers have multiple blades mounted under large decks to mow wide patterns and finish large areas quickly.

It is interesting to note that the blade under a rotary mower deck is not sharp across its entire length. In fact, the very tip of the blade is all that needs to be sharp to cut grass with a rotary mower at slow speed. High speed mowers may need slightly more blade sharpened because the rotational speed of the blade may not be fast enough to cut with only the tip when the mower is moving forward. Under normal circumstances, the blade is rotating so fast and the forward speed of the mower is slow enough that only the tip of the mower blade touches unmown grass as the mower moves forward. For the same reason, you can adjust the mower deck one step lower in the front than in the back to drop the mowing height approximately half a normal step. You can do that because the cutting action occurs at the tip of the blade as the mower moves forward. If the front of the deck is lower, the grass is cut normally but if the back of the deck is lower, the grass is cut twice, once in the front at a higher height and once in the back at a lower height. Not only will the mower have to work harder to cut grass when the back of the deck is lower, the grass will be injured twice with each pass.

Many rotary mowers are mulching mowers with a blade design that keeps the clippings under the deck for multiple cuts. Therefore you need to sharpen more than just the tip of the blade but the initial cutting action still occurs at the tip.

You might also notice that when you are using a rotary mower with multiple blades under a wide deck and you are traveling too fast while you turn, the outside blade cuts more raggedly than the inside blade and the area looks uneven. For one reason, the physical forces of the turn cause pressure on the outside of the deck pushing it down lifting the inside of the deck up so the outside blade is cutting lower and the inside blade is cutting higher. In addition, the outside blade has to travel farther and cut more grass than the inside blade. Both blades have the same cutting (rotational) speed. Consequently, the outside blade is expected to cut more grass at the same rotational speed and may not be able to keep up when the forward speed of the mower is excessive. Therefore, in order to get an even finish, the appearance of the turns determines how fast you can drive.

responsible for the work of others. Take a quick word of advice from a person who was an industrial supervisor for many years before becoming a professor. Your people are important. They are the only means you have of completing your assigned responsibilities. The saying "take care of your people and your people will take care of you" is just as true today as it was when it was first spoken by someone many years ago.

Your operators should be trained on the safe and proper use of mowing equipment. That will eliminate a lot of agronomic problems. They should be trained to look for signs of mower wear and improper operation. They should be alert to conditions in the field that relate to mower performance and to overall turfgrass health. Each piece of equipment should have a checklist assigned to it (Fig. 5.5). The checklist should contain the pre-use, mid-use and post-use responsibilities of the operator, and should be quickly reviewed by the operator before each use and at the completion of the assignment.

Mowing equipment has to be kept in good repair and mower blades have to be kept sharp. Otherwise agronomic problems will occur. A cut in your finger caused by a knife heals faster with less trauma than a ragged cut made by a sharp stone. Turfgrass cut with a sharp blade heals more rapidly than turfgrass cut with a dull blade and causes the least disruption to photosynthesis and the least desiccation. A sharp cut covers less surface area than a ragged one, so fewer cells are affected by the injury (Fig. 5.6). Consequently, a sharp cut exposes less surface area to air, causing less desiccation and less browning of the leaf tip. Mowing when the soil is wet may cause scalping as the mower sinks into the turf. Mowing during extremely dry periods should be followed by irrigation if possible to reduce leaf tip desiccation. Remain within recommended mowing heights for the species being mowed. Refer to the textbooks suggested at the end of Chapter 1 for recommended mowing height by species. See the suggested reading at the end of this chapter for websites of major equipment manufacturers, where considerable equipment information can be found.

Remember that mowing is a stress. The lower you mow the more stress that occurs. Low-mown turfgrass requires more management and care than higher mown turf within the range of species adaptation (Toler *et al.*, 2007). Low-mown turf is more susceptible to desiccation because it has a shorter root system. It is more susceptible to disease and insect invasion because it has to be mowed more often, creating chronic wounding and because it is under more stress and is therefore more vulnerable. Low mowing requires that irrigation be provided and that irrigation be more frequent because root systems are weaker. For that same reason, low mown turf usually needs more nutrients and more frequent nutrient applications. Finally, low mown turf requires a good manager; a manager with agronomic knowledge and the ability to think through problems.

Another factor to consider in mowing management is the design of landscape features. This is a time and labor decision rather than an agronomic decision. This topic is discussed in Box. 5.3.

5.3 Thatch and Thatch Management

Although our definition of a turfgrass does not include thatch, thatch is unique to grasses and thatch control is unique to turfgrasses. Thatch is an important consideration for any turfgrass management plan. Many people seem to believe that thatch has no benefit when, in fact, it is very beneficial to soil conditions in a long-term biological system. The most fertile soils are those that have sustained grass for long periods of time, meaning hundreds or thousands of years. Turfgrass managers tend to have a negative concept of thatch because many of our turfgrass management practices lead to excessive thatch layers that have to be controlled. However, thatch is mostly a positive influence in the natural environment and also provides some benefits to managed turf.

Thatch is a combination of dead and living plant material deposited by grass between the grass shoots and the underlying soil. It consists of stems and roots and, rarely, leaves. As long as mowing frequency is practiced according to the $\frac{1}{3}$ rule, clippings degrade quickly and do not add to the thatch layer (Soper *et al.*, 1988). Thatch is actually dead material, but living material has to grow through or across the thatch to reach the soil and consequently becomes a part of it. For that reason, any definition of thatch should include living plant material. The type and amount of plant material growing in the thatch differs depending on how thick the thatch layer becomes. If the layer of dead plant material becomes thick enough it can house a substantial amount of roots, stolons and sometimes even crowns and rhizomes, and begins to cause problems with turfgrass health (Ledeboer and Skogley, 1967). In natural situations, this rarely

Equipment Instruction Sheet

Equipment Name: _____

Equipment Number: _____

Safety equipment required for operation of this machine:

- Hardhat
- Safety glasses
- Ear protection
- Pesticide personal protective gear

Pre-use procedures inside:

- Inspect for obvious damage, report to mechanic if damage observed
- Visually inspect tires for sufficient air
- Check the fuel level
- Check the oil level
- Check hydraulic fluid level if applicable
- Quickly inspect cables for signs of wear
- Quickly inspect hoses for signs of wear
- Sign out on equipment log

Pre-use procedures outside:

- Check where machine was parked for oil leaks or other problems
- Evaluate engine operation
- Does machine move forward and backward without any problem
- Check operation of blades, booms, buckets, bed lifts, as applicable

During operation:

- Operate safely according to your training on this machine
- Be alert for broken hoses, cables or equipment malfunctions no matter how minor
- Note any equipment problems both major and minor, return to shop if necessary

Post-use procedures:

- Quickly inspect cables for signs of wear
- Quickly inspect hoses for signs of wear
- Report major or minor malfunctions to mechanic
- Clean equipment according to training
- Park the machine in its assigned location
- Sign in on equipment log

Equipment Log

Equipment Name: _____

Equipment Number: _____

Date	Operator name	Out		In		Equipment status
		Time	Odometer miles/hours	Time	Odometer miles/hours	

Fig. 5.5. A sample checklist of requirements and operations that an operator should review and perform before and after using a piece of equipment.

(a)  (b)

Fig. 5.6. A shredded leaf tip exposes more surface area to air immediately following mowing than a cleanly cut leaf tip. The shredded leaf tip usually sustains severe damage. (a) This leaf tip was cut cleanly with a sharp rotary mower blade. (b) This leaf tip was cut using a rotary mower with a dull blade.

occurs, but managed turfgrass systems commonly require forms of thatch management to keep thatch layers thin enough to be beneficial rather than detrimental.

Thatch is naturally degraded by saprophytic microorganisms, so an environment that encourages microbial activity speeds the degradation of thatch. Theoretically, it should be possible to accelerate thatch degradation by inoculating turfgrass with microorganisms that feed on thatch. Research performed many years ago suggested that inoculation is a possibility for reducing thatch in turfgrass (Sartain and Volk, 1984). However, to date, a successful program or commercial product that can accelerate thatch degradation consistently has not been found (McCarty *et al.*, 2007).

Turfgrass species, even cultivars, differ in their propensity to produce thatch based on anatomical differences that affect the speed of microbial breakdown (Stiff and Powell, 1974). In addition, the more aggressive grasses are often the ones that thatch up most quickly (Shearman *et al.*, 1980). Consequently, species and cultivar selection for your region is a viable method for maintaining thatch at desirable levels. Management procedures that encourage aggressive turfgrass growth tend to result in rapid thatch accumulation. If the environment favors rapid plant growth and slow microbial activity, thatch increases rapidly. This all means that the faster your grass is growing, the more likely you are to have a thatch problem. The thickest, greenest grass is probably going to require

Box 5.3. Design landscape features to facilitate mowing.

By designing landscape features that are easy to mow around, you can save yourself a considerable amount of time and labor. Even in a situation as small as a single home lawn, reducing your mowing and trimming time by 10 minutes per event results in about a 5-hour time saving each year if you mow 30 times per year. Consider the examples below:

All that you have to do to save yourself considerable trimming time is to design your bed-to-lawn interface using your mower. Once you have determined the shape that you would like to install in the landscape feature, lay out the edges of the feature using paint or a garden hose then paint and mow along the edge, maintaining easy turning radii that roughly follow the pattern. The mower determines the finished edge and you will save substantial mowing and trimming time with every mowing event.

---

cultural management for thatch control. So let us review some of the lessons that you have probably learned about thatch in earlier classes, previous reading, or on-the-job-training.

A reasonably thin thatch layer of approximately 0.5 inches (13 mm) on most turfgrass areas, or about 0.3 inches (8 mm) on a golf course putting green, helps to cushion turfgrass crowns and roots from traffic damage (Fig. 5.7). Thatch heats and cools rapidly, but as long as plant roots and crowns are growing in the soil instead of the thatch, the thatch layer actually provides an extra layer of insulation that helps prevent the rapid heating and cooling that is stressful to biological systems. Thatch is hydrophobic (water repellent) when it is dry but it is also porous (Hurto *et al.*, 1980). As a result, a thin layer of thatch is not particularly detrimental to water infiltration and is unlikely to cause a significant increase in water runoff. Once thatch is partially degraded to humus, it becomes a somewhat stable part of the soil system and provides exceptional benefits.

The richest, most fertile soils appear dark brown, even black, as a result of high levels of

Fig. 5.7. The soil profile of a sand putting green. The thatch on this green is very thin, but the mat layer, the mixture of thatch and soil below it, is very thick. During summer, the roots of this grass probably do not extend below the mat. If you were responsible for managing this green what changes would you make to the previous management plan?

Management Practices

organic material that has been degraded to humus, a semi-stable complex of organic compounds in colloidal form. Soils high in organic material that was once thatch have high cation exchange capacities, meaning that they are capable of holding a lot of nutrients (Carrow et al., 2001). These soils also hold a considerable amount of plant-available water, yet are highly aggregated to provide larger air-filled pores (Brady, 1990). It is the thatch and other organic compounds in various levels of degradation that bind clay soil particles together to form aggregates. These aggregates are the basic units of soil structure. They contain small pores that hold plant-available water and are separated by larger pores that drain readily to provide soil air. The oxygen in this air is used by plant roots in the respiration process. Respiration provides energy for water uptake. This water provides the cooling medium for the transpiration mechanism and provides nutrients for plant growth and development.

Although a relatively thin thatch layer has many benefits, an unreasonably thick thatch layer, approximately 1 inch (25 mm), is highly detrimental (Fig. 5.7). That is why thatch generally is viewed as a negative influence by turfgrass managers. In truth, before its degradation, the thatch layer has not much to offer a turfgrass plant other than a little cushion from traffic and some insulating value for crowns and roots. The primary advantages of water and nutrient retention, along with soil structuring, do not take place until the thatch has been or is well into the process of degradation to humus. As degradation occurs, a layer of partially degraded thatch mixed with soil called mat forms below the thatch layer (Williams and McCarty, 2005). The mat layer has much higher nutrient and water retention properties than those that occur in thatch. Consequently turfgrass managers tend to view mat more positively than thatch. They might even call it beneficial.

In our business, mat layers are more likely to be man-made than natural. Natural forces such as earthworm activity, burrowing insects and water or air deposition result in the mixing of dust, soil or water-borne silt deposits with thatch, resulting in mat. However, turfgrass managers also cause mat formation by adding topdressing materials to turf. By spreading a thin layer of soil or sand that closely matches the soil at the site into the thatch, the manager can encourage the formation of mat from thatch. Mat has all of the advantages of thatch and more. It is more resilient than thatch so it provides more protection for turfgrass crowns and roots, and mat retains more water and nutrients than thatch. Because water is slower to evaporate from mat than from thatch and because mat is less water repellent, mat is likely to retain plant-available water. Because water is slower to heat and cool, mat has better insulating properties than thatch. Most importantly, degradation occurs more rapidly in mat than in thatch.

As mentioned earlier, management methods that increase microbial activity reduce thatch by encouraging degradation of mat to humus (Berndt, 2008). It stands to reason that management methods that encourage aggressive turfgrass growth would lead to more plant biomass, hence more dead biomass, hence more thatch. High applications of nitrogen fertilizer have been linked to increases in thatch (Soper et al., 1988). However, multiple research projects have demonstrated that nitrogen rate and nitrogen source do not affect thatch (Dunn et al., 1981; Carrow et al., 1987; Johnson et al., 1987; Hollingsworth et al., 2005). From these conflicting results, it appears that nitrogen can have an effect on thatch accumulation but, as usual, it is not the only factor that matters. Turfgrass management and turfgrass growth and development rarely have simple relationships. There are nearly always multiple factors involved. In this case, the acidity of the soil is also a factor that influences thatch development (Sartain, 1985). Soil that is too acidic or too basic can slow microbial activity. Soil and thatch that is too wet or too dry affects microbial respiration in the same way that it affects root respiration. Too little oxygen or too little water slows degradation. Temperature affects microbial activity as well. As you learned earlier, the speed of chemical reactions increases geometrically with increasing temperature. The increases in microbial activity occur at about the same rate as increases in chemical reaction rates up to about 100 °F (38 °C). Consequently, there are times when grasses are growing exceptionally well but microbial activity is slow, and times when microbial activity is rapid but turfgrass growth is slow. Cool-season grasses, for instance, grow best at about 65 °F–80 °F (18 °C–27 °C). At those temperatures microbial activity is relatively slow, but at 100 °F (38 °C) microbial activity is rapid and cool-season grass growth is slow or nonexistent.

Turfgrass managers often feel that they are practicing aerification and topdressing to prevent thatch from occurring. That is certainly not the case.

Aerification and topdressing are practiced to increase thatch degradation, not decrease its occurrence. In situations where turf is managed for high expectations, like golf greens, bowling greens or professional athletic fields, thatch is a part of the process. It can't be helped. If you are managing the turf for high aesthetic and functional value, including good health and rapid recovery from injury, thatching is going to happen. A certain amount of thatch will occur and thatch management will have to be a part of your management program. Turf that is poorly maintained or not growing well does not produce excessive thatch. Only turf that is managed for good health and rapid growth produces excessive thatch. If you are practicing quality thatch management, but the turf is still producing excessive thatch, it is time that you consider your fertilization, irrigation and other management practices to slow the production of thatch. However, under a high-quality maintenance program, thatch will be produced but controlled through practices that encourage rapid microbial degradation.

As you probably know, there are three practices that we use to manage thatch: dethatching, which is most often called vertical mowing, aerification and topdressing. Dethatching is a process of thatch removal but also has some influence on microbial activity. Aerification is a process that encourages microbial activity but also has some thatch removal characteristics, and it is used to relieve soil compaction and may also improve water infiltration (Murphy et al., 1993a; McCarty et al., 2007). Topdressing is primarily used to promote microbial activity, but it is also used to help relieve compaction following aerification, to smooth and level the turf/soil surface, or to modify soil characteristics. Dethatching is accomplished using either a vertical mower or a power rake (Fig. 5.8). The only real difference between these is that vertical mowers have rigid blades and power rake blades work on hinges. You might even see a dethatching device that uses vertical chains or something similar to dig out thatch. The purpose is the same regardless of the equipment used, and that is to remove existing thatch. In doing so, the mowing device has to cut into the top of the soil and consequently may positively affect aeration and water infiltration. Vertical mowing, as this process is most commonly called, regardless of the equipment used, was once the most commonly recommended process for thatch control. Deep vertical mowing for thatch removal, however, is extremely destructive. The

Fig. 5.8. Vertical mowers have rigid blades something like saw blades. This power rake has blades mounted on hinges so that it can cut deeper with less chance of injury to the operator or the equipment if a blade hits a rock or a similar hard object.

destructive nature of vertical mowing for thatch removal is too destructive to be performed frequently (Hollingsworth et al., 2005).

Research suggests that core aerification may be more or equally effective for thatch management as vertical mowing and it is usually less destructive (Murray and Juska, 1977). Core aerification improves the environment for microbial growth and reproduction in the same manner that it improves the environment for root growth. Aerification allows the carbon dioxide produced by plants and microbes during respiration to escape the soil and encourages the entry of fresh air containing oxygen. Consequently, thatch and mat degradation improve. When core aerification is used on a golf course putting green or other highly managed area the soil cores are usually removed and the holes in the soil are usually filled with topdressing material that is applied in an amount slightly greater than that required to fill the holes. This topdressing procedure effectively mixes topdressing material with thatch, forming mat. Topdressing appears to be the best of the thatch management techniques (Carrow et al., 1987; Johnson et al., 1988), and four applications per year are better than one (White and Dickens, 1984). In home lawns, parks and other medium-maintenance turf, cores are usually left on-site and allowed to reincorporate naturally. Leaving the cores to incorporate naturally is a slightly less effective form of topdressing. Topdressing is believed to provide an environment more conducive for microbial thatch degradation than a typical thatch layer. Although topdressing alone is a fairly effective form of thatch control, the combination of light vertical mowing, aerification and topdressing has proved most effective for managing thatch on golf course putting greens (McCarty et al., 2007).

5.4 Issues Pertaining to Soil Compaction

Turfgrass and turfgrass soils usually have to sustain moderate-to-intense traffic. Turf most certainly has to sustain a certain amount of mowing traffic, but it is usually also required to sustain considerable foot traffic and sometimes vehicle traffic. This traffic results in soil compaction, a compression of the soil that destroys its structure and compresses large air-containing pores, converting them to smaller water-containing pores. Compaction has a detrimental effect on root growth by reducing root respiration. As respiration declines, transpiration is affected. A compacted soil is also dense, making it difficult for roots to grow through it. The concept of constructing playing fields of specialized sand systems resulted from the poor turf performance and poor drainage of compacted soils. Pure sand has no structure and, unless it is extremely fine, is unlikely to compact to the point that turfgrass growth is affected.

Aerification is a term that refers to the multiple processes used to help relieve soil compaction. Aerification includes coring, spiking, slicing, water injection and other procedures that are used to fracture the soil, causing large channels that contain air and drain water. Most soil compaction on turfgrass systems is surface compaction that usually does not extend deeper than 2–3 inches (6–8 cm). For that reason, the most common aerification practices are designed to primarily affect shallow soil layers. Aerification procedures lower soil bulk density, improve subsoil drainage and promote air circulation, which admits oxygen and vents carbon dioxide. Aerification is also effective for disrupting soil layers that interfere with water percolation and air exchange.

As soil is compressed and its pores become smaller, the number of soil pores that contain water increase and the number that contain air decrease. Small soil pores hold capillary water, meaning that they bind water against gravity. You learned about capillary action in Chapter 1. Larger pores are needed to drain the soil and it is the large pores that hold the oxygen for root respiration. With fewer and smaller pores holding air, there is limited soil air exchange with the atmosphere, resulting in the same effect on root respiration that a thick boundary layer resulting from poor air movement causes for leaf photosynthesis. As roots and soil microbes respire, oxygen in the soil air is depleted and the carbon dioxide concentration increases. High carbon dioxide concentrations in soil air can cause reductions in root length and mass (Bunnell et al., 2002). As normal air is about 21% oxygen, it is not likely that oxygen concentration would become low enough in the soil air to affect respiration. However, compaction results in more capillary water and considerably less soil air. Although the concentration of oxygen in the air probably remains sufficient, the availability of air in general could become a limiting factor for root and microbial respiration. That would not only affect root growth, it would affect also microbial activity, resulting in less thatch degradation. Multiple research studies have indicated that turfgrass water-use rates decline in compacted soil (O'Neil and Carrow, 1982, 1983; Agnew and Carrow, 1985).

5.5 Managing Wear Caused by Traffic

Traffic on turfgrass systems not only causes compaction, it also causes damage due to wear. Research on creeping bentgrass (*Agrostis stolonifera*) and velvet bentgrass (*Agrostis canina*) suggested that wear from traffic on those grasses caused more damage than soil compaction (Samaranayake et al., 2008). However, this research was performed on sandy loam soil. Had it been done on clay loam, the compaction from traffic may have been the major cause of decline. The amount of damage that occurs from wear and compaction is not only influenced by the amount of traffic at the site, it is also affected by environmental conditions that determine plant and soil status at the site. Saturated soil, for instance, does not hold up well to traffic because the soil is easily shifted, causing roots and stems to fracture or break. Plant succulence increases with high fertilization and in shade. Succulent plants are easily damaged because water does not compress and is likely to push out through plant tissue when the tissue is compressed. Although wear tolerance is affected by the major plant processes, photosynthesis, respiration and transpiration, species selection and simple traffic management are also important considerations for managing turf under traffic.

Research has suggested that Kentucky bluegrasses (*Poa pratensis*) that have more vertical shoot orientation, are less succulent and have stronger cell walls are more tolerant to wear (Brosnan *et al.*, 2005). Although this work indicates that cultivars differ in wear tolerance, it is equally useful for selecting management practices to help relieve traffic stress. Alternating mowing patterns to encourage upright growth, and applying only enough nitrogen fertilizer and water to encourage recovery is good management to reduce traffic damage. It is important that the grass chosen for the site is well adapted for the environment. A poorly adapted grass will not survive traffic (Dunn *et al.*, 1994). Both wear tolerance and recuperative ability are important considerations when choosing a species for a site where intense traffic is likely. However, it would be unusual to find a turfgrass species that is both highly wear tolerant and rapidly recovers from damage. Cool-season grasses with reasonably good wear tolerance are perennial ryegrass (*Lolium perenne*), tall fescue (*Festuca arundinacea*) and Kentucky bluegrass (Shearman and Beard 1975a,b; Minner and Valverde, 2005a). Of the three, Kentucky bluegrass recovers most rapidly so generally holds up best in high traffic situations. Among the warm-season grasses, zoysiagrass (*Zoysia* spp.) has outstanding wear tolerance, but most zoysiagrasses tend to be slow to recover from damage. Bermudagrass (*Cynodon dactylon*) is usually the most effective warm-season grass for high traffic areas (Duble, 1989).

Traffic on turfgrass may cause reductions in leaf water content, turf density and leaf chlorophyll content (Han *et al.*, 2008). These responses suggest declines in photosynthesis, transpiration and possibly root respiration. High shoot density, a product of good photosynthesis, transpiration and respiration management has been linked to wear tolerance (Trenholm *et al.*, 1999). Obviously, management practices that encourage photosynthesis, transpiration and respiration will affect the ability of a turf to recover from traffic damage, so managing to encourage these practices should improve performance. Turfgrasses should also be allowed sufficient time to recover between damaging events. A traffic simulation study indicated that intense traffic applied to Kentucky bluegrass weekly with 7 days recovery time between damaging events caused less turfgrass decline than the same amount of weekly traffic spread out over 3 days per week with only a day or two of recovery time between events (Minner and Valverde, 2005b). Managing traffic patterns to permit the longest period between damaging events is probably the most important maintenance practice for improving turfgrass quality under traffic.

5.6 Chapter Summary

Over the years, turfgrass researchers and managers have developed cultural practices that help turfgrasses to survive and flourish in spite of chronic stress. Species selected for turfgrasses are those that tolerate close mowing and can survive soil compaction and wear. Our expectations for these grasses are high. The grasses are highly adapted for most of the environments in which we use them, but they seldom meet our expectations unless we provide management. The development of cultural practices and other management techniques have made it possible for us to grow grasses that meet our aesthetic and functional expectations even in situations for which they are poorly adapted. The turfgrass must be healthy to tolerate regular mowing and it must be healthy to resist and recover from traffic. Proper turfgrass selection and cultural management practices make it possible for turf to survive our expectations. We aerify, topdress and vertical mow to improve water infiltration, which encourages transpiration, and we apply soil aerification, which encourages root respiration. We adjust mowing height and frequency to allow for adequate photosynthesis. In the future, it is likely that we will have to further adjust our practices to conserve more water and to further reduce fertilizer and pesticide applications. Techniques we develop for meeting our aesthetic and functional expectations, while reducing water, fertilizer and pesticide use, will only be effective if they encourage photosynthesis, respiration and transpiration.

Suggested Reading

Beard, J.B. and the United States Golf Association (2002) *Turf Management for Golf Courses*. Ann Arbor Press, Chelsea, Michigan.

Carrow, R.N. and Petrovic, A.M. (1992) Effects of traffic on turfgrass. In: Waddington, D.V., Carrow, R.N. and Shearman, R.C. (eds) *Turfgrass*. ASA-CSSA-ASSA (American Society of Agronomy-Crop Science Society of America-Soil Science Society of America), Madison, Wisconsin, pp. 285–330.

McCarty, L.B. (2001) *Best Golf Course Management Practices*. Prentice-Hall, Upper Saddle River, New Jersey.

McCarty, L.B. and Miller, G.L. (2002) *Managing Bermudagrass Turf: Selection, Construction, Cultural Practices and Pest Management Strategies*. Ann Arbor Press, Chelsea, Michigan.

Puhalla, J., Krans, J. and Goatley, M. (1999) *Sports Fields: A Manual for Design, Construction and Maintenance*. Ann Arbor Press, Chelsea, Michigan.

Rieke, P.E. and Murphy, J.A. (1989) Advances in turf cultivation. *International Turfgrass Society Research Journal* 8, 49–54.

Waddington, D.V. (1992) Soils, soil mixtures, and soil amendments. In: Waddington, D.V., Carrow, R.N. and Shearman, R.C. (eds) *Turfgrass*. ASA-CSSA-ASSA (American Society of Agronomy-Crop Science Society of America-Soil Science Society of America), Madison, Wisconsin, pp. 331–384.

White, C.B. (2000) *Turf Managers' Handbook for Golf Course Construction, Renovation and Grow-In*. Ann Arbor Press, Chelsea, Michigan.

Witteveen, G. and Bavier, M. (1998) *Practical Golf Course Maintenance: The Magic of Greenkeeping*. Ann Arbor Press, Chelsea, Michigan.

Suggested Websites

Some major mower manufacturers

Gravely (2009) Mowers and some other turf management equipment. Available at: www.gravely.com/ (accessed 5 June 2009).

Husqvarna (2009) Mowers and most other turf management equipment. Available at: www.husqvarna.com/ (accessed 5 June 2009).

Jacobsen (2009) Mowers and most other turf management equipment. Available at: www.jacobsen.com/ (accessed 5 June 2009).

John Deere (2009) Turfgrass, crop production, construction, irrigation and other equipment (includes video instruction). Available at: www.deere.com/ (accessed 5 June 2009).

Toro (2009) Mowers, irrigation and most other turf management equipment (includes video instruction). Available at: www.toro.com/ (accessed 5 June 2009).

Wikco.com (2009) Mowers, including flail mowers and most other turf management equipment (includes video instruction). Available at: www.wikco.com/ (accessed 5 June 2009)

6 The Importance of Light and Managing Shade

> **Key Terms**
>
> **Irradiance** is the combination of direct, diffuse and reflected sunlight (solar radiance) striking an object at any particular time.
> **Direct solar radiance** is the portion of sunlight striking an object that arrived on a straight line from the sun. Direct sunlight constitutes the largest portion of irradiance.
> **Diffuse solar radiance** is the portion of sunlight that has been scattered or diffused by atmospheric particles. Blue light suffers the most diffusion, hence the sky is blue.
> **Reflected solar radiance** is the sun's energy that bounces from one object to another without being absorbed. When we look at an object, it is the light reflected from that object that we see.
> An **aerosol** is a suspension of fine solid or liquid particles in a gas. Aerosols are the cause of atmospheric diffusion.
> **Light quality** is not the same as light quantity. Light quality is the spectral distribution of light striking an object. It could be a very small quantity or a very large quantity but both quantities, large and small, could have the same spectral distribution. In this chapter we are concerned about the proportions of blue, green, red and far red that make up the irradiance that penetrates shade.
> **Light quantity** is a measure of the amount of light that strikes an object. We are mostly interested in the amount of light that strikes an area of turf, especially in the shade.
> **Reflectance** is the portion of irradiance striking an object that is reflected from the object. The remainder of the irradiance striking the object is either absorbed or transmitted. In our case, reflectance is the amount of irradiance reflected from a turfgrass canopy divided by the amount of irradiance that is striking the canopy. Because reflectance, absorbance and transmittance are all fractions of the total amount of irradiance available, they are often expressed as a percentage of irradiance or a percentage of full sun.
> **Absorbance** is the portion of irradiance striking an object that is absorbed by the object.
> **Transmittance** is the portion of irradiance striking an object that passes through the object.
> **Morphological** plant characteristics are those characteristics such as size and color that we can see. Some of the most important turfgrass morphological characteristics are color, texture, density or cover, and uniformity. Density and cover are not characteristics of turfgrass plants but they are characteristics of turf.
> **Physiological** characteristics are characteristics that we can not see. It is the physiological processes that determine morphological characteristics. You already know the most important physiological processes. Those would be photosynthesis, respiration and transpiration.

6.1 The Practice of Making Adjustments to Improve Photosynthesis in Shade

Light provides biological energy. Without light there is no life as we know it. Plants must have sufficient light to provide the energy that fuels metabolism. Some plants grow proficiently in low light but most do not. Although some turfgrasses tolerate low light better than others, it is a rare environmental situation when a particular turfgrass grows better in shade than it does in full sun. Under nearly all circumstances, turfgrasses prefer full sun to shade and often will not grow satisfactorily in shade. In Chapter 2 you learned the basics of light absorption and conversion to chemical energy. In this chapter we will expand on that knowledge and help you understand how to use what you have learned.

For the purpose of plant study and management, the solar spectrum is divided into six categories called bands, ultraviolet, blue, green, red, near infrared (NIR) and infrared. The ultraviolet band is high-energy light that can be destructive to plants and to us. Ultraviolet radiation causes genetic mutations and tissue destruction but there is little that we can do to affect that, at least on our plants. Actually, some of our top-performing grasses are the result of natural or artificial genetic mutation caused by radiation. Otherwise the ultraviolet band is of little concern to us and we need not discuss it further.

The infrared band has some potential for nondestructive assessment of tissue and/or soil water in turf but, as yet, has not been fully developed for that purpose. As a source of energy it is important to us, but not in a positive manner. Infrared radiation degrades rapidly to heat. Much of the heat that builds in turfgrass leaves on bright summer days is a result of infrared radiation. Fortunately, most of the infrared radiation that strikes a green plant is reflected and, therefore, is inconsequential. That infrared radiation which is absorbed quickly degrades to heat and is dissipated mostly through active transpiration, and partially by conduction and convection.

The light bands blue, green and red are important spectral bands for us to consider. The NIR, which, for the most part, is not considered light because we can't see it, is also an important band. The NIR is generally considered to extend from a wavelength of 700 nm to a wavelength of 1350 nm (see Chapter 2 to refresh your memory concerning wavelengths). However, there is a narrower portion of the NIR from 700 nm to 800 nm that is more useful for plant study. We call the band from 700 nm to 800 nm the far red band and we can refer to it as light. Depending on the authority you would like to believe, we can see wavelengths of up to either 750 nm or 780 nm (Starr, 2000). Consequently, we can see half or more of the far red band, so we can loosely refer to it as light. The portion of the far red that we can see appears to us as dark to very dark red. It is the blue, green, red and far red bands that directly affect our turf. The blue, green and red bands provide energy through photosynthesis, and the red and far red bands are absorbed by phytochrome, the pigment that helps to release or control many plant functions (Grant, 1997; Nagy and Schäfer, 2002).

6.2 The Influence of Irradiance

Irradiance is the sum of direct, diffuse and reflected solar radiance (sunlight) striking a particular object at any one time. Those terms may be new to you and require explanation. If you were to look directly at the sun, not a good idea, your eyes would be absorbing direct radiation. Direct radiation is very intense. However, it does not exist outside a fairly straight line between an object, in this case you, and the sun. If you were to stand directly behind a tall building when the sun was on the other side of it, there would be no direct radiation in that area. All of the direct radiation would be blocked by the building. However, you would still be able to see. Why can you still see?

The reason that you can see when you are on the side of a building opposite the sun is because there are two other kinds of irradiance at work. When you look at the sky on a clear day, it looks blue. It looks blue because there are aerosols – suspended fine particles or molecules – in the air that scatter blue light (Gates, 1966). A portion of the blue light that is projected directly from the sun is detoured by diffusion throughout the atmosphere (Brine and Iqbal, 1983). Consequently, the atmosphere looks blue. Diffuse blue light strikes the earth from all the directions of the sky's hemisphere (Smith, 1982). It can strike an object in the open from any direction. If you are standing on a lawn in deep shade behind the building that we mentioned, the turfgrass that you are standing on is absorbing blue light from any direction in which you can see sky (Fig. 6.1). Diffuse blue light is one of many reasons that some turfgrasses can grow reasonably well in shade. Reflected light is another, and this is the main reason that you can see on the shaded side of a building.

White or shiny objects reflect nearly all of the light that strikes them. We have trouble looking at those objects in bright sunlight because they hurt our eyes. Direct sunlight reflected from a pure white object is almost as intense as direct sunlight itself. When we look at such an object, all that we see is blinding white light. Such an object is very rare. We usually see off-white objects and multiple colors of objects. Pure black objects, another rare occurrence, reflect no light at all. We see them as a hole in a field of colors. In other words, we don't really see pure black objects, we see where they are based on the position of surrounding and background objects that reflect light. We see most objects because, other than pure black ones, they all reflect light. The types or wavelengths of light that objects reflect determine what color they are to us. Turfgrass is green. Turfgrass is green to us

Fig. 6.1. Only a small portion of the direct rays of the sun can pass through a tree canopy. Consequently, there is shade on the side of this tree opposite the sun. However, notice that there is an entire hemisphere of sky surrounding this tree and consider that diffuse blue light radiates from all points in the hemisphere. That diffuse blue light penetrates the shade and helps to stimulate plant growth.

because it reflects more green light than any other color that we can see and because our eyes are more sensitive to green than to any other color. As you learned in Chapter 2, red and blue light are highly absorbed by chlorophyll. Some of the blue and red light striking turfgrass is also reflected, but usually only a small portion. Turfgrass reflects a much larger portion of the green light that strikes it than the red or blue light, hence the plants appear green. The high concentration of chlorophyll in turfgrass leaves masks the color of other pigments. The xanthophylls make up the second largest concentration of pigments in turfgrass leaves (Turgeon and Lester, 1976; McElroy *et al.*, 2006). Consequently, when turf experiences stress and chlorophyll begins to degrade, the plants turn yellow. Yellow is the color of the xanthophyll pigments. When our turf begins to yellow, we take notice. We know that when we begin to see the color of the xanthophylls, the plant's chlorophyll is degrading and its photosynthesis is slowing. Something needs to be fixed.

The greenness of turf has long been the factor most commonly used to assess its health (Box 6.1).

Turfgrass color has always been and still is a valuable indicator of plant health. Experienced professionals know when their grass is too green and when it is not green enough. In addition to its use as a discriminatory tool, reflected green and other colors of reflected light are sources of light for photosynthesis. Reflected light is capable of penetrating shade because, like diffuse light, reflection can occur from any direction. Shade occurs when the direct rays of the sun are intercepted by an object. However, direct irradiance is not the only light available to turfgrass. In shade there may still be sufficient diffuse and reflected light to maintain a small amount of photosynthesis. For some plants diffuse and reflected light is sufficient for growth. Not so for turf: turfgrasses, even shade-tolerant ones, require daily periods of full sunlight to maintain growth. You will not find turfgrasses growing directly under large trees or in the north inside section of U-shaped buildings. There is not enough light.

The amount of irradiance available in shade

Three things can happen to solar radiation when it strikes a plant leaf or any other object. The irradiance may be reflected, it may be absorbed, or it may be transmitted. The reflectance of an object (in this case a turfgrass canopy) is the portion of irradiance striking it that is reflected from it; the remainder is either absorbed or transmitted. The absorbance is the portion of irradiance striking the object that is absorbed by it, and the transmittance is the portion of irradiance striking the object that passes through it. Very little light is transmitted by a plant canopy such as that of a tree (Fig. 6.2). Although substantially more far red is transmitted by a vegetation canopy than are the other light bands, the proportion of the far red transmitted compared with the portion that strikes the canopy is very small (Bell *et al.*, 2000). Very little of the blue, green and red bands, and little of the far red band, penetrate right through a plant canopy. No light penetrates right through a building or other substantial structure. Consequently, shade from buildings, and for the most part shade from vegetation, are substantially devoid of direct sunlight. Turfgrass growing under shade conditions must survive on diffuse and reflected light until a time during the day when it can receive direct sunlight. If the period of direct sunlight is long enough, the turf will survive. If the period of sunlight is longer

Importance of Light and Managing Shade

> **Box 6.1. The role of green light in photosynthesis.**
>
> We generally refer to green light as light occurring in a band from a wavelength of 500 nm to a wavelength of 600 nm. That band is part of what we call photosynthetically active radiation (PAR). However, because plants are green and obviously reflect a substantial amount of green light, there is a common misconception that plants do not use green light for photosynthesis. That is simply not true.
>
> Plants appear green because they reflect more green than any other color. Chlorophyll has peak absorption points in the blue light range at approximately 410, 430, 455 and 460 nm, and in the red light range at approximately 640 and 660 nm (French, 1961). Therefore, chlorophyll absorbs more blue light and red light than it does green light. Consequently, the amount of reflected green light is much greater than that of any other color that we can see, so plants appear green. Green, however, is also quite active for photosynthesis in grasses.
>
> Measures of plant reflectance can be used to estimate plant health based on greenness (Bell and Xiong, 2008). That concept was introduced in Chapter 1, as were vegetation indices. A vegetation index is a combination of two or more bands reflected from plants that when combined in a specific way provide an accurate measure of plant greenness. The most common vegetation index used is the normalized difference vegetation index (Rouse et al., 1973), which is abbreviated NDVI. The NDVI is calculated as the difference between near infrared reflectance (R_{NIR}) and red reflectance (R_{red}) divided by their sum:
>
> $NDVI = (R_{NIR} - R_{red})/(R_{NIR} + R_{red})$
>
> Using this calculation, as the amount of red reflectance increases or the amount of near infrared reflectance decreases, the NDVI gets smaller. A large NDVI indicates a green plant and a small NDVI indicates a yellow or brown plant. The same is true of GNDVI, an abbreviation for another vegetation index called the green normalized difference vegetation index (Shanahan et al., 2001). The GNDVI is exactly the same calculation as the NDVI except that a green light band is substituted for the red light band:
>
> $GNDVI = (R_{NIR} - R_{green})/(R_{NIR} + R_{green})$
>
> When the amount of green reflectance increases, the GNDVI declines. A high GNDVI indicates a green plant and a low GNDVI indicates a yellow or brown plant. If chlorophyll reflected more green light than it absorbed, the GNDVI would increase when NDVI decreased, but it does not. A low red reflectance means that substantial red light is being absorbed and NDVI increases, a low green reflectance means that substantial green light is being absorbed and the GNDVI increases. Both a high NDVI and a high GNDVI indicate very green plants. If more green light was being reflected from a green plant than was being absorbed, the GNDVI would not work in the same way as the NDVI. The GNDVI has been thoroughly tested on wheat (Moges et al., 2004), corn (Dellinger et al., 2008), creeping bentgrass (Bell et al., 2004) and bermudagrass (Xiong et al., 2007).
>
> More green light is absorbed by chlorophyll than is reflected. Green light helps our plants assimilate carbon as an active component of the photosynthesis process.

than that minimally required for growth of the species at the site, the turfgrass will prosper and spread. Because little direct sunlight can penetrate shade, trimming a tree canopy to allow more light to penetrate is not nearly as effective for turfgrass shade management as is strategically removing individual trees to extend the period of direct sunlight (Bell and Danneberger, 1999).

We know that blue and red light is highly absorbed by turfgrass leaves because they contain high concentrations of chlorophyll. What you may not know is that blue and red light is also reflected from turfgrass. The bluish color of buffalograss (*Buchloe dactyloides*) and slight blue tint of Kentucky bluegrass (*Poa pratensis*) is caused by the reflection of blue light. Blue light is highly reflected from objects that appear blue or white, and red is highly reflected from objects that appear red or white. For that reason, turfgrass may grow slightly better in the shade of a building that is painted blue, red or white than a building painted a different color. I would not count on that as a management practice though unless all other options were exhausted. Then again, turfgrass areas of athletic fields painted blue or red, especially red, may decline more rapidly than turfgrass areas painted other colors. In that case, too much of the most important photosynthetic light is being reflected and is not available for absorption. If marking paints specially formulated for application

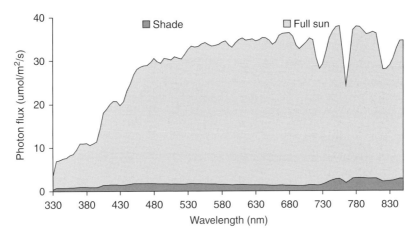

Fig. 6.2. A tree canopy such as the one under which these data were collected substantially restricts the penetration of direct sunlight.

to turfgrass are used, however, blue and red colors should not be a major problem. Green light is highly reflected from turf, but a substantial amount of it is also absorbed for photosynthesis. Because of that, reflected green light can be used as a measure of chlorophyll concentration and potential photosynthesis. Contrary to popular opinion, the greenest turf absorbs the most green light. Green reflectance actually declines as turfgrass gets more green (Shanahan et al., 2001; Bell et al., 2004; Moges et al., 2004; Xiong et al., 2007; Dellinger et al., 2008). Reflectance is not an amount of reflection, it is the proportion of available irradiance that is reflected, so although the turf turns greener owing to reflected light, it is also absorbing a larger portion of the total green light available. The turfgrass turns more green because the grass is also absorbing a greater portion of the available blue and red light, allowing the green reflection to be more dominant. In simple terms, the more green your turf, the more light that is being absorbed.

The far red band is transmitted through vegetation at a higher rate than the other light bands and it is also highly reflected (Bell et al., 2000). Substantially more far red is reflected than blue, green or red, and little far red is absorbed. The reflection of far red light presumably has little or nothing to do with pigments, and is believed to be strongly influenced by cell structure (Knipling, 1970). It has been suggested that the smaller the cells in the leaf blades, the higher the proportion of cell walls compared with cell contents and the higher the reflection of far red that occurs (Gausman et al., 1969; Maas and Dunlap, 1989). As our vision is not very sensitive to most of the NIR band, its reflection has little influence on the color of turfgrass. However, its ability to penetrate vegetation canopies at a greater rate than blue, green or red light is a point of major concern. We will spend considerable time on that phenomenon during the discussion on light quality. For now, remember that phytochrome absorbs irradiance in a band centered near 730 nm in the far red and another band centered near 660 nm in the red (Grant, 1997).

The quality of irradiance available in shade

With a little instruction, light quality is reasonably easy to understand. Turfgrass does not grow well in shade because there is not enough light for it to perform adequate photosynthesis. However, that is not the only light-related factor that causes turfgrass to decline in shade. Light quality is also a consideration.

Light quality is nothing more than the amount of each spectral band of interest in a given amount of irradiance (Fig. 6.3). As turfgrass managers, the light bands of interest are four: blue, green, red and far red. Actually, these bands account for less than half the total photon flux in sunlight, but they are the only bands known to have a major influence, except for genetic mutations and heat, on plant growth (to refresh your memory on photon flux, see Chapter 2).

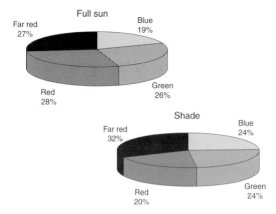

Fig. 6.3. The spectral distribution of irradiance determines its light quality. Irradiance in full sun, for instance, contains more red light than it does in shade from vegetation. The chart on the left depicts irradiance containing only blue, green, red and far red spectral bands, and their typical proportions in full sun. The chart on the right displays the light quality typical of shade from vegetation.

Wilkinson, Beard and Krans conducted landmark shade research and identified many of the morphological and physiological responses of turfgrass to shade (Wilkinson and Beard, 1975; Wilkinson et al., 1975). Some of these effects are caused by low light but many are caused by changes in light quality. These and other turfgrass shade responses are summarized in Table 6.1. In truth, physiological and morphological effects can not be separated as easily as this table implies. Morphological effects are the result of physiological effects which are the result of chemical reactions caused by plant response to environmental stimuli predetermined by genetic code. However, separating physiological effects and morphological effects into things that we can see and things that we cannot see often makes plant study easier to understand. The things that we can see, the morphological effects, are important indicators of the things that we cannot see, the physiological effects. Turfgrass management has historically, and may always, be based on what our plants look like at any given time combined with our knowledge of recent environmental conditions and our knowledge and experience of typical plant responses to those conditions. In short, if our turfgrass looks good morphologically, it is probably in good shape physiologically.

Many of the responses of turfgrass to shade result from phytochrome activity (Box 6.2). As you probably recall, phytochrome is a pigment that absorbs red and far red light (Furuya, 2004). Another pigment called cryptochrome absorbs blue light (Lin and Shalitin, 2003). Cryptochrome is a relatively recent discovery compared with phytochrome and its activities are not well known. It is unlikely that cryptochrome influences turfgrass growth in shade, but we do know that the cryptochrome or some other blue light-absorbing pigment must be present for stomates to open (Shimazaki et al., 2007). Cryptochrome has also been linked to phototropism, and both phytochrome and cryptochrome have been linked to photoperiodism and circadian rhythms in plants (Christie, 2007; Bae and Choi, 2008).

The ratio of red to far red light in sun and shade influences phytochrome status. With the absorption of red light, phytochrome enters an activated state called Pfr. The Pfr state is sensitive to far red light and enters the inactivated state called Pr when it absorbs far red light. At any given time within a single turfgrass plant, a certain portion of the plant's phytochrome is in an inactived state (Pr) and the rest is in the activated state (Pfr). Under normal conditions, phytochrome reverts to the inactive (Pr) state at night when light is not present and is inactivated during the day in proportions roughly equal to the proportion of far red to red light in the irradiance striking the plant (Fig. 6.4). The amount of light present during the day is not particularly important for phytochrome-mediated reactions, as phytochrome-mediated reactions are primarily influenced by the balance between red and far red irradiance, a characteristic of light quality rather than the amount of light present (light quantity).

The diurnal qualities of phytochrome enable plant processes, such as flowering, that respond to photoperiod. Phytochrome enables the regulation of other processes, such as shade responses, based

Table 6.1. Turfgrass morphological and physiological responses to the low light and poor light quality in shade from vegetation.

Physiological effects	Morphological effects
Decreased photosynthesis	Thinner, more delicate leaves
Decreased carbohydrate reserves	Increased succulence
Decrease transpiration	Vertical shoot growth
Changes in pigment concentrations	Reduced root growth

> **Box 6.2. The difference between shade from a building and shade from vegetation.**
>
> If you have ever seen bermudagrass (*Cynodon dactylon*) growing on the north side of a building where it only gets full sun for 1 to 2 hours in the morning and 1 to 2 hours in the evening you might have asked how it could possibly grow in that much shade.
>
> It is possible because the grass is able to allocate resources better in the shade of a building than it can in the shade of a tree. Some of the far red light striking the tree penetrates the canopy but none can penetrate the building. For that reason more phytochrome is inactivated in the shade of a tree than is inactivated in the shade of a building. The turf in the shade of the tree is encouraged by phytochrome activity to grow upright quickly and expends a large proportion of its energy to produce rapid shoot growth at the expense of root growth. In the absence of mowing that reaction would be a positive one. In the presence of mowing, however, it is a negative reaction. Because of mowing, the turfgrass cannot grow the taller shoots that would be more satisfactory for performing additional photosynthesis. As soon as the shoots begin to reach a more satisfactory height, we mow them off. Therefore, the plants in the shade of a building have an advantage because the light quality is different enough for less energy to be allocated to upright growth and so more is available to grow roots and stems.

on the amount of phytochrome inactivated during daylight. In other words, phytochrome is an indicator of light quality. You learned earlier in this chapter that shade, especially shade from vegetation, has a higher proportion of far red light and a lower proportion of red light than full sunlight. Consequently, phytochrome tells our turf when shade is present. The turf responds by partitioning a higher portion of its total energy into shoot growth with less going to root and stem growth. Shoots are stimulated to grow rapidly and more upright than normal as if they were trying to out-compete their neighbors and reach the sun. Photosynthesis is severely limited in shade, consequently less than normal energy is available. The situation results in thinner leaf blades and thinner cuticles as the turf distributes most of its energy to rapid upright growth. This shade response is common in plants. For instance, a tree, grown in full sun will be shorter and have a thicker trunk than the same tree grown in shade. Under natural conditions, this redistribution of energy is a beneficial response. Ecologically speaking, the strongest, in this case the most shade-resistant individuals, will garner the most sunlight and perpetuate the species. Theoretically, if some individuals survive, the species will become more shade tolerant. However, this naturally beneficial shade response is highly detrimental to mown turf.

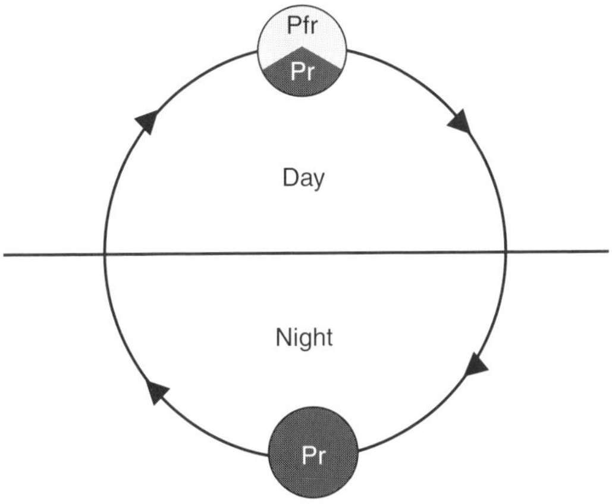

6.3 Managing Turf in Shade

Consider how the redistribution of energy in response to shade works against a grass that is mown regularly (Fig. 6.5). As the plant is attempting to grow shoots rapidly, we are removing leaf material by mowing. Not only does our grass have to regrow that shoot material removed by mowing, it must also direct energy to heal the mowing injury. The period immediately following mowing is usually accompanied by a period of shock that results in very low metabolism. The shock period is followed by a period of slow growth as the plants repair the mowing injury. Following this period of

Fig. 6.4. Most of the phytochrome in a plant reverts to the inactive form (Pr) overnight. During the day, a portion of the phytochrome is then activated to Pfr according to the balance of far red and red irradiance striking the plant.

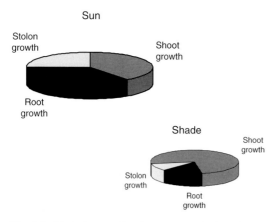

Fig. 6.5. Phytochrome signals grasses and other plants to redistribute energy to encourage rapid vertical shoot growth when the plants are exposed to shade. This naturally beneficial response is detrimental to mown turf because each time meaningful shoot growth occurs, we mow it off and the plants have to start all over again. In this situation, the small amount of photosynthetic energy available in shade is perpetually partitioned to shoot growth, causing a rapid decline in stems and roots.

slow growth, the plants can again direct most of their energy into shoot growth. However, previous growth progress has been removed, and the plants are, in essence, beginning over again.

Mowing in shade

In a natural environment, where mowing did not occur, the individual grass plants best adapted to shade would eventually out-compete their neighbors for available light. As the neighbors declined, more space and nutrients would be available to the best adapted plants. The tallest, best adapted plants would further shade their neighbors and as their neighbors died, more light would be available to the best adapted plants. In addition, the leaf area that the adapted plants accumulated during rapid upright growth would increase their ability to gather light and perform photosynthesis. Consequently, they would eventually have more energy available to partition into root growth and stem growth. Eventually, their energy distribution would return to one more conducive to overall growth. That distribution would be more similar to the distribution usually present in full sun. When we mow, we remove that intraspecies competition. Under mown conditions, the plants best suited for rapid vertical shoot growth no longer have a natural advantage.

Consequently, we must provide an artificial advantage to all of our plants if we expect them to survive and grow reasonably well under shade and mowing. One of the most effective means of increasing shade tolerance in turfgrasses is to increase their photosynthetic leaf area by raising the mowing height.

When we increase mowing height, we increase the potential for sufficient photosynthesis to occur. Increasing the mowing height even a small amount can increase photosynthesis substantially (Fig. 6.6). Raising the mowing height on a golf course putting green by as little as 0.031 inches (0.8 mm), from 0.125 inches (3.2 mm) to 0.156 inches (4.0 mm) increases photosynthetic capacity by 25%. Raising the mowing height on a home lawn from 2.0 inches (50 mm) to 3.0 inches (75 mm), increases its photosynthetic capacity by 50%. If we mow shaded turf at the same height that we mow the same species in full sun, we are not giving our shade turf much chance of survival.

The interesting effect of shade versus sunny conditions on the speed of a putting green is outlined in Box 6.3.

Managing air movement in shade

The detrimental effects of restricted air movement have already been discussed in Chapter 2. However, because restricted air movement and shade commonly exist together, air movement deserves another mention here. The effects of restricted air movement and its accompanying boundary layer of oxygen-rich, carbon dioxide-poor air can be as detrimental to turfgrass growth as shade (Koh et al., 2003). When restricted air movement and

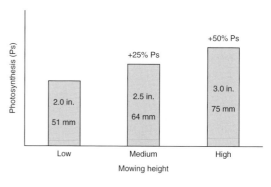

Fig. 6.6. A comparison of mowing heights. Notice that even a slight increase in mowing height results in a substantial increase in potential photosynthesis.

> **Box 6.3. Which putting green is faster, one in full sun or one in shade?**
>
> Most people would probably say that a putting green in shade is slower than one in full sun because the one in shade stays wetter than the one in full sun. That is not correct. The green in shade is usually wetter, but it is not slower. In fact, the green in shade is faster and you can prove that to yourself with a stimpmeter, an instrument specifically designed to measure putting green speed, and multiple trials on greens that do not differ except for their location in either shade or sun.
>
> The leaf blades on a shaded green are growing more upright than those in full sun. Consequently, as a golf ball rolls across the surface it is touching more leaf tips and fewer leaf blades. The tips have less surface area than the blades so they cause less resistance to the motion of the ball. In addition, the grass is nearly always less dense in shade, which causes even less resistance to the motion of the ball. Therefore, shaded greens are usually faster than sunny greens, and a slight increase in mowing height does not affect their playability in comparison with the other greens on the course.

shade exist together, the combination causes an environment sufficiently detrimental to turfgrass growth that it is very difficult to overcome, even with shade-resistant grasses.

The buildings and vegetation that cause shade often cause restricted air movement as well. Reducing or eliminating either the shade component or the air-restriction component is sometimes sufficient to promote turfgrass growth. As it is usually easier to encourage air movement rather than to remove a shade source, the recommendations for managing air movement presented in Chapter 2 should always be followed on shaded turfgrass sites. Start by trimming tree branches to a height of at least 10 feet (3m) from the ground. Remove all low-growing vegetation, including ornamental plants. In wooded and other areas where debris tends to accumulate, try to keep the surface of the ground free of this litter to allow good air flow. Pay particular attention to both the leeward and windward sides of the site that correspond with the regions of predominant wind direction. Finally, remove or level hills and valleys that affect air flow. Good air movement encourages all three of our most important processes: photosynthesis, respiration and transpiration.

Managing irrigation in shade

Irrigation management is as much an art form as it is a science. There are so many variables to consider with irrigation scheduling that it often becomes a trial and error pursuit. Irrigation scheduling is affected by many factors, including shade. Plants in shade grow more slowly and need less water. In addition, shade sites are cooler and evaporation is slower. Consequently, less water is needed. Turfgrass root systems in shade are usually shorter and may need more frequent irrigation, but a lower amount of water is required compared with the same turfgrass in full sun.

Although a turfgrass requires less water in shade than in full sun, it may have to be watered more under trees or other vegetation. Turfgrass in shade is often competing for water with the trees or other vegetation creating the shade. Then again, if the shade is caused by a building or some other structure, root competition does not occur. Considering all of these possibilities, and how they interact with each other and with climate and weather, causes great difficulty when trying to determine how much water is required by your turf on any particular shaded site. There is no standard requirement and these choices are far from easy.

Let us consider the needs of the turf in a shaded situation. Shade affects all three of our most important functions. Photosynthesis is reduced in shade compared with full sun because less light is available. The shade environment is cooler than that of full sun, consequently, respiration is slower. Because a shaded site is cooler and often includes air restriction, evaporation and transpiration decline. However, because photosynthesis and respiration proceed more slowly, creating less heat, and because less light is present to create heat, less transpiration is needed, and because transpiration is by far the greatest use of turfgrass water, less water is needed. How much less water is needed, however, is a very difficult question to answer. I suppose that if we were dealing with similar soils, similar climate, similar weather, the same turfgrass species and the same type of shade at most turfgrass sites, we could study the problem and find a reasonably good irrigation determination based on the amount of time that the turf is in full sun and in shade during each day. However, these conditions are rarely, if ever, the same and hardly ever similar. For that reason, there are no valid recommendations for irrigating turfgrasses in shade. Turfgrass plants in shade require less water, but because of root competition from other plants they often require more irrigation than the same plants in full sun. In many cases, we have to irrigate trees or other vegetation as well as our turf. Otherwise our turf will not get enough water.

Healthy turfgrasses have excellent root systems. Compared with overall plant size, most turfgrass root systems are stronger than those of most trees. The plains in North America, South America and Africa, for instance, are covered with grasses but do not have sufficient water for trees except in lowland areas or near rivers and streams. Trees are huge plants, however, and a shaded turf has difficulty competing with the shear size of the root system of such a large plant. Because turfgrasses are best adapted to full sun, they have a difficult time competing with trees for water in shade. If we hope to grow turf in the shade of a tree, we have to provide sufficient water to satisfy the tree before we can provide sufficient water for the turf. Again, there is no standard calculation to determine how much water is needed. It depends on the species of turf, the amount of shade, the type and size of the tree and all of the other factors mentioned in the previous paragraph. Regular observations and probably some trial and error will be required to determine the irrigation needs of such a site.

Irrigation frequency is also a consideration on shaded sites. Because shoot growth is favored in shade, turfgrass root systems are usually weak. The root system of a tall fescue plant (*Festuca arundinacea*) mowed at 3.0 inches (76 mm) may have a root system that is 12 inches (30 cm) deep in full sun, but that root system may be less than 6 inches (15 cm) in shade (Wherley *et al.*, 2005). The shorter turfgrass root systems generally found in shade require more frequent irrigation than the deeper root systems that are usually present in full sun. In most cases, it is advisable to allow the top 1 to 3 inches (25–76 mm) of soil to dry between irrigation events. If a grass plant is healthy and growing conditions are conducive, a more extensive root system will form if the plant experiences a slight drought stress between irrigation events. A plant in shade, though, does not have the energy to chase water deeper into the soil profile. This plant is under stress. It requires more care than a plant in full sun and it is highly susceptible to drought stress in its weakened condition. Consequently, long periods between irrigation events will harm the plant rather than encourage it to grow a more extensive root system. Plants with weak root systems require frequent irrigation. In shade, it is not advisable to allow more than the top 1 inch (25 mm) of soil to dry before the next irrigation event. Because shaded sites exhibit poor drying conditions, this may take 3 days or more, whereas in full sun, it would normally occur in 1 day. Severe root competition, however, will shorten this period considerably.

When attempting to grow turfgrasses on shaded sites it is often necessary to irrigate lightly but often. Tree root competition is an important factor in determining how often and how much irrigation is required. Tree roots can usually be restricted from important turf sites using a technique called root pruning. Root pruning is a simple process of cutting tree roots between the tree and the turf you intend to protect. It is normally accomplished by digging a ditch, but in some cases can be accomplished using machines with thin knives or saws designed to cut through tree roots in the upper 2 to 3 feet (60–90 cm) of the soil. As long as the pruning takes place no closer that 10 feet (3 m) from the trunk of the tree, and on only one side of the tree, the tree will not sustain lingering damage. In order to be effective, this process must be repeated once every 3 years or so as the tree roots tend to grow back fairly quickly. Because this process is time-consuming and expensive, it is usually only used on areas of high-value turfgrass, such as bowling greens, tennis courts, or golf course greens and tees.

Growing turfgrass in shade takes knowledge and care. Irrigating that turf requires redundant observation. Turfgrass quality and soil moisture should be checked daily during dry weather until the manager is confident that the situation has been properly assessed and that an acceptable program has been established.

Managing fertilization in shade

Turfgrass grown in shade requires very little fertilizer (Baldwin *et al.*, 2009). In fact, it is probably better to not fertilize at all than it is to fertilize the same as you would in full sun. I once executed a simple study in which I covered a creeping bentgrass (*Agrostis stolonifera*) putting green with shade cloth and allowed it to grow under that cloth for two full growing seasons. During that period, I applied a normal amount of nitrogen, phosphorus and potassium to some plots, and normal phosphorus, potassium and no nitrogen to others. At the end of two seasons, all of the turf was poor in quality because of the shade stress. However, the plots that had not received nitrogen fertilization were significantly better than the plots that received a normal rate of nitrogen. As you know, nitrogen is our most important fertilizer nutrient for turfgrass management. However, in this case, nitrogen was detrimental. Let me explain why.

As I stated earlier in this chapter, the light quality present in shade causes responses that encourage rapid vertical shoot growth. Nitrogen fertilization tends to do the same thing (Stanford *et al.*, 2005). In a different study, in 1998, we fertilized plots of a creeping bentgrass putting green at an exceptionally high rate (2 lb/1000 sf (sq. ft.); 97 kg ha^{-1}) of urea nitrogen during the summer and irrigated immediately (Bell *et al.*, 2004). A week later the plots had turned exceptionally green and looked beautiful to most people. To a turf manager, however, they looked too green to be healthy. The plots remained that beautiful green color for about 10 days until they died. They died because their root systems all but disappeared. The putting green was irrigated daily but the root systems on our highly fertilized plots were so weak that they could not sustain water uptake through the day. Had the plants not been under summer stress they probably would have survived the excessive nitrogen application and the reduction in root growth. However, the combination of stressful summer conditions, photorespiration and excessive fertilization was enough to affect the root system so severely that the plants died.

Nitrogen fertilization in shade at normal full-sun rates has a similar combination effect. The balance between shoot growth, root growth and stem growth has already been upset by the shade conditions. Consequently, even a fertilizer program typical of full sun can cause turfgrass to decline in shade. The plant's photosynthesis, respiration and transpiration rates have slowed considerably owing to the shaded environment. Because of that, the plants do not need much fertilizer. In addition, nitrogen fertilization tends to encourage rapid vertical shoot growth, as does light quality and perhaps light quantity in shade (Tan and Qian, 2003). Consequently, fertilizer in an amount greater than about half what is normally applied in full sun, can be quite detrimental to root growth. Nitrogen fertilization in shade should be practiced in low amounts and monitored closely for detrimental effects. In some situations, nitrogen fertilizer may not be required to maintain acceptable turf.

Factors that affect root growth in shade

With the exception of St. Augustinegrass (*Stenotaphrum secundatum*) and to a lesser extent, some zoysiagrasses (Japanese lawngrass, *Zoysia japonica*) warm-season (C_4) turfgrasses have poor shade tolerance. However, warm-season grasses, in general, form deeper root systems than cool-season (C_3) grasses and appear to have stronger root systems in shade. In addition, the warm summer months are normally very conducive to warm-season turfgrass growth, so it is easier for one of the rare shade-tolerant warm-season grasses to maintain an acceptable root system even though the shade environment encourages rapid vertical shoot growth instead. Cool-season grasses are detrimentally affected by high summer temperatures, high-intensity summer light and long photoperiods, and root decline during the summer is a common occurrence in these grasses. Consequently, shade tends to make an already poor situation worse.

Cool-season grasses usually grow better in full sun than in shade, but most will tolerate light shade and some will tolerate moderate shade. Tall fescue and the four fine fescue species, creeping red fescue (*Festuca rubra*), sheep fescue (*Festuca ovina*), hard fescue (*Festuca trachyphylla*) and Chewings fescue (*Festuca rubra* ssp. *fallax*) are probably the most shade tolerant of the cool-season grasses (Gardner and Taylor, 2002). However, even the fescue species suffer weaker than normal root systems in shade (Wherley *et al.*, 2005). Because these are cool-season grasses, root growth can be encouraged without enhancing shoot growth during cool periods. In shade caused by deciduous trees, some root growth also occurs after leaf fall when sunlight is less restricted.

In cool-season grasses, shoot growth occurs best when the air temperature is between 60°F and 75°F (16°C and 24°C), but root growth occurs best when soil temperature is between 55°F and 65°F (13°C and 18°C) (Beard, 1973). For that reason, there are periods during the year when nitrogen fertilization encourages root growth without encouraging shoot growth (Fig. 6.7). Root growth will continue in the fall after shoot growth

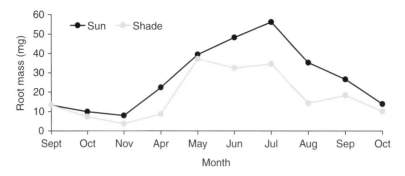

Fig. 6.7. Root growth of creeping bentgrass (*Agrostis stolonifera*) managed as a putting green over the course of a year from fall 1996 to fall 1997. Shade cloth on shaded sites was removed in the fall to coincide with deciduous leaf fall and replaced in the spring as deciduous leaves developed. Notice the amount of winter root growth from November to April in both shade and sun, and the difference in root mass between shade and full sun during the summer.

has slowed to almost nothing, and will continue to grow slowly throughout the winter and into the spring when the turf is exposed to light and the soil is not frozen. During late fall, winter and early spring, nitrogen fertilization encourages root growth rather than shoot growth. For that reason, low applications of nitrogen fertilizer in soluble form or slow-release isobutylidene diurea (IBDU) are recommended to encourage root growth in shade. Such a program is also effective in full sun.

6.4 Chapter Summary

Turfgrasses do not particularly like shade but some will tolerate light shade and some will tolerate more. We can't grow grass in a forest or under an opaque canopy, but we can often grow it beside a building or under a single tree or small group of trees. Whether or not your turf will grow in shade depends on many factors. However, the primary factors that determine shade tolerance are the adaptability of the species selected and the duration of shade each day. Even the most shade-tolerant turfgrasses will require at least a short period, usually 2 to 3 hours, of full sunlight each day to maintain acceptable density. Non-tolerant species such as bermudagrass (*Cynodon dactylon*) or buffalograss (*Buchloe dactyloides*) will require periods much longer than that. If you are responsible for growing grass in shade, and conditions and species are suitable for that purpose, there are management practices that will make the turf more presentable and improve its functional value.

There are three main factors that reduce the effectiveness of turfgrass growth in shade. There is not enough light to perform adequate photosynthesis, there is often insufficient air movement to refresh the boundary layer and encourage transpiration, and in many cases there is significant root competition with shade-producing vegetation. In addition, plants, including turfgrasses, are programmed to respond to shade by partitioning most of their available energy to rapid vertical shoot growth. The result is poor root growth, poor stem growth and thin, succulent vertical leaf blades with weak cuticles. These leaf blades are easily damaged and slow to heal because the low temperature and high humidity normally present in shade interferes with respiration of the limited carbohydrates available. For instance, turfgrasses rarely tolerate traffic in shade because they are too easily damaged and too slow to heal. Consequently, shade areas should be protected from foot and vehicle traffic. If used, the shaded areas need to have sufficient time to heal before they are used again. Damaged leaf blades are more attractive to feeding insects and more susceptible to disease. Because of this susceptibility, shaded turfgrass requires frequent observation and may need special pest protection.

Trimming tree canopies is probably not worth the effort; it may not be sufficient to significantly improve photosynthesis and if the trimming is successful, it will have to be repeated regularly, perhaps as often as every year. Tree removal or, better stated, specific tree removal may be required in order to grow turf on the site. The trees selected for removal should be those that will allow the longest periods of full sun to irradiate the site. However, most people are resistant to tree removal and it may be a battle that you can't win. In that case, you need to have other options available. Shade gardens, ground covers, decorative stone patios, or even simple mulch are examples. In the case of playing fields or golf courses where alternatives to turfgrass are not an option, the decision-makers will have to be educated about turfgrass and shade. Your arguments may be easier to support if you invite the opinion of an outside consultant.

Air movement is often restricted on shaded sites and should be encouraged by the removal of low-growing tree limbs, vegetation and debris. The simple act of removing a fence may be all that is needed to improve air flow and grow acceptable turf. Shaded turf requires little water unless it has to compete for water with other vegetation. Turfgrass grown on shaded sites has to be monitored closely and soil moisture sampled regularly until you are confident that the irrigation program is acceptable. Fertilization with nutrients other than nitrogen should be based on growth. If the shaded turf is growing about half as rapidly as the same species in full sun then fertilization at half the rate is a good starting point. Nitrogen fertilization that encourages root growth is desirable, nitrogen fertilization that encourages shoot growth is not desirable. Most importantly, raise the mowing height. That is often all that is required to grow good grass in shade. Managing fertilization, irrigation and other factors then become practices that improve the aesthetic and functional value of shaded turf rather than practices that improve its chances for survival.

Suggested Reading

Dudeck, A.E. and Peacock, C.H. (1992) Shade and turfgrass culture. In: Waddington, D.V., Carrow, R.N. and Shearman, R.C. (eds) *Turfgrass*. ASA-CSSA-ASSA (American Society of Agronomy-Crop Science Society of America-Soil Science Society of America), Madison, Wisconsin, pp. 269–284.

Puhalla, J., Krans, J. and Goatley, M. (1999) *Sports Fields: a Manual for Design, Construction and Maintenance*. Ann Arbor Press, Chelsea, Michigan.

Stier, J.C. and Gardner, D.S. (2008) Shade stress and management. In: Pessarakli, M. (ed.) *The Handbook of Turfgrass Management and Physiology*. CRC Press, Boca Raton, Florida, pp. 447–472.

Suggested Websites

NASA Goddard Space Flight Center. Electromagnetic Spectrum. Available at: http://imagine.gsfc.nasa.gov/docs/science/know_l1/emspectrum.html (accessed 6 April 2010).

7 Understanding and Prescribing Nutrition

Key Terms

The **basic plant nutrients** are carbon, hydrogen and oxygen.
The **macronutrients** consist of two groups of three nutrients each, the primary macronutrients and the secondary macronutrients.
The **primary macronutrients** are nitrogen, phosphorus and potassium.
The **secondary macronutrients** are calcium, magnesium and sulfur.
Turfgrasses need **micronutrients**, iron, copper, molybdenum, chlorine, zinc, boron and manganese in very small quantities.
Soil texture is the proportion of three minerals, sand, silt and clay that exist in a specific soil.
Soil structure occurs when clay particles combine with organic material to form soil aggregates.
Cation exchange capacity is a measure of a soil's ability to attract and hold nutrients.
Organic matter is dead plant material that exists in the soil in various stages of degradation. As you learned in Chapter 5, mature turfgrass sites nearly always have organic soil layers of thatch and mat.
Soil compaction is the compression of soil pores, usually caused by traffic, resulting in less soil air and greater soil water-holding capacity.
Humus is the fairly stable form of soil organic matter that remains after the major portions of the plant and animal matter have been decomposed. Humus is usually dark in color.
Soil aggregates form when clay combines with organic matter. Aggregates are as large or larger than sand particles enabling them to form large drainage pores between aggregates and small capillary pores within aggregates.
Macropores are the large pores formed between sand particles or soil aggregates. These pores drain quickly following soil saturation, providing air for root respiration.
Micropores are very small pores capable of holding water against gravity by capillary action. Although water drains readily from soil macropores, it remains in micropores and provides the water for plant metabolism and transpiration.
Compounds that **volatilize** (i.e. are volatile), like nitrogen fertilizer, have a propensity to change from a solid to gaseous state under normal atmospheric conditions.

7.1 Manage Your Fertilization to Match Your Turfgrass Needs and Local Conditions

On some of the old science fiction television shows and movies it was common for aliens to refer to humans as carbon-based life forms. That is an apt description. In fact, all life on earth, including turfgrass, is carbon based. Biochemistry is a study in carbon combinations. Consequently, it is curious that we tend to forget that carbon is our most basic nutrient. Oxygen and hydrogen are also basic nutrients. The ranking of turfgrass nutrients by importance depends on the amounts that are found in turfgrass plants. Carbon, hydrogen and oxygen are called basic nutrients because they are the most common elements found in animals and plants, including turfgrasses. The second category of turfgrass nutrients is the macronutrients which consist of the primary macronutrients nitrogen, phosphorus and potassium, and the secondary macronutrients calcium, magnesium and sulfur. Turfgrasses contain more primary macronutrients than secondary macronutrients and more secondary macronutrients than the third category of nutrients, called micronutrients. The micronutrients are made up mostly of metals and our grasses only require trace amounts of them. Nonetheless, a missing micronutrient can make the difference between acceptable turf quality and an unacceptable stand. You can find standard fertilizer recommendations for each turfgrass species in books, but most scientists and managers are reluctant to make general statements concerning fertilization. Different situations, climates and overall environments require custom fertilization practices to grow the best turf possible. For that reason, standard recommendations are a good place to start but not a good place to end. As turfgrass managers, you are required to develop the best fertilization practices for your sites and to be flexible enough to alter those practices as the environment dictates. This chapter will help you to make those adjustments.

In order to determine our most effective fertilizer plan, we need to consider a number of factors and variables. The most important factor is probably soil. Everything about the soil affects nutrition. A basic understanding of soil pH, soil texture, soil structure and other soil attributes is not just important, it is absolutely necessary to designing and implementing a successful custom fertilization program. Factors such as climate, rainfall, temperature and seasonal growth rates are also necessary considerations when designing a program. If these variables are not considered, adjusting your program to your situation is nothing more than trial and error management, which will probably fail, if not this year, next year or some future year. Managing a fertilization program for turfgrass is not easy. It requires knowledge of the soil and plant, consideration of the climate and weather, and specific knowledge of the site. Experience also helps a lot, so, in addition to reading this book, ask questions of experienced teachers and managers until you have acquired additional knowledge of various practices and procedures that work.

7.2 Basic Soil Attributes

Plants have to have a medium in which their roots can find the nutrients and water that enable them to grow. In hydroponics, this medium is simply aerated water. Container plants are grown in various types of soilless medium, but turfgrass people, except for some experimental growth of sod on plastic, still use either natural soil or an imported sand mix. For that reason, basic soil properties are very important to us. We not only need to manage our plants, in most cases, we need to manage our soil to achieve our best plant growth.

Cation exchange capacity

Natural soils are made up primarily of three minerals, sand, silt and clay. These mineral classifications are based on size. Sand is the largest of the mineral particles, followed by silt and then clay (Fig. 7.1). Sand particles are huge compared with clay particles. Large particles have great volume but small particles have high surface area to volume ratios. A beach ball, for instance, has a much greater volume than a golf ball. However, the golf ball has a greater surface to volume ratio than the beach ball. If you were to fill a large dump truck with beach balls it would probably only accept 20 to 30 balls before it was full. However, if you were to fill it with golf balls, it would accept hundreds of them before it was full. The golf balls have lower volume so more of them fit into the truck, but the sum of the surface area surrounding those golf balls is much greater than the total surface area surrounding the beach balls that fit in the same truck (Fig. 7.2). This surface area property and the highly weathered condition of clay make it very much more reactive than sand. In other words, clay will attract considerably more nutrients than sand.

As you probably recall, when some chemical compounds, primarily salts, are added to water, they disassociate to form cations and anions. Table salt (sodium chloride, NaCl) disassociates to form Na^+, a cation, and Cl^-, an anion. It is the molecular bond between the positive cation and the negative anion that hold the two together to form table salt. However, when they are added to water, the attraction of the polar water molecules is strong enough

Fig. 7.1. These three cylinders represent the size differences between sand, silt and clay. This drawing is not to scale; there is a much greater difference in size among these particles than is depicted here.

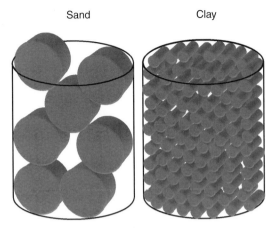

Fig. 7.2. It takes a much larger number of clay particles to fill a container than it does sand particles. However, the clay particles that fill the container on the right (not to scale) constitute a much greater total surface area than the sand particles on the left.

to pull the cation and anion in opposite directions and separate them. We say that salt dissolves in or forms a solution in water. Once in solution, if a dissolved cation encounters a stronger negative force than the water molecules it will bind to the stronger force. Clay particles provide a stronger force. We call clay particles reactive because they are highly charged. Although the surface of clay has some positive charges it is mostly covered with negative charges that attract cations more strongly than does water. The charges on the surface of clay particles are numerous and are called exchange sites. The number of exchange sites in a given amount of soil determines the soil's cation exchange capacity, commonly called CEC.

There are six nutrient anions: phosphate, sulfate, nitrate, borate, molybdate and chloride. These six negative nutrient-containing anions either bond weakly to the positive sites of soil particles or enter complex reactions that bind them to the soil in other ways. The rest of the plant nutrients are cations. These cations are attracted to the negative exchange sites on clay particles. The number of negative exchange sites in a soil, the CEC, is a measure of its potential fertility. Cation exchange capacity is measured in units called milliequivalents (meq) per 100 grams of soil or centimoles (cmol) per kilogram. One meq per 100g is equal to one cmol per kg. I am sure that you would understand these units if I explained them to you but understanding them is really not necessary. You simply need to know the difference between a high CEC and a low CEC. Sandy soil has a very low CEC and clay loam has a very high CEC. The CEC of soil organic matter that has been mostly degraded is extremely high (Table 7.1). Consequently, organic matter such as peat, a highly degraded material, is usually added to pure sand systems for athletic fields and golf courses to provide some much needed nutrient and water retention characteristics.

The highly degraded organic material and clay particles in a soil are called soil colloids. Do you remember soil colloids from Chapter 5? Colloids are fine particles of one substance somewhat evenly distributed throughout another. When a soil scientist or instructor refers to soil colloids, he/she is referring to the highly reactive portion of a soil, the portion that determines the CEC.

The influence of soil texture and soil structure

In the preceding paragraphs I introduced the terms "loam" and "clay loam" assuming that you knew what they were. These are terms that refer to soil texture, and soil texture is the proportionate combination of sand, silt and clay that exists in a particular soil. Soil texture is a description of the size of the particles in the soil and has a major influence on soil properties. You are probably familiar with the soil texture triangle of Fig. 7.3. Soil structure is a property unique to combinations of clay and organic matter. Consequently, it is influenced in part by soil texture and also by soil compaction.

Soil texture

Soil compaction is a very important problem associated with growing turfgrass on common areas. It is rare for a turfgrass area to be displayed and not used. Most turfgrass stands have

Table 7.1. General approximations of cation exchange capacities (CECs) common to certain types of soils.

Soil texture	CEC (cm kg^{-1})
Sand	2
Sandy loam	7
Loam	25
Clay loam	35
Clay	100
Organic matter	250

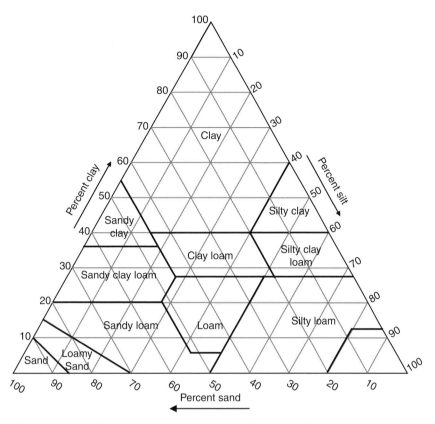

Fig. 7.3. The soil triangle figuratively describes the soil texture classifications and is found in most introductory soils texts.

to sustain a certain amount of traffic which results in compaction. Compaction will be discussed more thoroughly in Chapter 10. For now, understand that soil compaction is the compression of soil particles and results in reduced soil air space and increased soil water-holding capacity. Because air space containing oxygen is an important requirement for root respiration, compaction can be a major problem on turfgrass sites. Compaction does not influence soil texture but soil texture determines how easy soil is to compact.

Sandy soil is not easily compacted and pure sand cannot be compacted enough to cause a decline in turfgrass growth. That said, however, there are other factors such as natural accumulations of dust, sediment and organic material that can contaminate pure sand enough to encourage compaction sufficient to restrict root growth and water uptake (Murphy *et al.*, 1993b). In that case, the original pure sand is no longer the same texture. It now contains silt and clay and has a different soil texture classification.

Soil texture affects many important soil factors, such as bulk density, hydraulic conductivity and others, which you can study in soil textbooks or soil classes. For purposes of this text, air-holding capacity, water-holding capacity and nutrient retention of soils are the important factors that satisfy our purposes.

In the preceding section, you learned that sand has very poor nutrient retention compared with clay. We refer to sand as inert and to clay as reactive. Silt is also inert and organic material in the form of humus (highly degraded organic material) is extremely reactive (Brady, 1990). Clay and humus by themselves hold lots of water, very little air and lots of nutrients. Silt also holds lots of water and very little air, but few nutrients. Sand holds lots of air, little water and few nutrients. The properties of silt are mostly negative (Table 7.2). It is easily compacted, holds very little air and few

Table 7.2. Some of the advantages and disadvantages of the mineral fractions that determine soil texture.

Mineral	Advantages	Disadvantages
Sand	High compaction resistance Rapid drainage High oxygen retention	Poor nutrient retention Poor water-holding capacity
Silt	High water-holding capacity	Poor nutrient retention Poor oxygen retention Poor compaction resistance Poor drainage
Clay	High nutrient retention High water-holding capacity Good drainage if not compacted	Poor drainage if compacted Poor oxygen retention Poor compaction resistance

nutrients. Sand and clay, however, have advantages depending on the expectations of the site. Sand does not compact substantially and provides superior aeration. Clay and humus hold high amounts of water and nutrients and are desirable for that reason. We will discuss these properties further but before we do we need to review soil structure.

Soil structure

Soil structure occurs when clay and organic matter, primarily humus, combine to form large particles called aggregates (Fig. 7.4). These aggregates are the units that combine to form soil structure. There are at least seven different types and subtypes of aggregates that form from various combinations of materials and determine the various forms of soil structure (Brady, 1990). There are normally at least two types of structure present in a mineral soil. However, that is not our basic concern. We need to know how structure affects turfgrass and how we can sustain it and make it work for us.

Plants are facilitators of soil structure. Root growth helps to break up large compacted units of soil and helps to form smaller, more useful aggregates. Plant roots, mostly root hairs, secrete organic materials that help to cement soil aggregates together (Brady, 1990). Theoretically then, a soil with a high number of plants and plant roots, such as a turfgrass stand, maintains soil structure better than a soil containing few or larger plants with less extensive rooting. However, turfgrass areas tend to sustain high levels of pedestrian and sometimes vehicle traffic that negatively affect structure. As pedestrian and/or vehicle traffic moves across turfgrass sites, the upper 1 to 2 inches (2.5–5.1 cm) or more of the soil are compressed, destroying structure. This compaction also helps to seal the soil surface, thus interfering with water infiltration and encouraging water runoff. For that reason, turfgrass managers regularly aerify, the process introduced in Chapter 5 primarily for thatch control. Aerification helps to relieve soil compaction and promote water infiltration (McCarty, 2001). Although aerification does not necessarily promote soil structure, it does promote root growth, which leads to better structure.

Refer to the representation of soil structure in Fig. 7.4. Notice that the aggregates are large complexes that consist of much smaller particles. These complexes are too large to fit tightly together, much

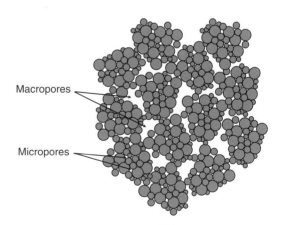

Fig. 7.4. Soil structure occurs when clay and organic material combine to form particles as large or larger than sand called aggregates. Aggregates are the building blocks of soil structure. Large pores between the aggregates (macropores) hold air and small pores within the aggregates (micropores) hold water.

like the cylinders that represent the sand particles in Fig. 7.2. Large rounded objects cannot physically fit together no matter how tightly packed without leaving relatively large gaps or pores between them. They can only fit tightly together along flat surfaces. Small rounded particles cannot fit tightly together either, but there are much smaller gaps between them (Fig. 7.2). The large gaps that form between sand particles, between soil aggregates and between sand particles and soil aggregates are called soil macropores, and the small gaps that form between clay particles and within soil aggregates are called soil micropores.

Soil aggregates are large enough to have the attributes of sand as well as the attributes of clay. I remember my introductory soils instructor asking the class a question many years ago. The question was "Which soil drains fastest, a sandy loam or a highly aggregated clay loam?". Of course, we all answered sandy loam but, according to him, a highly respected soil scientist, we were wrong. A highly structured clay loam forms aggregates as large or larger than sand. Consequently, the soil assumes the drainage properties of sand. The large particles cannot fit together tightly and macropores are left between them. Yet, the aggregates are composed of clay particles and organic matter that maintain very small pores. These micropores hold water against gravity because of capillary action, a physical phenomenon that was described in Chapter 1 and again in Chapter 4. Hence, a highly aggregated clay loam has both the drainage advantages of sand and the water and nutrient retention capabilities of clay. Unfortunately, the clay loam does not maintain the sand characteristic of resistance to compaction and its structure is easily destroyed by traffic. If the site has to maintain regular traffic, the compression of the structure in the upper soil layers cannot be avoided and aerification must be performed regularly to help relieve the inevitable soil compaction.

7.3 Plant Nutrition

In order for a plant to assimilate a nutrient, the nutrient must be soluble. The air, for instance, is about 78% nitrogen, but that nitrogen is not available to plants because it is not soluble. The soil in my area of the USA is often red, indicating high iron content. That is mostly iron oxide, however, and it is neither soluble nor plant available. Consequently, applying fertilizer to turf requires more thought than action. You not only have to select the right product, determine how much to apply, calibrate your equipment and select the proper time, you also have to take precautions that the fertilizer is not going to volatilize, runoff or leach before the plant has a chance to use it. Incidentally, how do you apply carbon, hydrogen and oxygen, and what do you do if your soil pH is too high or too low?

Managing the basic nutrients

Can we manage carbon, hydrogen and oxygen? Carbon, yes; oxygen, yes; and hydrogen does not matter. Hydrogen is ubiquitous. It is the most available element in nature and it is everywhere. Our plants can get hydrogen from water, from air and from nearly any compound that they assimilate. We don't have to manage hydrogen but we do have to manage carbon and oxygen if we expect our turf to look and perform at its optimum. The three most important plant physiological functions are photosynthesis, respiration and transpiration, and these require carbon dioxide, oxygen and water, respectively. If we cannot manage carbon and oxygen then all we can do is mow and hope for the best.

Carbon

How do grasses accumulate carbon? Grasses accumulate carbon by assimilating carbon dioxide during photosynthesis. Consequently, management procedures that encourage photosynthesis also encourage carbon nutrition. We may not be able to control photosynthesis but we can certainly affect it.

In review, photosynthesis requires light, water and carbon dioxide. Light is reasonably easy to manage in most situations. There is not much we can do about buildings or other shade-producing structures, but we can affect trees and other vegetation. In Chapter 6, you learned some techniques of shade management. There is a limit to the amount of shade that a turfgrass can stand. That limit has a lot to do with you, or your customer's expectations. The better you want the turf to look and the lower that you want to mow it, the more shade you will have to eliminate. You can often grow a sparse turfgrass cover in shade but in most cases a sparse cover is not acceptable. If trees need to be removed in order to grow good grass, it is up to you to educate your customer on his/her alternatives. It is also up to you to know which trees need to be removed and which trees can remain. In the case of tree removal your role is advisor. Tree removal decisions have to be made by your customer. Otherwise, you may be jeopardizing your job. However, the customer must realize that growing acceptable turf in the shade currently present is simply not possible. The decision is to save the tree or save the grass. In some cases, saving both is not possible.

As you learned in earlier chapters, we can also provide more carbon dioxide to our turf by promoting air movement. Unrestricted airflow refreshes the boundary layer, replacing air that is rich in oxygen and water and low in carbon dioxide with air containing normal gaseous proportions. Good airflow also encourages transpiration. You learned in Chapter 6 that you can help to mitigate low photosynthesis in shade by enhancing airflow. Sometimes, enhancing air movement on shaded sites is sufficient to provide adequate carbon for acceptable turfgrass nutrition.

Mowing height has a very noticeable effect on carbon accumulation in stressful situations. The lower you mow the turf, the more efficient it has to be to accumulate carbon. A long leaf provides substantially more photosynthetic potential than a short one. Finally, grasses have to have water to accumulate carbon. If yellow to brown grass is not acceptable during extended dry periods, irrigation must be provided. When turfgrasses do not have sufficient water to transpire or to perform photosynthesis, they enter a period of quiescence that we call dormancy, or they die. Some species can sustain long periods of dormancy, others cannot. Although we do not list water as a nutrient, it is obviously important for all plant functions.

Yes, you can affect carbon nutrition and oxygen nutrition, but the processes required are not nearly as simple as a fertilizer application. Most management practices are designed to improve carbon and oxygen assimilation. It is your responsibility to assess environmental factors and choose the correct management for your situation(s). Managing the basic nutrients requires evaluation, planning and effective timing of the practices required. Do not forget that your decisions not only enhance the assimilation of carbon and oxygen, they can also cause a decline.

Oxygen

As photosynthesis proceeds, oxygen is produced. Normal air also holds a substantial amount of oxygen. Consequently, oxygen is not limiting for respiration in aboveground plant organs. However, it can become limiting in the soil. As you learned earlier, oxygen is necessary for root respiration and root respiration is necessary for water and nutrient uptake. Roots do not function unless they respire. Hence, saturated soil around turfgrass roots shuts down root uptake. Turfgrass managers must provide adequate drainage for their turf. There may be situations where adequate drainage is not possible but those situations are extremely rare. Like trying to grow grass in near perpetual shade, you will not be successful consistently growing turfgrass in areas that cannot be drained properly. During any rainy season, your turf will struggle or die.

Soil compaction is another major factor contributing to poor soil oxygen levels. Compaction is similar to poor drainage in that the soil holds too much water and too little air. In the case of compaction, however, better drainage will not be adequate to completely relieve the problem. Compaction results in smaller pores that hold water more tightly. In effect, compaction turns macropores into micropores. So following compaction, many of the macropores that previously held air become micropores that hold water. If these pores give up water to transpiration or evaporation and become filled with air, there is sufficient oxygen present for respiration, but there is no longer water present for uptake. Aerification and traffic control are the two primary practices that can help control compaction. We have already discussed those procedures.

Soils high in salt also have a restrictive effect on soil oxygen. Salts, especially sodium, destroy soil structure and attract and hold excess water. Similar to compacted soils, sodic soils hold too much water and too little air (Harivandi et al., 1992). Sodium is a large element that is larger when water is available. When water is present, sodium is surrounded by it and holds it tightly. Consequently, where large complexes of sodium and water are bound to the soil CEC, the colloids cannot move close enough to each other to form aggregates. Hence, soil structure is destroyed and few macropores are formed. With no macropores to hold air, the effects of the salt are similar to those of compaction. However, even if oxygen was present for respiration, much of the water in the soil would not be plant available because the turfgrass roots would not be able to create osmotic potentials sufficient to break the strong bonds between water and salt. We will discuss these phenomena further in Chapter 10.

For now, remember that the best management practice for salty soil is applications of gypsum – a common name for calcium sulfate.

What to do with soil pH

Soil pH can have a detrimental affect on nutrient availability (Carrow et al., 2001). Most turfgrasses grow adequately in a wide range of pH, but all appear to grow adequately at a soil pH of about 6.5 (Waddington, 1992). A pH that is too high affects the solubility of micronutrients such as iron, copper, zinc, boron and manganese. Nitrogen becomes less available at a pH of over about pH 8.0 or less than about pH 6.0. Phosphorus has a more narrow range than nitrogen, with decline in availability beginning at about pH 7.5, and a sharp decline at about pH 6.5. Acidic soils of pH below 6.0 tend to limit the availability of important nutrients such as potassium, calcium, sulfur and magnesium, and the micronutrient molybdenum. Consequently, managing pH is an important practice.

Agricultural sulfur (elemental sulfur, S_2) is used to lower pH. Microbial breakdown of S_2 results in sulfuric acid, which eventually lowers the pH of the soil. Sulfur applications can burn turfgrass so they should be limited to less than 10 pounds per 1000 square feet (488 kg ha^{-1}) of turf per application and only applied in spring and fall when the weather is cool. Be careful not to overlap passes, resulting in strips of double applications. Applications of iron sulfate or aluminum sulfate can also help to acidify the soil. Iron is an important micronutrient of turfgrass and is preferable to aluminum, which rarely adds anything positive to the soil.

Soils are highly buffered, meaning that they strongly resist changes in pH. A downward, more acidic change in pH is more difficult to accomplish than an upward one. On some highly buffered soils, it could take a lifetime of sulfur applications to make even a minor change in pH. Under those circumstances it is not worth the effort, although sometimes, the application of acidifying fertilizers such as ammonium nitrate, ammonium sulfate or potassium nitrate, particularly ammonium sulfate, are enough to make sufficient nutrients available.

Lime ($CaCO_3$) is applied to raise soil pH. It is usually possible to raise the pH of acidic soil but it often requires a substantial amount of lime. Lime is safer on turfgrass than sulfur and can be applied at rates of up to about 40 pounds per 1000 square feet (1950 kg ha^{-1}) in a single application. Again, it is best to apply lime in the spring and the fall, and avoid applications in the summer when the weather is hot. A soil test is required to determine how much sulfur or how much lime will be required to change soil pH. It may be an incredibly large amount. If sulfur/lime applications are not practical, foliar feeding will be necessary. Foliar feeding is simply the application of nutrients in light, frequent fine sprays. The objective is to make applications in amounts small enough and at times that are conducive for the plants to absorb the nutrients through the foliage instead of through the roots. Foliar feeding is effective, but it is an expensive, labor-intensive practice.

Soil test reports

Fertilizer decisions should be based on soil test reports. Over-application of nutrients such as nitrogen, and particularly phosphorus, can be an environmental hazard and should be avoided (Soldat and Petrovic, 2008). Although P is bound tightly to the soil, fertilizer P applications are soluble and run off easily into surface water, thereby creating biological hazards (Bell and Moss, 2008).

Not all soil reports are created equal. There are several different methods available for determining the soil content of major nutrients. Particular soils are better tested with one method than they are with others. Consequently, it is best to get your soil tested at a laboratory in your general region. A laboratory close by is most likely to use the best testing methods for your location and should be familiar with the specific soils in your area. A local laboratory will also be familiar with your turfgrass species and be able to make competent recommendations for the amount of fertilizer to apply. Unless you are very familiar with soil science, it is best for you to at least start with laboratory recommendations rather than trying to make your own. With experience and enough recommendations, you will probably learn how to determine fertilizer requirements for your site, or at least the requirements for macronutrient fertilization.

Some laboratories report soil nitrate nitrogen (Fig. 7.5), and some report total soil nitrogen. For purposes of turfgrass management, however, measures of soil nitrate or total soil nitrogen have no real value. Soil N changes daily and is affected by numerous factors. Consequently, we have to apply the proper amount of N fertilizer based on knowledge and practical experience. Soil reports

OKLAHOMA COOPERATIVE EXTENSION SERVICE

SOIL, WATER & FORAGE ANALYTICAL LABORATORY

Division of Agricultural Sciences and Natural Resources • Oklahoma State University
Plant and Soil Sciences • 048 Agricultural Hall • Stillwater, OK 74078
Email: soils_lab@agr.okstate.edu
Website: http://clay.agr.okstate.edu/extensio/swfal/intro.htm

SOIL TEST REPORT

DR GREG BELL
HORTICULTURE
360 AG HALL

744-6424

Name:
Location:

Lab I.D. No.:
Customer Code:
Sample No: 15
Received: 03/22/00
Report Date: 03/24/00

TEST RESULTS

— Soil Reaction —

pH: 7.4
Buffer Index:

— NO3-N (lbs./acre) —

Surface: 3
Subsoil:

— Test Index —

P: 64
K: 119

———— Secondary nutrients ————

Surface SO4-S (lbs./acre): 13
Subsoil SO4-S (lbs./acre):

Ca (lbs./acre):
Mg (lbs./acre): 95

— Micronutrients —

Fe (ppm): 18.5
Zn (ppm): 1.83
B (ppm):

INTERPRETATIONS AND REQUIREMENTS FOR *No Crop Provided*

— Test —   — Interpretation —   ——— Requirement ———   ——— Recommendations and Comments ———

Signature

Oklahoma State University, U.S. Department of Agriculture, state, and local governments cooperating. Oklahoma Cooperative Extension Service offers its programs to all eligible persons regardless of race, color, national origin, religion, sex, age or disability and is an Equal Opportunity Employer.

Fig. 7.5. A typical soil report. Notice that this laboratory reports soil nitrate and other nutrients in parts per million (ppm) or pounds per acre (lbs./acre).

are generally meaningless for determining applications of N fertilizer to turfgrass. Soil reports for P and K, however, are accurate and useful. Reports of calcium, magnesium and sulfur are useful for pointing out potential deficiencies, and micronutrient tests may lead you to the answer for a difficult-to-diagnose problem.

While some laboratories report nutrient content per unit of soil in parts per million and/or pounds per acre (Fig. 7.5), some may use kilograms per hectare. Others report nutrient content as percent base saturation. Still others issue reports that include both nutrient content per unit of soil and nutrient content in percent base saturation (Fig. 7.6). There are general guidelines for nutrient content per unit soil for both cool- and warm-season grasses but many specific circumstances apply and it is my opinion that you are best served to ask the lab for recommendations. If you don't completely understand the report, call the laboratory and ask questions. Your success may depend on knowing the answer.

If your laboratory reports base saturation, they are telling you the proportion of soil exchange sites that are theoretically occupied by each nutrient. That is very useful information because it ignores soil fertility and reports nutrient retention in comparison with other nutrients. In most soils, calcium occupies the majority of exchange sites and there is a high percentage of magnesium. If that is not the case in your soil, it indicates problems that need to be addressed. Other deficiencies may also be apparent. A quality soil laboratory knows what your base saturation should look like for growing turfgrass and will make recommendations to correct existing problems.

Choose a laboratory before you sample your soil. A laboratory is generally chosen based on reputation and the recommendations of other turfgrass managers and horticultural or crop production specialists. Nearby turfgrass managers are your best source of information. Call the laboratory before you collect samples. Ask them for advice on how to proceed with sampling. Ask for recommendations concerning the proper size of sampling areas and the number of samples to collect on each area. Also ask for directions on sampling depth; this is important and differs by laboratory. In addition, ask for any other advice that the laboratory might feel is important.

Many soil laboratories also do tissue testing, and many turfgrass managers are convinced that tissue testing is superior to soil testing. Tissue testing, however, is exceedingly more complicated than soil testing and requires a thorough knowledge and/or history of what nutrient levels are necessary to maintain your expectations. It gets complicated because adequate tissue levels differ with season, just as response to fertilizer differs with season. If you are willing to keep good records and thoroughly assess responses to fertilizer input, tissue testing may be right for you. Regardless of the method you use to assess your nutrient requirements, you will be money ahead and exceedingly more effective if guesswork is mostly eliminated.

The primary macronutrients

On any bag of consumer fertilizer you will normally find a formulation. The formulation describes the amount of the three primary macronutrients, nitrogen (N), phosphorus (P) and potassium (K), in the bag. These amounts refer to percentages and are presented in a form consistent with N-P-K. The formulation 14-3-6 means that the bag contains 14% pure N, 3% P_2O_5, and 6% K_2O. The 3% P_2O_5 in that bag is not the same as 3% P. In fact, P_2O_5 is only 44% P and K_2O is 84% K, which results from a simple comparison of the molecular weights of P, O and K. So the bag really contains 14% N, 3% × 0.44 = 1.32% P and 6% × 0.84 = 5.04% K. Turfgrass managers apply fertilizers based on pure N applied, so we need to know these calculations so that we can choose formulations that result in the proper application of P and K when we apply the expected amount of N. That sounds complicated, but it really is not; it is just simple mathematics as we use every day.

Nitrogen

Nitrogen is the most important nutrient for turfgrass. If water is present and a minimal amount of the other important nutrients are present, an application of N will encourage turfgrass growth and improve green color. Adequate N fertilization will improve turfgrass overall health, and increase turfgrass cover and density. However, N fertilization at a rate above optimum is detrimental to turfgrass health and can result in increased traffic damage, pest susceptibility, decreased root growth and even death.

Turfgrass gets nitrogen from soil organisms that remove it from the air. It can also get nitrogen from

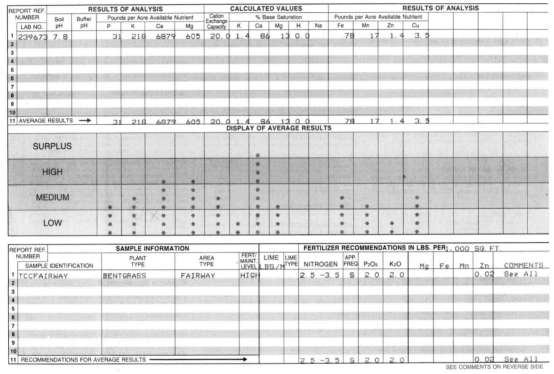

Fig. 7.6. A typical soil report from a laboratory that reports base saturation as well as soil nutrient content in pounds per acre. Notice the recommendations at the bottom of the report.

organic material as microorganisms degrade it. Nitrogen is a part of rainfall, especially during thunderstorms, when lightning causes N in the air to combine with precipitation. Turfgrasses rarely need N in addition to that supplied by nature to survive. However, they normally do need additional N to look and function the way that people want them to. We can make a huge difference in turfgrass by adding N fertilizer at the right time and in the right amount.

You can find turfgrass fertilization guidelines on university websites in your region based on species and climate. Introductory turfgrass textbooks also include general guidelines for turfgrass fertilization. You can often find guidelines in trade magazines and you can surely get advice from other turfgrass managers in your area. However, you must remember that these are just general guidelines, they are not always the best programs for your site and they should be adjusted according to recent weather conditions. Fertilizer recommendations, like species selection, are based on climate. Climate, however, is substantially different from weather. Climate is the average of weather conditions over an extended period. In the USA, the period is 30 years. Weather changes substantially on an annual basis and weather determines the conditions under which your turfgrass must thrive, not climate. When we talk about ecology in later chapters, you will learn that disturbances, sometimes caused by weather, can change the species composition of an ecosystem. Consequently, weather can certainly influence the health of a turfgrass stand and the amount of fertilizer required to sustain acceptable growth and development. You must be able to adjust to a changing environment to grow your best turf.

One of the most important weather factors that influences fertilization is temperature. Temperature influences all three important physiological processes, photosynthesis, respiration and transpiration, therefore it affects the amount of nutrients required for plant maintenance. In Chapter 6, we discussed how applying fertilizer at the proper time of year could encourage root growth in shaded cool-season (C_3) grasses. This same procedure is useful for cool-season grasses in full sun (Wehner et al., 1988). Fertilization, especially nitrogen fertilization, tends to encourage shoot growth, sometimes at the expense of root growth (Hull, 1992; Bowman, 2003). If we hope to use N fertilizer to encourage root growth in a cool-season grass, the best time to affect that process is when it is too cool for shoot growth but warm enough for root growth (Moore et al., 1996; Grossi et al., 2005). As we discussed in Chapter 6, those conditions occur in late fall, winter and early spring (Wehner and Haley, 1993).

Summer is a poor time to fertilize cool-season grasses. During summer, high temperatures encourage rapid photosynthesis, respiration and transpiration, increasing the need for nutrients. In C_3 grasses, however, high temperatures also encourage photorespiration. As photorespiration increases during summer (see Chapter 3), plant energy declines. During this period, cool-season grasses will use most of the available energy to maintain shoot growth, and partition less available energy into root growth (Hull, 1992). Consequently, although some new roots are formed many roots die and are not replaced. This process results in a gradual loss of the root system of a cool-season grass during summer (Huang and Liu, 2003). If the root system was not adequate when the summer began, the plant will probably die. If the root system was healthy when the summer began, the plant will survive. When nitrogen is added, the balance of shoot growth to root growth shifts farther toward shoot growth, causing an increase in stress on plant roots (Bushoven and Hull, 2005). As the root system declines the plant's ability to transpire also declines. Eventually, the root system may become so weak that it can no longer keep the plant hydrated during hot, dry days and the plant dies. Fertilizer timing of cool-season grasses is important and should be based more on the times when root growth can be encouraged as well as or in place of shoot growth. A nitrogen rate that is typically healthy in the fall can damage a cool-season grass in the summer. The best time to fertilize cool-season grasses to encourage the root growth that will help them survive summer heat is after the weather cools to the point that the shoots stop growing.

The timing of warm-season (C_4) grass fertilization differs from cool-season grass fertilization. Warm-season grasses are better adapted to summer conditions and apparently have been naturally selected for deep root systems (Christians, 2007). Although, similar to cool-season grasses, the natural tendency of warm-season grasses is to favor shoot growth over root growth, there appears to be a greater energy partition to root growth in C_4 grasses than there is in C_3 grasses. For that reason, and because photorespiration does not occur in the C_4 photosynthetic pathway, warm-season grass fertilization is more simple. Growth slows in warm-season grasses when daytime high temperatures exceed 95 °F (35 °C) for extended periods. However, warm-season grasses can be fertilized to match seasonal growth rates or can be fertilized on a regular periodic basis throughout the growing season with few detrimental effects. Most managers caution against fall N fertilization of warm-season grasses. Nitrogen fertilization tends to make plant leaves more succulent. In view of this, and because warm-season grasses are very sensitive to cold temperatures, it is believed that damaging ice crystals might form in succulent warm-season plants at higher temperatures than in drier plants. Although this reasoning is sound, scientific tests have not demonstrated such an occurrence (Reeves et al., 1970; Goatley et al., 1994). In spite of that, even scientists are cautious about recommending fall N fertilization of warm-season grasses.

Rainfall is also a factor for consideration during N fertilization. Nitrogen fertilizers may be lost to volatilization or to leaching, but the most likely loss of N fertilizer is in surface runoff (Petrovic, 1990). Runoff occurs when rainfall exceeds the infiltration rate of the soil. Once the soil pores fill with water, surface runoff begins. If the soil is saturated or partially saturated before rainfall begins, runoff occurs more rapidly than it would if the soil was dry. Consequently, fertilizer on saturated or partially saturated soil may be wasted if rainfall occurs within the next 24 to 48 hours. In addition, periods of rainy weather are usually accompanied by long periods of cloud cover which reduce photosynthesis. The temperature is usually cooler than normal as well, and as long as the turfgrass is wet it stays cool. Therefore, respiration slows and transpiration is not required. Less nutrition is required too. Periods of wetter than normal weather require adjustments to your fertilizer program. Periods of exceptionally dry weather also require alterations to your fertilizer program.

If irrigation is available, fertilization should continue normally during periods of dry weather. However, roots do not take up fertilizer unless it is dissolved in water, so fertilizing non-irrigated turfgrass when the weather is dry is a waste of time and money. During dry weather, a large amount of N fertilizer may volatilize before rainfall occurs and soil water becomes available. Other fertilizers (P, K, etc.) will also degrade, but degradation will occur more slowly. Humidity is another factor that should be considered before fertilization.

High humidity slows transpiration and also encourages fungal diseases. When N fertilizer is added to turfgrass the grass becomes greener. The grass greens because N fertilizer encourages chlorophyll synthesis. As chlorophyll levels increase, turfgrasses absorb more light energy and have to dissipate more heat. If transpiration is slow because humidity is high, the turf becomes too hot and its overall health declines. Therefore, N fertilization should be adjusted during periods of higher than normal humidity. If disease occurs, depending on the pathogen present, N fertilization may make the disease worse.

Nitrogen fertilizer and water are the two major inputs that make the most difference between managed and unmanaged turfgrass. Nitrogen fertilizer makes turfgrass green. It also encourages turfgrass to grow and spread, and increases the overall health of the plant. However, there are occasions when N fertilization should be reduced in comparison with the normal amounts applied. Periods of high temperature, high rainfall and high humidity call for less than normal fertilizer. As in late-fall fertilizer applications to cool-season turf, sometimes the timing of the application is as important as the amount applied. A turfgrass manager should always know what weather is expected for the next 3 or 4 days following a fertilizer application. A severe storm within 24 to 48 hours of a fertilizer application could result in substantial losses to runoff (Shuman, 2002).

NITROGEN FERTILIZERS Nitrogen fertilizers for turfgrass are separated into two categories, quick release and slow release. Quick-release fertilizers are soluble in water, relatively inexpensive, can be applied in spray or granular form, can be applied in any season and provide a quick turfgrass response. Slow-release N fertilizers are not readily soluble, are moderately to highly expensive, with one or two exceptions can only be applied in granular form and provide a slow fairly constant N release over a long period. Slow-release N provides a consistent turfgrass response over time, but with one exception, IBDU (isobutylidine diurea), has little effect when applied in cool weather (Volk and Horn, 1975). Turfgrass managers tend to be loyal to the types of fertilizer they use. Not just to the quick-release, slow-release type of fertilizer used, but often to the manufacturer and formulation. These managers trust the products that they use but, even more important, they trust their ability to accurately determine how best and when to apply them. With these particular products, they are confident that they can determine how their turfgrass will respond under different environmental conditions and to different rates of application. That should give you some idea of how important they deem N fertilization to be in a management scheme for providing outstanding turf (Box 7.1).

There are several types of quick-release and slow-release N sources to choose from. Some of the most common forms are listed in Table 7.3. These and other N sources are often mixed to provide specific formulations and combinations that include both slow-release and quick-release N in a bag of fertilizer. Explanations of how these fertilizers are manufactured and combined, and the special properties of each, can be found in many sources (Carrow et al., 2001; Christians, 2007; Turgeon, 2008).

One of the most difficult aspects of N fertilization is its propensity for N being lost to the atmosphere or to runoff or soil water. Nitrogen fertilizer volatilizes easily, meaning it is lost to the atmosphere as a gas (Petrovic, 1990). The warmer the temperature, the more likely that volatilization will occur. Losses of N to the atmosphere are more likely to occur from quick-release N sources than from slow-release sources (Torello et al., 1983). Because quick-release N is highly soluble, it is also more likely to be lost to surface or soil water. A major rainfall event that is sufficient to produce runoff will carry surface-applied soluble fertilizers in solution off the target site and into surface water (Cole et al., 1997). However, irrigating with approximately 0.25 inches (6mm) of water immediately following fertilizer application will prevent most volatilization and runoff losses (Titko et al., 1987; Shuman, 2002).

Although quick-release N has advantages, it also has the disadvantage of being easily lost to volatility or water movement. Quick-release N is less expensive than slow-release N but must be applied

> **Box. 7.1. Nitrogen fertilizer sources: believe it or not.**
>
> Turfgrass managers have all kinds of opinions concerning fertilizer sources. I have heard experienced golf course superintendents claim that quick-release granular N fertilizer is more likely to burn creeping bentgrass (*Agrostis stolonifera*) putting greens than spray applications of the same source. That does not seem to make sense, but when more than one experienced superintendent makes that statement you have to believe that there might be something to it. I asked one experienced superintendent, one of my old students, if he believed that a granular slow-release fertilizer would burn his greens. He said "probably not but why take the chance". I have heard many managers claim that quick-release N sources, when applied properly, result in a slightly greener, slightly denser canopy than an equally good slow-release source. There is no scientific basis for that opinion. Nonetheless, I also have that opinion. But I also believe that a turfgrass fertilized regularly (monthly or so) with a slow-release fertilizer is healthier than one fertilized properly with a quick-release source. Although that opinion may have some basis, it is not scientifically sound and is just an opinion.
>
> Although these opinions have no scientific basis and may seem unreasonable, there is something to be said for having confidence in the product that you use and less confidence in those that you do not. Considerable experience with a good fertilizer product that you believe in is substantially better than using a supposedly more reliable product in which you have little experience and little faith. Knowing your product intimately is a huge advantage when environmental conditions warrant a change in your normal fertilization plan. Unsupported opinions are usually destructive, but there are rare times when they can be advantageous. In this case, there are a number of good fertilizer products and companies seem to be making them better and better. Choose the one that works best for you, learn its performance characteristics under a variety of conditions and use it with confidence.

Table 7.3. Common quick-release and slow-release sources of nitrogen for turfgrass fertilization.

Quick-release N sources	Slow-release N sources
Ammonium nitrate	Natural organics (manure, etc.)
Ammonium sulfate	Methylene urea
Potassium nitrate	Sulfur-coated urea
Calcium nitrate	Polymer-coated urea
	Isobutylidene diurea (IBDU)

at lower rates more frequently. Slow release N is more expensive but is less labor intensive because it can be applied less frequently. Turfgrass managers often choose the type of fertilizer they use because they like the turfgrass response that occurs because of it. However, those on a tight budget must also consider the economics of the fertilizer they choose, including the labor expense required to apply it. If your labor is expensive, slow-release sources may be more economical than quick-release sources. If labor is relatively inexpensive, then quick-release sources are likely to be cheaper. Regardless, you can save expensive fertilizer movement off the site by handling the product properly. Once fertilizer is watered into the soil it is not likely to volatilize and once in the soil it will not run off. Some fertilizer may be lost to leaching after watering in, but most will bind to the soil CEC strongly enough to resist leaching. As soils high in sand have low CEC, leaching is more probable in sand than in soil (Petrovic, 1990). However, if a moderate amount of soluble fertilizer is watered in with about ¼ inch of irrigation or rainfall, as mentioned earlier, it will be assimilated by the plants quickly. Consequently, after 2 to 3 days, there will be little fertilizer left to leach.

Quick- and slow-release N sources cause slightly different color/growth responses from turf (Fig. 7.7). Quick-release sources elicit strong color and growth responses that reach a peak in about 10 to 14 days, depending on environmental conditions, and then decline at about the same rate. A slow-release N source elicits more gradual color and growth responses that are sustained over a longer period (Waddington and Duich, 1976). Consequently, monthly applications of slow-release fertilizer result in a consistent growth rate and a more consistent color response compared with monthly applications of quick-release sources (Zhang *et al.*, 1998). For that reason, quick-release sources need to be applied at lower rates more frequently than slow-release sources to elicit consistent turfgrass responses. A steady growth rate is preferable and considered healthier for turfgrass than a series of peaks and

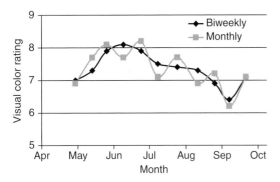

Fig. 7.7. Actual color response of bermudagrass (*Cynodon dactylon*) fertilized with urea nitrogen, a quick-release N source every 2 weeks (biweekly) at 0.3 pounds N/1000 square feet (15 kg ha^{-1}) averaged over two seasons in Stillwater, Oklahoma (black line). The gray line represents what the response curve would probably look like (estimated) if the fertilizer was applied at 0.6 pounds N/1000 square feet monthly. Notice the peaks and valleys as the monthly applications reach maximum color then decline until the next application cycle. A slow-release N source at the same rate would closely approximate the smooth response curve of the biweekly applications. Color was rated visually on a scale of 1–9, where 1 = brown and 9 = dark green.

valleys that occur with each fertilizer application. However, it must be remembered that each slow-release product has slightly different attributes, and some research among suppliers and producers is required to determine the product that best meets your needs (Carrow, 1997).

It should also be remembered that most slow-release products are affected by microbial activity and therefore by temperature. Release of all slow-release fertilizers except for IBDU is delayed by cold temperatures (Wilkinson, 1977). The N release of IBDU is controlled by water and determined by particle size; the larger the particle, the slower the release. Microbial activity need not be present. However, N release from natural organics such as manure, or from methylene urea, is controlled by microbial activity and does not occur in cold weather (Wilkinson, 1977). Release of N from sulfur-coated or resin-coated urea is partially controlled by imperfections in the particle coating and partially by microbial breakdown. Therefore release slows but does not stop in the winter. The same is true of polymer-coated urea particles that work by diffusion. As water cools it becomes more viscous, and the diffusion of urea in and out of the polymer-coated particle is affected. Another factor that should not be forgotten is that if the coating on a coated urea particle is damaged, it becomes a quick-release fertilizer. Mowing, aerification and other cultural practices can damage coated particles of urea, and the N application that you expected to release over a long period may release much quicker. Coated particles need to be watered in before mowing and fertilization should not be followed by other cultural practices that are likely to cause damage to the particle. Both quick-release and slow-release N sources provide satisfactory results for the N fertilization of turfgrass. However, application timing and application methods vary and need to be seriously considered when developing a fertilizer program.

Clippings are a primary source of slow-release N. When the temperature is warm, clippings are broken down fairly rapidly by microbial activity and nutrients are returned to the soil and plants (Starr and DeRoo, 1981). Research has indicated that the N recovered from clippings of Kentucky bluegrass (*Poa pratensis*), perennial ryegrass (*Lolium perenne*) and tall fescue (*Festuca arundinacea*) was often greater than the N fertilizer applied (Liu and Hull, 2006). It has been estimated that N fertilizer could be reduced by 50% on a mixed stand of Kentucky bluegrass, perennial ryegrass and creeping red fescue (*Festuca rubra*) if clippings were returned instead of collected (Kopp and Guillard, 2002). Unless clippings must be collected to promote the playability of a surface, or the aesthetic value of the turf requires collection, clippings should be left in place; they are a great source of fertilizer.

Phosphorus

Phosphorus (P) is the least important of the primary macronutrients for turfgrass fertilization. A primary use for plant P is in cellular energy molecules such as ATP (adenosine triphosphate) that are in high demand during periods of rapid growth or reproduction. Because turfgrasses are normally mowed below the heights at which they flower and seed, P is not as important for a reproductive energy boost in turfgrass as it is in most flowering plants. In addition, turfgrasses have very extensive root systems for their size and are capable of sequestering soil P better than most plants (Christians, 2007). Research has demonstrated that turfgrass roots actually seek out P and will grow toward the greatest concentration of it (Lyons *et al.*, 2008).

Soil tests determine soil P availability and whether or not P fertilization is required. A very general estimate of the P subsistence level for adequate turfgrass growth is around 12 ppm (parts per million). Phosphorus levels recommended by local laboratories are usually accurate. However, subsistence levels can be affected by many factors, and a little experimentation is sometimes required. If your turfgrass is not responding to fertilization as you think it should, at least one nutrient may be deficient. Soil tests and laboratory recommendations may often lead you to the missing nutrient. Apply fertilizer containing the suspected nutrient to some small areas at the rate used previously with unacceptable results, and fertilizer containing a high rate of the nutrient to other small areas and compare responses. That procedure will usually demonstrate whether or not the suspected nutrient is deficient in the fertilizer mix. If P is deficient it is often because of turfgrass competition with trees or other flowering plants for soil P or because soil pH is below 5.5 or above 7.5.

There was a time when we recommended that turfgrass managers determine an adequate annual P to apply based on turfgrass use and soil P present, and make one or two applications per year to apply sufficient nutrient. Because of the dangerous environmental effects of excess P in surface water mentioned earlier, we no longer recommend that procedure. Now we recommend that P be applied in regular small applications and watered in immediately with about ¼ inch (6.4 mm) of irrigation whenever possible. We also recommended applications of P during establishment, but this too has changed. It is true that seedling turfgrass has a difficult time assimilating P in its early stages of root growth. However, the establishment period is also the best time for P runoff to occur because the soil is practically bare and the turfgrass root system is not strong enough to hold the soil together. After seedlings have begun to grow, it may be wise to apply a light application of P for support, but the old recommendations of starter fertilizers high in P are no longer valid. We have not experienced a decline in growth or delay of turfgrass establishment substantial enough to warrant a P application at establishment. Soils low in P should be adjusted before establishment by tilling P into the topsoil. As long as adequate P is present in the soil, germination and seedling growth should proceed as desired. Sometimes additional P at establishment makes no difference at all (Frank *et al.*, 2002). Soil tests for P are quite accurate, inexpensive fertilizers are reasonably easy to find and mature turfgrasses are very good at finding P when they need it. Consequently, planning a P fertilization program is relatively easy.

Potassium

Maintaining adequate potassium (K) levels on fine textured soil is fairly easy, but it can be more difficult on soils with very high sand contents. On sand, potassium is not tightly bound and leaches easily. Potassium soil analysis and laboratory recommendations are accurate and it is widely believed that excess K is not toxic to turfgrass, nor does it have detrimental environmental effects. Consequently, a little too much doesn't hurt, or so we believe.

Potassium fertilization has been linked to partial relief from numerous turfgrass stresses (Turner and Hummel, 1992). As yet, we do not know exactly why potassium has such desirable stress relief qualities, but we know it works. Potassium is second to nitrogen in its concentration in turfgrass plants (Turner and Hummel, 1992). It is not a constituent of important biological molecules, but it is involved in many metabolic reactions and in the maintenance of cell turgor, and is often responsible for maintaining electrical gradients across membranes. Potassium is sometimes applied at rates equal to those of N, especially on sandy soil. However, that is typically more than is needed.

Potassium is sometimes applied to warm-season turfgrasses as they enter dormancy to help protect them from cold and desiccation during the winter. However, there have been mixed reports on whether or not such a program is actually effective. Research in Massachusetts determined that late-fall applications of K reduced winter damage on perennial ryegrass, a cool-season grass, but results in North Carolina on hybrid bermudagrass (*Cynodon dactylon* × *Cynodon transvaalensis*) suggest that late-season K applications had no effect on winter hardiness (Peacock *et al.*, 1997; Webster and Ebdon, 2005).

Fertilization of K at about half the rate of N applied is usually adequate. However, K fertilization should be adjusted for soil tests and other site conditions based on observation and site history.

The secondary macronutrients and the micronutrients

Soil tests from respectable laboratories are typically very accurate. However, most laboratory analyses, including those involving soil nutrients, become less accurate with decreasing concentrations. Reported amounts of the primary and secondary macronutrients, except perhaps for sulfur, are highly accurate, but reports of micronutrients are harder to evaluate and may have marginal accuracy. That is not to say that micronutrient reports are not useful. More often than not, a soil test report will identify micronutrient deficiencies. If a micronutrient deficiency is suspected, test areas (as described earlier) are a good method for determining whether or not the nutrient should be included in the fertilizer program. A small amount of a safe-for-turf chemical containing the nutrient of interest can be applied with normal fertilizer to some areas and not to others, and the turfgrass response evaluated.

Calcium, magnesium and sulfur

Calcium (Ca) is a primary constituent of most soils but not all Ca is plant available. At low soil pH, Ca is often present in insoluble forms that turfgrass cannot assimilate. However, as you should recall, lime ($CaCO_3$) is the product best used to increase soil pH and it also contains soluble Ca. If the soil pH is not too low but Ca is still not available, gypsum ($CaSO_4$), the same product as used to remove soil salts, is the product of choice for Ca application. Gypsum is considered to be pH neutral.

Magnesium (Mg) may also be deficient at low soil pH. Epsom salts ($MgSO_4$), available at most drug stores, is a good product to use for testing of Mg deficiencies. However, it is probably not the best product to use for treating Mg deficiencies and may not be legal for direct soil application in your area. A Mg-containing fertilizer is a better choice once a Mg deficiency is identified. Sulfur (S) can also be unavailable at low soil pH and this should be tested using agricultural sulfur (S_2). Agricultural sulfur or an S-containing fertilizer can be used to correct deficiencies.

Iron and other micronutrients

Micronutrient deficiencies are rare in native soils but can occur. Most commercial fertilizer producers now include micronutrient fertilizers in their product lines. That is primarily because there has been such a strong industry movement toward sand-based sports fields and putting greens that are considerably more likely to be micronutrient deficient than native soil. Native soils, especially those polluted by human activities, are as likely to have toxic levels of micronutrients as they are to be deficient.

Iron (Fe) is the most important micronutrient for turfgrass growth and maintenance. Iron directly promotes chlorophyll synthesis, so its addition to turfgrass almost always results in a nearly instant color change. The turf color usually turns blackish at first, but changes to green after about 24 hours. Some scientists believe that the black response is actually a staining response rather than a response to increasing chlorophyll or heme synthesis. Whether or not a staining effect is involved, Fe clearly encourages chlorophyll synthesis and will nearly always add green color to turf (Lee et al., 1996). For that reason, light applications of chelated or soluble Fe may be used to cause turf to green without encouraging shoot growth. Deficiency testing for Fe by spraying some plots and not others is not productive because a change in chlorophyll concentration and color nearly always occurs. A high pH will usually result in an Fe deficiency, and the missing Fe causes the plant to yellow noticeably as a result of chlorophyll deprivation. Many turfgrasses, especially those that prefer acid soils, need supplemental Fe in very small amounts. If your turf is reasonably healthy but yellow for no apparent reason, it may need Fe. Centipedegrass (*Eremochloa ophiuroides*) is notorious for needing supplemental Fe (Carrow et al., 1988). However, excess Fe can cause slight growth declines in creeping bentgrass (*Agrostis stolonifera*), annual bluegrass (*Poa annua*) and possibly other grasses (Xu and Mancino, 2001). Growth may also slow following an Fe application when air temperatures are high (Schmidt and Snyder, 1984). Light Fe applications, however, are relatively safe, and if growth slows, it does not have a severe negative effect on the turf.

Manganese (Mn) is very important for the splitting of water molecules during photosynthesis. However, it is rarely deficient and is one of those nutrients that is typically difficult to assess by soil testing. A Mn-containing fertilizer should be tested before investing in a potentially expensive program of Mn fertilization. Deficiencies in plant available

Mn may be related to high soil pH. In that case, lowering soil pH to below 7.0 will probably solve the Mn problem and also positively affect the availability of other nutrients (Snyder et al., 1979). Copper, zinc and boron are rarely deficient, but may be unavailable at high pH. Many micronutrient fertilizers contain trace amounts of these elements. Test for deficiencies before investing in a micronutrient program. Soil tests for micronutrients are inexpensive and have value, but are not always accurate (Faust and Christians, 2000). If a nutrient deficiency or toxicity is observed through soil testing, or suspected, simple experiments should be designed for verification before applying micronutrients on a large scale. If tissue tests are used to determine micronutrient content, it should be remembered that micronutrients, like macronutrients exist in plant tissue in different concentrations depending on season and environment (Belesky and Jung, 1982).

Nutrient interactions

Nutrient uptake and concentrations within turfgrass plants are affected by each other (Christians, 1993). Turfgrass, for instance, may not take up K if N is deficient (Petrovic et al., 2005). In fact, interaction between K and N for plant maintenance is common. Potassium is needed for a number of physiological processes to occur and also for other purposes within a turfgrass plant. Those purposes include helping to maintain cell turgor pressure and osmotic potential, and helping to preserve electrical gradients across cell membranes (Carroll et al., 1994). These functions are extremely important, and if K is not present to perform them plants will use sodium (Na) as a substitute. Plant use of Na in place of K can lead to a decline in turfgrass quality (Snyder and Cisar, 2005). Therefore, K fertilization is extremely important if you expect to manage turfgrass at optimum levels.

When sufficient K is part of an N fertilizer program, the turf will use N more efficiently (Fitzpatrick and Guillard, 2004). For that reason, less N may be needed in the program and it can be adjusted to match your demands. As far as we know, K is not damaging to the natural environment as excessive N and P can be. However, it can reach levels detrimental to plants. It would not really be accurate to say that K can reach toxic levels, because that would suggest that excessive K could kill plants; that is highly unlikely and would require a huge amount of K in the soil. However, excessive K can cause deficiencies of other macronutrients (Miller, 1999). Potassium competes with Ca, Mg and other nutrient cations for exchange sites on the soil CEC. If K is excessive it can cause soil deficiencies of Ca and Mg (Woods et al., 2005).

Earlier, I recommended that a K fertilization program should be approximately half that of the N applied. Based on research and observation that is a good starting point for your fertilizer program (Christians et al., 1979; Snyder and Cisar, 2000). In fact, your entire fertilizer program should be based on the amount of N applied. Basically, nothing happens in turf without nitrogen. If, for instance, your soil is deficient in S or Mg your turf may not take up those nutrients if N is not also present (Goss et al., 1979). If green turfgrass is an important goal for your site, application of mixtures of Fe and N will increase green color better than applications of either nutrient alone (Yust et al., 1984). Iron is the nutrient that has to be present for the heme synthesis pathway, the pathway that also produces chlorophyll, to occur. Nitrogen is an important part of the chlorophyll molecule and also encourages chlorophyll synthesis when present. The nutrients in combination are more likely to encourage chlorophyll synthesis together than either would alone. You may be able to grow green grass with less N input if Fe is also applied. However, N must always be present for Fe to be most useful. That is the case for nearly all turfgrass activity. Nitrogen must be present in adequate supply for maximum turfgrass performance. Otherwise, the remaining nutrients will not encourage turfgrass performance as efficiently. Begin your fertilization program based on the amount of N needed to meet expectations, then adjust additional nutrients to combine with the program most effectively.

7.4 Chapter Summary

Carbon, hydrogen and oxygen make up most of the dry matter in a turfgrass plant. To supply those nutrients we must encourage photosynthesis, respiration and transpiration. Sufficient light must be available for photosynthesis to occur. Once carbon dioxide is removed from the air surrounding a turfgrass canopy, fresh air must be available to replace it. As transpiration cools our plants, the humidity in the air around them increases and transpiration slows unless fresh air is available. Oxygen must be present for respiration to occur. Air is limited in some soils and fresh air infiltration may also be limited. As oxygen is used by turfgrass roots, fresh soil air must be available to replace the air that has become oxygen deficient. Soil drainage, aerification and other soil cultivation and management practices are required to encourage adequate oxygen for sufficient root respiration.

Nitrogen fertilizer is required to synthesize chlorophyll molecules, and high N availability tends to encourage chlorophyll synthesis. Hence fertilized turfgrass becomes more green. Nitrogen is also a part of every enzyme that catalyzes metabolic reactions and a part of every protein. It is the most conspicuous nutrient in turfgrass plants (Jones, 1980). Properly applied N fertilizer programs encourage both shoot growth and root growth. They also encourage high turfgrass density and a nice green color. A good N fertilization program helps turfgrass out-compete weeds and resist pathogen and insect invasion. Too much N, however, can encourage certain diseases and other problems. Too much N causes plants to retain more water, becoming more succulent and more easily damaged by traffic. Excessive N also encourages rapid shoot growth, usually at the expense of root growth and that can result in a very weak root system. A typical N fertilization plan should be a part of the maintenance schedule, but it should be flexible and adjusted for weather conditions such as temperature, rainfall and humidity.

If sufficient N is present, turfgrass plants will respond to P and K fertilization when needed. Phosphorus is important for the formation of cellular energy and for DNA and RNA synthesis. Potassium helps to maintain electrical gradients across cell membranes and cell turgor. It also has to be present for many metabolic reactions to occur. Sometimes Ca, Mg or S are required and occasionally micronutrients are missing from the soil solution. A regular program of soil testing helps to determine the fertilizer program required to manage your turf at acceptable levels. Proper planning, execution and adjustment of that program results in healthy turfgrass and satisfied customers.

Suggested Reading

Brady, N.C. and Weil, R.R. (2008) *The Nature and Properties of Soils*, 14th edn. Prentice Hall, Upper Saddle River, New Jersey.

Carrow, R.N., Waddington, D.V. and Rieke, P.E. (2001) *Turfgrass Soil Fertility and Chemical Problems: Assessment and Management*. Ann Arbor Press, Chelsea, Michigan.

Hull, R.J. (1992) Energy relations and carbohydrate partitioning in turfgrasses. In: Waddington, D.V., Carrow, R.N. and Shearman, R.C. (eds) *Turfgrass*. ASA-CSSA-SSSA (American Society of Agronomy-Crop Science Society of America-Soil Science Society of America), Madison, Wisconsin, pp. 175–206.

Turner, T.R. and Hummel, N.W. Jr. (1992) Nutritional requirements and fertilization. In: Waddington, D.V., Carrow, R.N. and Shearman, R.C. (eds) *Turfgrass*. ASA-CSSA-SSSA (American Society of Agronomy-Crop Science Society of America-Soil Science Society of America), Madison, Wisconsin, pp. 385–440.

Suggested Websites

Duble, R.L. (2010) Turfgrass Fertilization. Texas A&M University. Available at: http://aggie-horticulture.tamu.edu/plantanswers/turf/publications/fertiliz.html (accessed 12 March 2010).

Georgia Turf (2010) A Balanced Soil Fertility Program is Essential in Turfgrass Management. Available at: http://mulch.cropsoil.uga.edu/fertility/ (accessed 3 August 2010).

Mississippi State University Extension Service (2010) Home Lawn and Turf in Mississippi. Available at: http://msucares.com/lawn/lawn/fertilization/index.html (accessed 12 March 2010).

Penn State Cooperative Extension, Berks County (2010) Turfgrass – Fertilization. Available at: http://berks.extension.psu.edu/mg/hgargd/turffert.html (accessed 12 March 2010).

8 Irrigation and Water Management

> **Key Terms**
>
> A **water window** is the length of time that you have available to irrigate your site each day. It is not the length of time required, but the length of time available.
>
> The term **precipitation** is normally used to describe natural rainfall. It may also be used to refer to other naturally occurring climatic forms of water such as hail, snow, sleet and even dew. We also use it to refer to irrigation and other forms of simulated rainfall.
>
> **Static pressure** refers to the pressure in a closed irrigation system when no water is moving. Although static pressure is constant throughout the system, it is really only useful for irrigation design when it is considered at the water source. Once the water starts moving, friction becomes a factor and the pressure declines.
>
> **Dynamic pressure**, also called working pressure, is the pressure that exists at a given point in an irrigation system when it is operating. Dynamic pressure changes throughout the system, declining as a result of friction loss as it travels from the source and increasing or decreasing with elevation. A system must be designed to have adequate dynamic pressure at each sprinkler head.
>
> A **backflow preventer** is a sophisticated check valve that prevents the siphoning of water from an irrigation system backwards into a potable water supply. Without a backflow preventer, pesticides, fertilizers and any number of other materials could flow into sprinkler heads and siphon back into a public water source.
>
> The **friction factor** is a calculated number that irrigation designers use to help them choose the proper pipe size required to supply the individual irrigation heads in a zone.
>
> **Evapotranspiration** (or **ET**) is the term used to describe the total water lost through a combination of evaporation and transpiration from a turfgrass or any other cropping system.
>
> The ET_o is the evapotranspiration reference standard determined by entering specific weather parameters into a mathematical model developed to calculate ET from a reference crop. The reference crop chosen for this calculation happens to be a grass.
>
> The K_c is a crop coefficient determined by scientific study of a particular species in a particular location.
>
> The ET_c is the estimated evapotranspiration of a particular crop on site or at a location nearby.
>
> **Precision irrigation management** is the mapping of a large turfgrass area for sectioning into management units based on irrigation requirements, followed by adjustment of the irrigation system to best accommodate the needs of each unit. Once the site is mapped and the units determined, similar units are combined and irrigated independently based on need.
>
> **Water use rate** is the amount of water a turfgrass species extracts from the soil over a given period during adequate precipitation.
>
> **Leaf fire** is a measure of leaf browning of a turfgrass caused by long periods without precipitation.
>
> **Drought tolerance** is the ability of a turfgrass plant to survive long periods without precipitation.

8.1 Manage Your Irrigation and Drainage for Site-specific Objectives

Over the course of my life, I have been fortunate to live in areas of the USA that have either too much or too little rainfall. Not long ago, I experienced a summer storm when over 2 inches (5.1 cm) of rain fell in less than 30 minutes. It was not a tornado, a hurricane or a cyclone, it was just a rain storm. It was incredible and, needless to say, drainage was a problem. Drainage was a particular problem because the event occurred in a relatively dry portion of the USA where drainage is often ignored.

It has been my experience that regions where large amounts of rainfall occur tend to have poor irrigation systems and regions where little rainfall occurs tend to have poor drainage systems. As a turfgrass manager, you need to be intimately familiar with both irrigation and drainage. If you are managing high-quality, high-value turf, both irrigation and drainage are major concerns for you. In most areas of the world, periods of dry weather and periods of wet weather occur annually regardless of overall climate. If you are not prepared for both, you are going to lose grass.

In this chapter, we will address both too much water and too little water. In nearly all regions, dry periods occur, and turfgrass requires irrigation to look and function best during those periods. Irrigation is not only a science but an art. It requires complete flexibility for weather changes and site-specific differences in soil, light and airflow environments. Plant competition and ecological systems have to be considered, and the limitations of water use or irrigation system design have to be part of irrigation planning. Drainage is more of a common-sense technology than irrigation, but in some situations it requires some creative ideas and problem solving. Although most turfgrasses can survive several weeks of drought and recover, it only takes 1 or 2 weeks to die under submerged or waterlogged conditions. Remember, even if you are dealing with an extremely well-drained soil, like pure sand in a dry climate, it is still going to rain occasionally. No matter how well drained your soil is, the water has to have someplace to go or you are going to be dealing with saturation or submersion. You have to have a plan, even if that plan is as simple as draining the area into a collection swale or cistern where excess water can be pumped off-site.

8.2 Irrigation Management

Students often ask why most turfgrass college curriculums include an irrigation class where they have to learn about the calculations required of irrigation design and where they might have to learn about methods involved in central pivot irrigation or other non-turf systems. The answer is this: you are not training or taking classes or reading this book so that you can be a technician. You are training to make decisions for technicians. No matter how deeply you respect a highly qualified technician, and you should, that is not your job. Knowing how to be an irrigation technician would make your job easier, but that knowledge is really not necessary. It is your job to know how irrigation systems work, not how to install or repair them. You need to know the design. You can learn the installation and repair later and in far less time.

In earlier times, I was hired as a supervisor into two different industries where I had absolutely no experience. However, in spite of the fact that I did not know how to make the products, I did know how to manage people, organize a work force and make intelligent decisions. Therefore, in both cases, I was very successful and once I earned their respect, my employees were happy to help me learn how to do their jobs. Knowing how to do their jobs made my job easier but it wasn't absolutely necessary.

I don't wish to offend our agricultural colleagues, but a central pivot irrigation system is the simplest of irrigation systems. If you don't understand how central pivot irrigation works, you can't possibly hope to understand a simple home lawn system and most certainly not a system for a golf course or athletic field complex. Those systems are far too complicated for you. Several of my students have commented to me after they entered management positions that they wished they had paid more attention in irrigation class. Now they have to design a system, make an addition to a system, or oversee a design and installation by an independent contractor. If you don't understand basic irrigation system design, you are not qualified for those tasks. In the opinion of many who might consider hiring you, being proficient in irrigation design is part of your job.

When it comes to irrigation systems, turfgrass people are the very best. We may lag behind the expertise of other agricultural industries in some plant management methods, but we are the very best in irrigation. This text is not going to provide you with an education in turfgrass irrigation systems. I hope that you will further your education in that area with more informative books and classes. However, we will discuss the basics of irrigation design because you need to know those basics to make good management decisions.

Basic irrigation design

Everything about an irrigation system is affected by its water source. The water source determines how much area you can irrigate, how large your pipe has to be, how many heads you can put on a zone, and how many zones you can run at one time. The whole idea of water source is based on a simple concept called the "water window".

The water window is the length of time available to irrigate everything that you need to irrigate each day. It is not the length of time required to irrigate your site, it is the length of time available to irrigate your site. In other words, if you want to be able to irrigate your entire athletic complex in one night and the complex closes at 20:00 and opens at 7:00 during the longest day of the year, your water window is 11 hours. If you only require the capability of irrigating half your complex each night, your water window is still 11 hours but you are allowing two windows (22 hours) to complete the job. If it requires 16 hours to pump all of the water that is required to irrigate your complex and you need the capability to irrigate it all in one night, then your window is still 11 hours but your system is too small. You need a new water source. Hence, your water window determines how large your water source has to be to irrigate your site. It is always best to have a larger water source than you need because you never know when you might lose your water or have to replace a pump or get into some other situation where your turf has not received water for some time, and you need to get it all irrigated as fast as you can. If your water source is a public water supply, you do not have much control over the rate of supply. However, if your water source is a pumping station from a lake or other water feature with a nearly unlimited supply of water, the supply rate is mostly up to you. In that case, the supply usually depends on how much you or your sponsors are willing to spend for the system. When making those decisions, the reliability of the pumping equipment and the rest of the supply system are of profound importance. In addition, a backup pumping system must be considered and included in the purchase.

You can find instruction manuals and tables at most irrigation supplier websites (see the list at the end of this chapter); these are very helpful for understanding irrigation principles and helping you learn to design and install an irrigation system. There are also books available on turfgrass irrigation systems and drainage which are listed at the end of this chapter. I will not attempt to instruct you on how to design or install an irrigation system, but you need a basic knowledge of irrigation systems to use them most efficiently. Certain basic principles and construction apply to all turfgrass irrigation systems, large or small. The water window determines how large your water source must be to fulfill your irrigation needs. You may use a public water source, an irrigation pond, a well or some other source. Assuming that there is enough water available at the source to satisfy all of your irrigation needs, the size of the pipe and the pressure in the pipe determine how much water you have to work with. Your public system or your pump must be able to fill that pipe continuously at the pressure required. Before you start buying pumps or pipe, however, you must first determine how much water is required to irrigate the target area. To keep this discussion simple, let us assume that you will be irrigating a home lawn with water from a public water source.

Just because you are using a public water source from a municipal system does not mean that you have an unlimited supply of water. Some systems are barely adequate to supply drinking water. You could be dealing with a system that has a small supply line on 30 pounds of pressure per square inch (psi) or 26 bar. That source will not supply very much water and you will have to plan accordingly. Assuming that you have a $\frac{5}{8}$-inch (16-mm) water meter on 30 psi, your source will theoretically supply about 4 gallons per minute (gpm) (or $0.908\,m^3/h$) from a 1-inch (25-mm) supply pipe (Hunter Irrigation Innovators, 2009a). Now let us assume that the lawn you intend to irrigate is 10,000 square feet (sf) ($929\,m^2$) and that during the hottest, driest periods, you will have to supply irrigation at about 2 inches (51 mm) per week. Under a worst-case scenario, the system would break down and the homeowner would be forced to apply all 2 inches of irrigation in 1 day as soon as the system was repaired to keep the grass alive. A few calculations bring you to the conclusion that at the maximum rate of supply, the lawn will require 12,467 gallons ($47\,m^3$) of water and 52 hours to irrigate completely with 2 inches of precipitation. Consequently, the homeowner cannot irrigate in 1 day, but if nothing unforeseen happens, the homeowner could potentially supply the amount of irrigation needed during the driest season by running the system all night for four nights per week. However, if the homeowner happens to prefer the capability to irrigate everything in 1 night (8 hours), a total irrigation time of 52 hours is not workable. This particular system, because of a poor water source, would require a booster pump and probably a storage tank to supply enough water to irrigate the entire site within an 8-hour water window. If you are not in control of the water supply, you have to adjust to the situation. Now, let us assume that you have the same system, but a new water source that can meet the 8-hour water window, and you will typically run all zones on the system for 4 hours, two nights per week during the driest time of the year.

Our new water source has a 1-inch (25 mm) water meter and 60 psi (52 bar) pressure on a 1.25 inch (32-mm) supply line. This particular source has a maximum output of about 26 gpm: over six times greater than the old source (Hunter Irrigation Innovators, 2009a). In that case, we can supply 12,467 gallons (47 m^3) of water to the lawn in 8 hours, which will meet the requirements that we determined for the water window.

Assume that, in this example, we used a measure of static pressure to estimate the water available to operate our system. However, static pressure is a measure of potential pressure, not actual pressure. Dynamic pressure, often called working pressure, is the pressure that you actually have to work with at any particular point in your system. If you were to check the pressure on the system when it was at rest with no irrigation running, that would be the static pressure of the system. The static pressure would be basically the same throughout the system, except for the effects of gravity when elevation changes. However, if you were to check the pressure at any point when the system was at rest, and then check the pressure at the same point when the system was operating, the pressure would be lower during operation because water is moving through the system. When the system is operating, the pressure is not fixed or static, it is moving or dynamic. During operation, you are not measuring static pressure, you are measuring dynamic pressure. Dynamic pressure, the actual working pressure, changes throughout the system based on elevation and friction (Fig. 8.1). Pressure decreases as water is pushed uphill but increases as water is pushed downhill. Any restrictions inside the pipe cause turbulence and friction, which result in pressure loss. Our theoretical home lawn application, for instance, will have to have a backflow preventer on the main line before any water flows to the irrigation heads. The backflow preventer is a special valve that only allows irrigation water to flow in one direction. Consequently, groundwater or surface water cannot flow backwards from the irrigation heads to the potable water system and contaminate it. Valves have a manufacturer's rating describing how much pressure is lost as water moves through them. A backflow preventer on a small system might cause a 5 to 10 psi pressure loss, so the pressure in the system could be 10 psi lower than that of the source after water passes

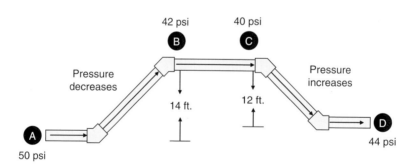

Pressure at point A = 50 psi (pounds/sq.in.)
Water passes through 2 fittings (−0.4 psi), 16 ft of pipe (−2.2 psi) and climbs 14 ft (−6.1 psi)
Pressure at point B = 42 psi
Water passes through 10 ft of pipe (−1.4 psi)
Pressure at point C = 40 psi
Water passes through 2 fittings (−0.4 psi), 12 ft of pipe (−1.7 psi) and falls 12 ft (+5.2 psi)
Pressure at point D = 44 psi

Fig. 8.1. Water flowing through pipe causes friction, resulting in pressure loss. Pressure is also lost when water is forced uphill but pressure increases as water flows downhill.

through the backflow preventer. Pipe fittings also cause friction, so each time water flows through a tee or an elbow, pressure is lost. Last, and possibly most important, the pipe itself causes friction. The amount of pressure lost is affected by the roughness of the pipe and the distance that the water has to travel through the pipe. The material used to make a pipe, plastic, steel, etc., differs in its roughness and in the amount of friction caused as water flows through it (Table 8.1). The size of the pipe also affects the amount of friction that occurs. A small pipe causes greater friction than a large pipe. Consequently, less pressure is lost in a 1.25 inch pipe than is lost in a 1.00 inch pipe. Therefore, pipe size is extremely important to the amount of water that can be supplied to the irrigation heads that are farthest from the source. Irrigation designers use a number they call "the friction factor" to help them determine what pipe size is required to adequately supply each head in the system (Box 8.1).

All but the simplest of irrigation systems are divided into sections called zones. Each zone is operated independently by a valve. The amount of water available from the source determines how many irrigation heads can be controlled by a single zone. Zones may also vary by the type of head used in the zone or by the type of plant material in the zone. Planting beds, for instance, should be on different zones from turf because bedding plants and turf are usually irrigated on different schedules. The friction factor helps to determine when pipe sizes can be reduced without seriously affecting the flow rate at the heads on a zone. Each head should distribute approximately the same amount of water uniformly on an area basis (e.g. gallons precipitation per square foot). How this is accomplished is not vital to you learning how to irrigate effectively, but it is interesting. Let us suppose that you have calculated a friction factor of 1.50 (see Box 8.1) and that you plan to use CL 200 PVC pipe. You can use a friction factor short cut chart (Table 8.2)

Table 8.1. The amount of friction as water flows through a pipe is influenced by the velocity of the water flowing, the size of the pipe and the type of pipe used. The velocity and the pipe size determine the flow rate. Irrigation supply companies provide friction loss charts that make it relatively easy to calculate pressure from one point in a pipe to another. A sample of a friction loss chart for comparison of polyvinylchloride (PVC) schedule 40 plastic pipe, polyethylene (PE) SDR (standard dimension ratio)-pressure rated tubing, schedule 40 standard steel pipe (steel) and type K copper tubing (copper) is given below. Notice that velocity does not change with the type of pipe used. That is because velocity is a function of flow rate and pipe size, and is not affected by friction. fps = feet per second; gpm = gallons per minute; psi = pressure per square inch.

Type of pipe	PVC		PE		Steel		Copper	
Size of pipe	1 in.		1 in.		1 in.		1 in.	
Flow (gpm)	Velocity (fps)	Pressure loss (psi)	Velocity (fps)	Pressure loss (psi)	Velocity (fps)	Pressure loss (psi)	Velocity (fps)	Pressure loss (psi)
1	0.37	0.03	0.37	0.04	0.37	0.07	0.37	0.05
2	0.74	0.12	0.74	0.14	0.74	0.26	0.74	0.18
3	1.11	0.26	1.11	0.29	1.11	0.55	1.11	0.38
4	1.48	0.44	1.48	0.50	1.48	0.93	1.48	0.65
5	1.85	0.66	1.85	0.76	1.85	1.41	1.85	0.98
6	2.22	0.93	2.22	1.06	2.22	1.97	2.22	1.37
7	2.59	1.24	2.59	1.41	2.59	2.63	2.59	1.82
8	2.96	1.59	2.96	1.80	2.96	3.36	2.96	2.33
9	3.33	1.97	3.33	2.24	3.33	4.18	3.33	2.90
10	3.70	2.40	3.70	2.73	3.70	5.08	3.70	3.53
11	4.07	2.86	4.07	3.25	4.07	6.07	4.07	4.21
12	4.44	3.36	4.44	3.82	4.44	7.13	4.44	4.94
14	5.19	4.47	5.19	5.08	5.19	9.48	5.19	6.57
16	5.93	5.73	5.93	6.51	5.93	12.14	5.93	8.42
18	6.67	7.13	6.67	8.10	6.67	15.10	6.67	10.47
20	7.41	8.66	7.41	9.84	7.41	18.35	7.41	12.73

Box 8.1. A friction factor helps to determine pipe size.

A friction factor is used to determine the size of lateral line pipes required to operate within a desired range of flow rates without exceeding a detrimental amount of pressure loss (Hunter Irrigation Innovators, 2009b). To maintain consistency and adequate coverage, all of the sprinklers on a zone should be operating within ±10% of and certainly no worse than ±20% of their desired operating pressure.

The friction factor is calculated by:

$$F_f = (P_o \times P_v)/L_c$$

Where:

F_f = the friction factor, which is the loss of pressure that you are willing to accept per 100 feet of pipe;

P_o = the sprinkler operating pressure suggested by the manufacturer;

P_v = the pressure variation between the valve and the last sprinkler head on the zone, not to exceed 10% or certainly no greater than 20%; and

L_c = the length of pipe from the control valve to the farthest head in hundreds of feet.

Let us assume that we have a zone of sprinklers that operate best at 30 psi (pounds per square inch) (P_o = 30 psi) and we want to operate all of the sprinklers on the zone within 20% of the recommended pressure (P_v = 0.20). The farthest sprinkler head is 400 feet from the valve (L_c = 400 ft./100 ft. = 4.0). Then the friction factor is:

$$F_f = (P_o \times P_v)/L_c = (30 \times 0.20)/4.0 = 1.5\,\text{psi}/100\,\text{ft.}$$

Once the friction factor has been determined, a friction factor short cut chart (Table 8.2) is used to determine the proper pipe size for each section of the zone (Fig. 8.2).

Table 8.2. A portion of a typical friction factor short cut chart. Once a friction factor is calculated and the flow rate that is needed for that particular section of a zone is known, the pipe size required can be estimated using the chart. A pipe that is too large will always work, but it is also more expensive. gpm = gallons per minute.

	Pipe	Friction factor				
		1.00	1.25	1.50	1.75	2.00
Size (in.)	Type	Max. gpm	Max. gpm	Max. gpm	Max. gpm	Max. gpm
0.50	CL 315 PVC	2.30	2.60	2.80	3.10	3.30
0.75	CL 200 PVC	4.50	5.10	5.70	6.10	6.60
1.00	CL 200 PVC	8.70	9.80	10.80	11.70	12.60
1.25	CL 200 PVC	16.00	18.10	19.90	21.70	23.30
1.50	CL 200 PVC	22.90	25.80	28.50	30.90	33.20
2.00	CL 200 PVC	41.00	46.30	51.10	55.50	59.70
2.50	CL 200 PVC	67.80	76.40	84.30	91.70	98.50

to determine the size of pipe required to supply an adequate amount of water to each section of a zone (Fig. 8.2). On a system with small zones, such as our lawn example, using the same pipe size throughout a zone is typical. However, on systems with large zones, such as that shown in Fig. 8.2, reducing pipe size in the farthest sections of the zone can save a considerable amount of money on installation.

Rotor-type sprinkler heads (the heads that rotate as they spray) should not be included on the same zone as common spray heads (those that do not rotate as they spray). Rotor heads work best at higher pressures than spray heads, usually at about 50 psi (43 bar) for residential applications. Spray heads usually work best at about 30 psi (26 bar). Rotor heads are tasked to cover much larger areas than spray heads, and although they use water at a much higher rate, they also take much longer to apply precipitation over their assigned area because the area is so much larger. Therefore, if spray heads and rotor heads are mixed in the same zone, the area covered by the spray heads always receives more water than the areas covered by the rotors and the turf is either too wet at the spray heads or too dry under the rotors.

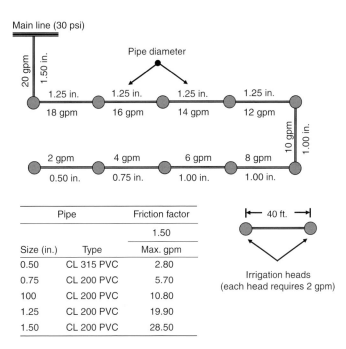

Fig. 8.2. A simple schematic of a typical zone in a large irrigation system. Irrigation heads are set 40 ft. apart making the total length of pipe from the main line to the last head 400 ft. Each section of pipe is labeled with the flow that it is required to sustain and the pipe size that is required to carry that flow based on a friction factor of 1.5 psi/100 ft. A friction factor table is included for reference. gpm = gallons per minute; psi = pounds per square inch.

Getting back to our home lawn irrigation system, it is now time to choose the types of irrigation heads that we need to accomplish the tasks in each zone. We will need both spray heads and rotors to complete the system because we have small lawns on each side of the house and large lawns at the front and back (Fig. 8.3). If we ignore the planting beds, we need at least two zones, one for spray heads and one for rotors. However, as we design our system, we will probably find that our water source is not sufficient to supply both the front yard and the back yard at the same time, and we will have to irrigate them individually on different zones.

We will use typical residential spray heads and rotors. Spray heads are adjustable by using nozzles that differ by radius and area of coverage, and adjustable for flow rate by adjusting a screw that moves to fill the supply tube and works much like an adjustable valve. As each spray head on a zone can potentially supply an equal amount of water, a head in the corner of a zone that sprays a 90° radius will supply twice as much water to its assigned area as a head on the side of a zone assigned to supply a 180° radius because the 90° head only has to cover half the area that the 180° head has to cover. Most irrigation companies sell nozzles for spray heads that vary in output so as to supply the same precipitation rate under the same pressure and flow regardless of radius. You can usually purchase those nozzles in fixed radii of 45°, 90°, 180°, 270° and 360°. However, not all spray nozzles are designed to deliver uniform flow at different radii, so you have to be aware of what you are buying. The most versatile spray nozzles are those that can be adjusted for radius from just a few degrees to 360°. On those nozzles, you sometimes have to adjust the flow rates when you adjust the radii to match your needs. When using rotors, you have to adjust nozzle size to accommodate uniform precipitation rates. A rotor tasked to cover a 180° arc, for instance, needs a nozzle with twice the output of a rotor tasked to cover a 90° arc on the same zone.

Rotors are highly adjustable for radius, spray angle, precipitation rate and distribution pattern. The radius on most rotors is adjustable over almost 360° and the spray angle that controls the height of

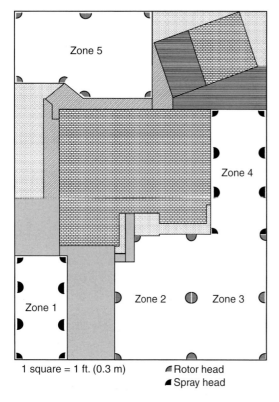

Fig. 8.3. A simple home lawn irrigation system designed to supply precipitation to the various lawn sections (white areas). Zones 1 and 4 contain spray heads and zones 2, 3 and 5 contain rotor heads. Additional zones would need to be added to supply precipitation to the planting beds (dotted areas).

will be some deterioration in the system and the water source over time.

Our next step should be to choose the heads that we will use and lay out the system based on the amount of water that we have available. We will use spray heads to apply precipitation in arcs up to 17 feet (5 m) from the heads and use rotors for larger arcs up to 40 feet (12 m). On a perfect system, the arc of one spray head or rotor should extend to the head(s) adjacent to it. We call that head-to-head spacing (Fig. 8.4). In that case, every part of the turfgrass area to be irrigated is covered by at least two sprinklers. In a rectangular area of turf that type of coverage is relatively easy to accomplish, but in most areas some creative placement will be necessary to achieve uniform coverage. It must also be remembered that even in the very best equipment the edges of the spray pattern supply slightly lighter precipitation than the inner portions of the pattern. The portion of the pattern closest to the head supplies the largest amount of precipitation and the pattern diminishes slightly as the distance from the head increases. For that reason, we can sometimes adjust the arc on opposing heads to provide uniform distribution when head-to-head coverage is not entirely possible.

Consider the diagram in Fig. 8.3. To adequately irrigate our turf, we need 16 spray heads to cover the side lawns, zones 1 and 4. Each 180° head in zone 1 must irrigate a radius of 12 feet (3.7 m) and according to manufacturer's specifications applies approximately 1.26 gpm (4.8 l/min) at 30 psi (26 bar). The heads in zone 4 must irrigate a 13 feet (4 m) arc and require 1.86 gpm (7 l/min) at 30 psi (26 bar). The 90° heads need half the water of the 180° heads, so the total supply required for zone 1 is 7.56 gpm (29 l/min) and for zone 4 is 11.16 gpm (42 l/min). As our source can supply 26 gpm, we could run all of these sprinklers on one zone, but our system will be more versatile if we split them into two zones. It is likely that these two sections of

the spray pattern above the ground is adjustable by using high- or low-angle nozzles. The precipitation rate can also be adjusted by the size of the nozzle chosen and the distribution pattern can be slightly altered by adjusting a deflector screw in the head. Some of the adjustments made to a spray head or rotor alter the amount of water that it needs to complete its assignment and in nearly every case this alteration can be estimated reasonably closely. By referring to manufacturer's specifications for nozzle output and summing the water requirements of each nozzle on the zone, you can estimate the amount of water needed to supply the entire zone and determine whether your water supply is sufficient to meet the needs of the zone or whether the zone will need to be divided into smaller zones. Some designers also build a comfort factor into each zone to account for the likelihood that there

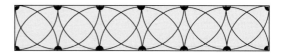

Fig. 8.4. In a well-designed turfgrass irrigation system, each spray head or rotor covers an area that ends at the head(s) adjacent to it. Arcs of 90° and 180° are normally used to accomplish this head-to-head coverage. Arcs of 45°, 270° and 360° are also common.

lawn will have different irrigation requirements, and by placing them on different zones we will be able to accommodate our water needs more efficiently. If we use small nozzles in the rotors on the front lawn, we can irrigate the entire lawn on one zone, zone 2, but we can probably increase water use efficiency by splitting it into two zones, zones 2 and 3. That also makes it possible to operate the system with less water flow in case our water source diminishes over time. We can also try long-radius nozzles or short-radius nozzles on the outside edges to get the best possible uniformity with the short-radius nozzles that we will use in the center.

Uniform precipitation rates are an important goal for an irrigation system and are the reason that we have discussed basic system design. Irrigation systems employ multiple heads to irrigate single areas to gain uniformity of precipitation. Consequently, when water is plentiful and natural rainfall is close to adequate, irrigation system deformities are rarely evident; however, they tend to stand out when environmental conditions are more severe. Extended dry periods usually define the inadequacies of a system rather clearly, but once they appear, it may be too late. At the very least, expensive hand watering will be required in the areas of turf where the system fails to meet expectations. If the system is not designed properly to provide uniform coverage, areas that do not receive enough irrigation will have to be hand watered, or the system will have to be run longer to provide enough precipitation to those areas – resulting in too much precipitation on the areas that are adequately covered. A turfgrass manager has to be intensely aware of the strengths and weaknesses of the irrigation system that his/her turfgrass depends on. Otherwise, normally small problems can become severe when conditions are extremely dry.

We all seem to realize that irrigation systems are likely to develop leaks over time. We realize that wires and heads may be broken, that valves will require repair or replacement and that pumping systems will require overhaul or replacement. However, we tend to overlook the smaller things that can create big problems. Nozzles, for instance, wear out. As they wear, the irrigation pattern becomes unbalanced and the heads use more water with less uniform coverage. Nozzles can also get plugged occasionally and so block distribution. Gears on rotors can wear out and stop working. When a rotor stops turning, the problem is usually identified easily by puddles in the direction the rotor is facing when it stops turning. However, if it turns intermittently the fault will not be noticed so easily and it will still create problems. Although portable sprinklers will probably always be more common than automatic irrigation systems on small sites, the days of manually operated permanent systems are long over. However, automatic systems still have to be monitored and inspected regularly or there will be problems that require immediate attention every time the weather gets dry.

Determining your irrigation needs

Irrigation efficiency is becoming increasingly important as demands for potable water become greater. For most people, turfgrass irrigation is a luxury. The importance of adequate water supplies for home and industry far exceed the importance of turfgrass irrigation. Irrigation for the production of food crops is also more important than turfgrass irrigation. However, there are alternatives to the use of potable water for irrigating turf (Thomas et al., 2006). Irrigating turf with used or partially treated wastewater is becoming more common and techniques for the desalination of seawater have improved and are becoming more cost-effective (Throssell et al., 2009). In many cases, effluent water can be used for turfgrass irrigation and provide acceptable turf without detrimental effects (Mancino and Pepper, 1992). However, irrigating with effluent has its own particular problems, especially during excessively dry periods (Hayes et al., 1990; Dean et al., 1996). When effluent water is used a system of soil, plant and atmospheric monitoring is a firm requirement of irrigation programming (Lockett et al., 2008). As you read and study this text you should come to the conclusion that a monitoring system should be practiced at all levels and for all components of plant management. However, irrigation is perhaps the most important of those components to be monitored. The more wisely that we use our water, the more likely that we are going to have enough water to satisfy our future needs. We have the means to manage irrigation precisely if we have the knowledge and experience to use it.

Evapotranspiration

The term evapotranspiration (ET) is a simple combination of two familiar words, evaporation and transpiration. Factors that affect evaporation affect ET and factors that affect transpiration also affect ET. Scientists like to use and study evapotranspiration because it can be calculated from a quantitative model. Most scientists, however, will be quick to explain that it is best not to take evapotranspiration too literally. If recent environmental conditions estimate evapotranspiration at 0.3 inches (7.6 mm) over the past 72 hours that does not necessarily mean that every turfgrass manager in the area needs to apply 0.3 inches of irrigation to relieve the deficit. In fact, if you are managing a large site, your irrigation needs may be for 0.35 inches in one area, 0.20 inches in another area and 0.30 inches on the remainder. Irrigation is as much an art as it is a science and you need to use history, knowledge and examination to help estimate how much water each area of your site really needs.

In most cases, ET does not have to be completely replaced to manage turfgrass at acceptable levels (Qian and Engelke, 1999). Most turfgrasses have root systems extensive enough to acquire deep or tightly held soil moisture remaining from earlier periods of rainfall that saturated the soil. For that reason, an intensively managed Kentucky bluegrass (*Poa pratensis*) fairway might require as little as 60% of ET replacement in areas where periods without rainfall last no longer than about 4 to 6 weeks (Shearman *et al.*, 2005). Relatively drought-tolerant species such as tall fescue (*Festuca arunidinacea*) and bermudagrass (*Cynodon dactylon*) generally do very well under irrigation practices of 60% ET in many locations (Fu *et al.*, 2004). Even poorly rooted species such as creeping bentgrass (*Agrostis stolonifera*), velvet bentgrass (*Agrostis canina*) and colonial bentgrass (*Agrostis capillaris*) may grow reasonably well at 60–80% ET during the summer, and can nearly always handle such deficits in the spring and fall (DaCosta and Huang, 2006b). Obviously, deficit irrigation can be practiced at higher levels and longer periods in areas with fairly frequent rainfall or during seasons when rainfall is reasonably frequent (DaCosta and Huang, 2006a). A good turfgrass manager adjusts the irrigation deficit by season and when conditions vary from normal.

The reference ET, called ET_o, is the basic unit of an ET calculation. It is most commonly derived from the Penman–Monteith scientific model (Monteith and Unsworth, 1990). The Penman–Monteith model is a combination of climatic factors that affect evapotranspiration. In order to calculate ET_o with accuracy, the altitude and latitude at the site and several weather factors have to be known. Air temperature and wind speed 2 m above the soil surface must be known, as well as humidity and solar radiation. These factors must be recorded and converted, if necessary, to the proper units of measure before being used to determine additional parameters, such as vapor pressure and soil heat flux, that are needed to complete the calculation. Obviously, the calculation of ET_o is complicated and best left to on-site weather station software or to local meteorologists or agronomists. However, once ET_o has been calculated the result is not complicated to use and can be an important part of an irrigation program.

The reference ET (ET_o) is the result of a meteorological calculation that is designed to calculate evapotranspiration from a standard vegetated surface. Fortunately for us, the standard vegetated surface happens to be a grass cover crop. However, ET differs according to grass species and other important factors. In fact, ET differs by any number of factors and that is why it is best used as a starting point and should be adjusted for your specific management needs.

Once ET_o is known, it can be adjusted for the crop, in our case the particular grass, on-site (Aronson *et al.*, 1987b). The adjustment factor for a particular crop is called its crop coefficient (K_c). The K_c is a simple multiplier that customizes the ET_o to the transpiration attributes of a particular species. The ET rates of grasses vary substantially by season (Carrow, 1995). In the case of a turfgrass, the determination of K_c may also differ by mowing height and other factors (Devitt *et al.*, 1992). The multiplication of ET_o by K_c results in the reference ET standard for a particular crop, which is called ET_c.

$$ET_o \times K_c = ET_c$$

If you are familiar with the process, you can calculate your own crop coefficients specific to your site or you can simply adjust ET_c, or even ET_o, based on history, knowledge and on-site experience. The crop reference evapotranspiration, ET_c, is a standard, like ET_o. The ET_c is the ET_o adjusted for a specific crop. However, it is based on the crop when all environmental conditions are excellent for plant growth and maintenance and no stress is present. The ET_c is defined as the evapotranspiration of a disease-free, well-fertilized crop grown in a large field under optimum soil water conditions and achieving full production under the given climatic conditions (Allen *et al.*, 2004). Obviously, conditions are rarely, if ever, perfect so the ET_c must be adjusted for weather and on-site conditions before a truly accurate value of evapotranspiration can be considered. We could continue this discussion to include the adjusted ET_c, appropriately called $ET_{c\text{-adj}}$, but instead, I will direct you to the references at the end of this chapter for further education in evapotranspiration and turn, in the next section, to a more appropriate subject, precision agriculture, to further identify locally adjusted irrigation needs. Once you obtain ET from your weather station or any number of meteorological sources, you can adjust it to determine irrigation need in small units of a large site using precision turfgrass management.

Evapotranspiration adjusted for plant need is a very useful irrigation tool. It is also a good model to study as a learning device because all of the factors that have a major effect on the irrigation needs of your plants are included in the model. There are several different methods for calculating ET and their accuracy among locations, conditions and plant species is controversial (Allen *et al.*, 2004). Rather than enter into a debate, I suggest that they are all relatively accurate and that you should use the one that is the most common in your region, and is the most readily available, and stick with it. Changing from one model to another will only make it more difficult to assess your site and to customize the calculated ET to best satisfy your specific needs. Once the model or source of information is selected, your site must be divided into individual units based on irrigation need.

Precision irrigation management

Irrigation need has to be constantly monitored. Individual areas should be identified and placed into management units that have common need (Carrow *et al.*, 2010) (Fig. 8.5). The darkest areas in Fig. 8.5 (designated ET 0.7), for instance, may generally require 70% of the calculated ET (either ET_o or ET_c) that you choose to use, while areas that have a designation of ET 1.1 require 110% of your calculated ET. Observation and record keeping are the activities most useful for identifying areas of similar irrigation need. Large areas may be mapped for management purposes using soil probes and/or reflectance sensors to estimate their

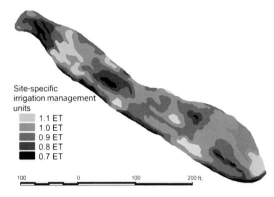

Fig. 8.5. A representation of a zoysiagrass (Japanese lawngrass, *Zoysia japonica*) golf course fairway mapped for turf greenness using a set of reflectance sensors mounted on a spray boom. These areas are believed to define differences in irrigation need, but further inspection is required to determine if that is truly the case. A mobile unit that measures soil moisture would be a better choice for mapping irrigation need. However, reflectance sensors are fast and can be used as long as site inspections to determine that water deficit is truly the source of the difference in greenness follows on from the spectral mapping. The darkest areas indicate the darkest green turf (the best watered turf) and the lightest areas indicate the lightest green turf. The abbreviation ET can refer to ET_o, the reference ET, or to ET_c, the crop-specific ET. The number designations represent irrigation need. For example, 0.7 ET identifies areas that require 70% of the calculated ET (ET_o or ET_c) that is consistently used at the site, while 1.1 ET refers to an irrigation need of 110% of the calculated ET.

irrigation requirements (Krum *et al.*, 2010). The maps are then used to section the area into individual management units. Further observation will facilitate the fine tuning of irrigation requirements over the entire site, resulting in a substantial conservation of water and an increase in positive turfgrass response.

Notice in Fig. 8.5 that individual irrigation patterns caused by poor irrigation head spacing or system malfunctions are not visible on this fairway. Poor irrigation patterns are generally obvious in precision mapping using spectral sensors. This particular system appears to be operating efficiently. It is unfortunate that areas of need are not distributed in such a manner that the irrigation system can be adjusted to irrigate each management unit independently. Although that is rarely the case, the irrigation system can normally be adjusted to provide more efficient irrigation of individual management units. Individual heads can be adjusted for arc or for nozzle size to accommodate more or less irrigation on areas where these needs have been identified. The adjustments will never be perfect, but the savings in water and the increase in turf uniformity and value can be substantial (Carrow et al., 2010).

Plant and environmental factors that affect evapotranspiration and water use

The environmental factors previously identified as affecting evapotranspiration, or ET – humidity, temperature, wind speed and radiation – are easily defined by their physical effects on evaporation. However, these factors and others also affect transpiration and turfgrass health. Because this text is primarily based on the physiological components of photosynthesis, respiration and transpiration, it is necessary to discuss these four environmental factors in more detail to understand how they affect these processes.

Humidity

Humidity is a measure of the amount of water vapor in the air and is usually reported as relative humidity, the amount of water vapor in the air compared with the amount of water vapor that the air can potentially hold at current conditions. Relative humidity is normally reported as a percentage (%). Consequently, if the relative humidity is 20% the air is dry, at 50% the air holds half the water it is capable of containing and at 90% the air is quite humid. Air temperature has a direct effect on relative humidity. As temperature increases, the air can hold more water. So when the air temperature is 50 °F (10 °C) and relative humidity is 50%, the air is holding less water vapor than when the air temperature is 90 °F (32 °C) and relative humidity is 50%. A combination of heat and humidity is a stressful condition for most turfgrasses, especially cool-season (C_3) grasses.

As warm air can hold more moisture than cold air, water vapor in the air condenses on cold objects. As warm, humid air comes in contact with a cold object, the air cools rapidly. As the air cools it can no longer hold all of the water vapor that it was holding when it came into contact with the cold object, so water condenses on the object, forming dew. Although dew formation is more complicated than the simple condensation of water vapor from warm air touching a cold object, the rapid cooling of the air is the primary principle behind the condensation. Dew on turf is mostly caused by warmer air containing sufficient water vapor coming in contact with colder grass leaves. Dew usually forms shortly before sunrise when the difference between the temperature of the lower air and the temperature of the turf is greatest, but it may form at any time during the night. Dew is important as a form of precipitation. If a turfgrass is in need of water, it will absorb dew through its leaves, thus helping to avoid a stressful drought situation. Lack of morning dew on turfgrass in a particular area when a general dew formation has occurred is often an indication of drought stress. A turfgrass manager should be observant for such a situation and investigate it if it occurs. Some species, zoysiagrasses (*Zoysia* spp.) in particular, are slow to form dew on leaves (Fig. 8.6). Consequently, in a mixed or side-by-side stand, one species may have dew on its leaves and another may not.

When relative humidity is high, evaporation is low and transpirational cooling is negatively affected. If the temperature is warm and sunny and humidity is high, photosynthesis is proceeding rapidly but cooling is inhibited by the reduction in transpiration. That creates a stressful situation for warm-season (C_4) grasses and an extremely stressful situation for cool-season grasses. Under windy conditions, high humidity is less of a concern because wind increases evaporation and helps to counteract the high humidity. As always, there are multiple factors that affect evaporation, hence transpiration. Humidity is only one.

The need for irrigation is less during humid periods. Declines in both evaporation and transpiration limit the need for additional water. The decline in irrigation need should be reflected in the reported ET. Consequently, the turfgrass manager has an accurate estimate and should adjust irrigation accordingly. However, the irrigation manager does not have a measure of heat buildup in the turfgrass due to high photosynthesis, high respiration and poor transpiration. An infrared canopy temperature device may be of use for comparing turfgrass temperature with the surrounding air temperature (Throssell et al., 1987; Jiang et al., 2009). Often at midday the canopy temperature will be higher than the air temperature and light misting or syringing may be required on important turfgrass in spite of the reduction in evaporation due to environmental conditions (Carrow, 1993).

Fig. 8.6. The small rectangular plots in the middle of the photograph and the stand of turf in the foreground are of zoysiagrass (Japanese lawngrass, *Zoysia japonica*). The square plots are bermudagrass (*Cynodon dactylon*). The bermudagrass appears a lighter color because it is covered with dew. The zoysiagrass is damp but the dew is not noticeable. The zoysiagrass is not drier than the bermudagrass, but it has a waxy coating on its leaves that discourages the formation of large water droplets.

Cooling is particularly important on golf course putting greens where localized dry spots, a condition often observed on sand systems, occur (Tucker *et al.*, 1990). The most effective means of cooling turfgrass is to provide a combination of syringing with fans (Guertal *et al.*, 2005). Misting and syringing increase humidity above the turf and add to the problems previously discussed. Therefore, water cooling by misting or syringing should only be practiced when it is imperative to sustain turf. A critical canopy temperature is normally considered to be 104°F (40°C), the temperature at which enzymes begin to denature. Misting and syringing are labor intensive, therefore expensive, and their cooling effects are short lived, less than an hour (DiPaola, 1984). Combining syringing with air movement caused by fans can extend the cooling period for up to an additional hour (Guertal *et al.*, 2005). Although their effects are short lived, misting or syringing may be necessary to help turfgrass sustain itself through especially difficult environmental conditions. As manager you should be able to determine when misting or syringing is required both by sight and by consideration of recent weather conditions.

Temperature

High temperature increases evaporation compared with low temperature. Transpiration will also increase during periods of high temperature unless it is restricted by water deficit or another plant stress. Although turfgrass plants may stop growing or turn yellow, they can generally sustain periods of above normal temperature as long as humidity is not particularly high and winds are not extremely high or extremely low. Sufficient soil water is normally enough to maintain turfgrass health through exceptionally warm weather (Perdomo *et al.*, 1996). However, when extended drought conditions occur, sunlight is intense and the air temperature is high, turf canopy temperatures may be as much as 40°F (22°C) above air temperatures (Steinke *et al.*, 2009).

Evaporative demand is high and irrigation need increases when the weather is exceptionally warm. The ET rate will increase as temperature rises, indicating the need for additional irrigation. Heat stress also makes it more likely that a plant will succumb to additional stresses such as traffic or disease. Sufficient irrigation is usually the key to resisting heat stress and related stress factors.

As you learned in earlier chapters, high temperatures increase molecular activity. Consequently, photosynthesis and respiration as well as all other plant metabolic processes increase when the temperature increases. This rapid rate of less than perfectly efficient plant metabolic processes produces substantial heat in addition to high air and soil temperatures. For that reason, transpiration must increase to lower plant temperature. The plant root system must be sufficient to handle this increased demand and water must be available. A grass that has been preconditioned to drought, meaning that it has already experienced mild drought, is more likely to remain healthy through a sustained drought period or a period of exceptionally high temperatures than one that has not been preconditioned (Jiang and Huang, 2000a). Preconditioning to drought encourages grasses to grow deeper root systems (Jiang and Huang, 2001). Therefore, over irrigation or frequent rainfall during the spring encourages failure during a hot, dry summer. Short periods of dry weather and no irrigation during the spring encourage drought and heat tolerance, especially in cool-season grasses, during the summer. You should be aware of these effects and manage spring irrigation accordingly.

Wind speed

Environments that are extremely detrimental to turfgrass growth include high wind combined with high temperature and low humidity. Under such conditions, evaporation and transpiration are so rapid that turfgrass plants may enter water deficit situations by late afternoon in spite of the presence of plentiful soil water early in the day. Plant wilting may occur because plant roots cannot absorb water fast enough to meet transpiration demand or because the soil water at the depth where the roots are plentiful has been exhausted. More than likely, transpiration demand is too fast for the roots to keep up. Consequently the plant cells near the stomates signal water deficit and the stomates close either fully or partially, depending on evaporative demand.

Assuming that light is available, the condition of wind-renewed continuous carbon dioxide availability is perfect for rapid photosynthesis. The warm conditions are perfect for rapid respiration of the carbon sequestered by photosynthesis and rapid plant growth should ultimately occur. However, sufficient transpiration cannot be attained to satisfy the system and the chance for rapid growth and development is lost. Because transpiration cannot meet evaporative demand, the stomates close, effectively suppressing carbon uptake, and photosynthesis slows. Unless substantial reserve carbohydrates are available, respiration and growth also slow. Although it is rarely mentioned in relation to irrigation, extremely calm weather has the same effect on carbon dioxide availability. Although the plant stomates may be open during calm periods, little carbon dioxide is available because the air is not refreshed, causing photosynthesis to slow and affecting respiration.

For a particular temperature and humidity, an increase in wind speed causes an increase in evaporation and transpiration. These weather phenomena cause an increase in ET. If high winds continue for extended periods it will be necessary to increase irrigation. However, it must be remembered that irrigating in high wind is not particularly effective. Irrigation patterns are partially destroyed, causing the precipitation to miss areas where it is needed. In addition, much of the irrigation evaporates before it reaches the ground. If the nights are relatively calm, as often occurs in some locations, irrigation can be performed in the early morning before the next windy day. However, if the nights are also windy, postponing irrigation is advisable until calmer weather occurs or until it can no longer wait. Such times may require hand watering of valuable turf. Hand watering reduces the amount of irrigation water lost to evaporation and allows a competent technician to visibly assess and satisfy irrigation need.

Radiation

Solar radiation is a form of energy just like heat. Consequently, intense solar radiation over a long day increases evaporation and transpiration. It also increases soil and plant warming, and encourages rapid and sustained photosynthesis. Consequently, the need for irrigation increases when solar radiation is high and days are long. Solar radiation is also part of the ET calculation. It may seem that the energy increase in heat during high solar radiation and long days would be sufficient to account for the increase in ET. However, such is not the case. Radiation is an important component of evaporation and transpiration; partly because of its important effect on photosynthesis and partly because solar energy, like heat, encourages chemical reactions.

Irrigation frequency

If a means of irrigation is available and water is available for it, supplying sufficient water to your turf is easy. Building a root system is not. Irrigation frequency is best determined by the seasonal health of the turf and by the current strength of its root system. There are several factors that have been discussed that not only affect ET, they affect irrigation frequency. However, the extent of the root system and the season of the year are the two most important factors that influence irrigation frequency. When ET is extremely high, and turfgrass wilt occurs almost daily, it will be necessary to increase irrigation frequency. In that case, the upper layer of soil, about 3 inches (7.6 cm), that houses most of the turfgrass root system is drying out too rapidly and irrigation frequency needs to be increased. That problem is likely to occur during periods of high transpiration. Frequency adjustments will normally be made on a seasonal basis according to climate. However, they will also be made based on the extent of the turfgrass root system and the plant's ability to grow new roots.

Like nearly all plants, turfgrasses lose cell turgor and wilt when they are seriously low in water. However, healthy grasses can survive a certain amount of wilt so it is not necessary to increase irrigation frequency when wilt is noticed late in the day on an occasional basis. Turfgrasses that are irrigated more frequently may maintain higher leaf turgor pressures, but they are not necessarily more healthy (Jordan et al., 2005). However, irrigation should be applied more frequently when wilt is seen regularly. Wilt manifests itself as a grayish or bluish coloring that, with some experience, is easily detected. The leaf blades loose their rigidity to wilt and do not spring back after traffic. Consequently, footprints are deeper and easier to distinguish when the grass is wilting. Turfgrass managers look for wilted turf late in the day to assess irrigation need. They also use other techniques to test soil moisture before turf wilting.

Some turfgrass managers are quite proficient at testing moisture in soils they are familiar with by using a knife blade inserted into the soil. The ease with which the knife blade penetrates the soil tells the manager approximately how much soil water is present. Other managers may cut off a golf shaft then cut a vertical sighting slot in it for use as a miniature soil probe. With experience, it is fairly easy to determine the moisture content of the soil by looking at its profile exposed in the vertical slot. More technical managers may use a soil moisture probe to acquire more exacting measurements. Regardless, a good irrigation program includes methods, means and protocols for continuous monitoring of irrigation need. It also includes records of the site by specific location over years of high and low rainfall and variable weather conditions. The turfgrass manager always knows which areas of the site are most likely to dry out quickly and have to be monitored closely. The manager also knows and has mapped the areas that tend to hold water, where traffic should be avoided following major precipitation events.

Although environmental conditions influence irrigation frequency, the most important factors are root depth, root mass and season. In general, deep, infrequent irrigation is the most effective practice for managing turfgrass health (Richie et al., 2002). Under most circumstances, poor root systems are the result of poor management preceding stressful periods. Frequent irrigation during active growth periods does not encourage adequate root growth. The daily irrigation often practiced on sand systems, for instance, is generally not required during the spring and fall. Daily irrigation results in shallow soil wetting and may remain in the thatch where it evaporates easily (Sass and Horgan, 2006). Although winter desiccation is not uncommon, the most important season for irrigation is summer. Warm-season grasses tend to have deeper root systems than cool-season grasses (Christians, 2007). They also are better at performing photosynthesis and use less water during hot weather (Feldhake et al., 1983). Consequently, they are better adapted to summer stress and water deficit. Cool-season grasses are more likely to experience water deficit stress over the summer. In fact, the management practices of the fall and spring preceding the summer may determine whether or not a cool-season grass will survive.

All grasses, both cool-season and warm-season, grow roots rapidly in the spring. Although warm-season root growth continues to occur slowly over the summer, cool-season grass roots diminish (Fig. 8.7). Warm-season grass roots may diminish slightly in the late summer into the fall but not at anywhere near the rate that cool-season roots diminish during the summer (Sartain, 2002). During the summer, cool-season roots die faster than they can be replaced. A cool-season root system cannot improve itself over the summer. Instead it declines during the period. Therefore, trying to encourage cool-season grasses to grow deeper roots during the summer only stresses the grass further. The depth and mass of the root system as the plant enters the summer often determines whether or not it will survive. Once the summer season begins, the root system continues to decline until the fall. It is

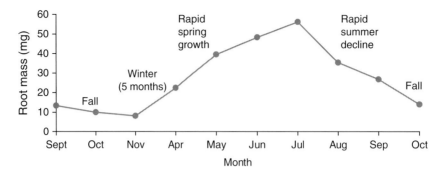

Fig. 8.7. Root mass of a 1.5-cubic inch (25-cm³) section of creeping bentgrass (*Agrostis stolonifera*) roots throughout a growing season in Columbus, Ohio. Observations were made on the 1st of each month.

highly unlikely that the turfgrass manager can do anything to prevent this decline. In fact, any management practice that causes damage to the turf results in decline of cool-season grasses during summer. If a plant's root system is poor in May, by September the plant may be dead.

Proper irrigation management during the spring and fall for cool-season grasses and during the spring and early summer for warm-season grasses can be used to promote root growth. During these periods when the respective grasses are healthy and growing, withholding irrigation temporarily will encourage root growth (Fu *et al.*, 2007). Frequent irrigation will discourage root growth. When turfgrass plants are healthy they respond to water deficit by extending their root systems in search of water. If sufficient water is always present, turfgrasses may respond by allocating more energy to shoot growth and stem growth and less energy to root growth. Consequently, if the plants have not experienced water stress before dry periods the plants will be poorly prepared to tolerate the dry conditions.

During periods of rapid growth and generally healthy environmental conditions, it is best to withhold irrigation until at least the top 2 inches (5 cm) of soil in a turfgrass stand is dry or until the first signs of wilt appear. During those periods, water deeply rather than frequently. A healthy turf will extend its root systems deeper into the soil to accumulate water. An unhealthy turf or a turf under stress, however, will not have the energy to grow new or deeper roots. An unhealthy or stressed turf will require close observation during dry periods. When necessary, adjust irrigation frequency to account for the depth of the root system. If your turfgrass roots are only 3 inches (7.6 cm) deep and the plants are under stress, it will do no good to water deeply. The deep water will be wasted. Stressed plants do not have sufficient energy to extend their root systems for water. As manager you must adjust your practices to irrigate frequently with less water to keep the short root systems supplied until conditions improve and the plants are healthy enough to grow more roots. An observant turfgrass manager knows that cool-season plants provided with frequent natural rainfall in the spring are more likely to need frequent irrigation during the summer. The manager also knows that these same plants will have to experience some light irrigation deficits in the fall and the following spring to help encourage the root expansion that will get them through the next summer.

Mowing and fertilization for low water use

Proper mowing and fertilization can help to reduce water use. From a common sense standpoint, it would seem that high mowing resulting in longer leaf blades would cause a turfgrass to transpire more water. Technically that is correct (Feldhake *et al.*, 1983). A high-mown turf has more biomass, a deeper root system and takes up more water (Madison and Hagan, 1962). The deeper root system of a higher mown turf, however, also helps it to avoid dehydration and wilt. Because its roots go deeper, there is more water available to it. As mentioned earlier, short-mown turfgrass, because it has a shorter root system than higher mown turf, requires more frequent irrigation. More frequent irrigation means more frequent evaporation of the irrigation and more likelihood that a small portion of the irrigation may run off or leach deeper into the soil profile than the short root system can reach. A low-mown turf may not use quite as much water as a high-mown turf but turf quality declines more rapidly between rainfall events (Feldhake et al., 1984). Therefore, the relatively rapid decline of low-mown turf may lead the manager to believe that more irrigation is needed than is actually required. The combination of all of these minor elements results in a larger irrigation requirement for low-mown turfgrass compared with the same turf at a higher height of cut. The lower mown plant uses slightly less water, but because of the multiple factors discussed requires more irrigation. Higher mown turf may use slightly more water, but will maintain turf quality longer between irrigation or rainfall events, thus reducing irrigation losses to factors other than plant uptake. Consequently, higher mown turf requires less irrigation.

In Chapter 7, you learned that high N fertilization leads to increased plant succulence. Increases in N fertilizer cause increases in leaf turgor, but increases in K fertilizer can cause decreases in leaf turgor (Carroll and Petrovic, 1991). Consequently, if the N fertilizer rate is relatively low and the K fertilizer rate is relatively high, plant water use may decline. However, if the N rate is relatively high and the K rate is relatively low, plant water use may increase. If you use effluent water for irrigation, N fertilizer rates will be likely to need adjusting for the N in the effluent water (Devitt et al., 2008). Ebdon et al. (1999) found that as N fertility increases, water use generally increases in Kentucky bluegrass; in the same study, they also found that when N and P were high, increasing K increased water use, but when N and P were normal, increasing K lowered water use. It has also been found that slow-release N, presumably because growth is more consistent, helps to lower water-use rate compared with quick-release N in both St. Augustinegrass (*Stenotaphrum secundatum*) and some ornamental plants (Saha et al., 2005).

Based on research and experience, the best mowing height for reducing turfgrass water use is the one that maximizes root depth. Therefore, the highest mowing height will produce the best results. The highest mowing height that meets your customer expectations for aesthetic value and/or playability is the best choice for minimizing irrigation needs. The best fertilizer program is a balanced one. Sufficient nutrition must be present for turfgrass to grow most efficiently using the least possible water. Within that realm of sufficiency, a program slightly low in N and slightly high in K will maximize water use efficiency.

8.3 Species Adaptation to Low Water Use

Any discussion of species adaptation to drought or water use must begin with a determination of objectives. Are your objectives to manage the turfgrass site to remain green all through the growing season or are you willing to accept some yellowing and possible dormancy during dry periods? Both cool-season grasses (Aronson et al., 1987a; Fry and Butler, 1989) and warm-season grasses (Beard and Sifers, 1997; Huang et al., 1997) and their cultivars differ in propensity for surviving drought and for resisting leaf fire during dry conditions. Scientists use leaf fire, a measure of the browning of leaves as they dry during periods without precipitation, as a measure of visual quality (Ebdon and Kopp, 2004). A plant's tolerance to leaf fire is not the same as its propensity to resist drought (Carrow, 1996). If your objectives are to irrigate to maintain visual quality, leaf fire resistance is most important to you. If you are looking for a species or cultivar that can survive long periods without irrigation, drought tolerance is your primary concern. When researching species and cultivars for drought or leaf fire resistance in your particular area, you must keep your objectives firmly in mind. You must also consider the adaptability of the species for your site. Although tall fescue is a drought tolerant cool-season grass it would not be considered more resistant to wilt than hybrid bermudagrass (*Cynodon dactylon* × *Cynodon transvaalensis*) or buffalograss (*Buchloe dactyloides*). However, research suggests that where tall fescue is better adapted to the environmental conditions than hybrid bermudagrass and buffalograss, it grows a deeper root system and is more tolerant of dry conditions (Qian et al., 1997). Consequently, a species or cultivar will not demonstrate its potential for drought or leaf fire resistance in a location where it is not particularly well adapted.

Water use rates, the speeds at which a turfgrass takes up water when sufficient water is present, are presumably a measure of the amount of water required by a particular turfgrass species. However, trying to determine the water use rate of any particular species is difficult. Again, it is likely that the best adapted species for the location will have an advantage in water use rate. Also, the water use rate of a particular species does not necessarily mean that it is or is not drought tolerant. For instance, perennial ryegrass (*Lolium perenne*), colonial bentgrass (*Agrostis capillaris*) and tall fescue were found to use approximately equal amounts of water in a study in Oregon, but tall fescue is clearly more drought tolerant than the other two species (Doty *et al.*, 1990). However, in a subirrigated study in Arizona, bermudagrass and zoysiagrass (Japanese lawngrass, *Zoysia japonica*) used less water than St. Augustinegrass and tall fescue, which is what would be expected at that location (Kneebone and Pepper, 1982). St. Augustinegrass is known to have poor drought tolerance, yet research in Florida suggested that it could be grown on loam soil with as little as 0.5 inches (13 mm) of precipitation every 6 days (Peacock and Dudeck, 1984). One reason that water use rates are so hard to define is because grasses are likely to use more water than necessary when more water is present. Bermudagrass, for instance, can survive on very little water, but it will use a substantial amount of water when enough water is present (Kneebone and Pepper, 1984). Consequently, the best choice of grasses are those that are best adapted to the conditions at your site and have the best drought tolerance or leaf fire resistance, depending on your objectives. Those are the grasses that you can grow best at the lowest water use.

8.4 Managing Water Overload

In most regions of the world, periods exist where the soil is saturated for a short time. Rainfall is sometimes sufficient to saturate the upper layers of soil and produce substantial runoff, perhaps even flooding, but these conditions rarely last for long. Turfgrass cannot function when submerged or when rooted in saturated soil. It has been found that saturated soil even 15 cm deep affects creeping bentgrass (Jiang and Wang, 2006). However, turfgrass is perfectly capable of tolerating such conditions for short periods. Depending on species, submersion in cool water or residing in saturated soil for a few days is not particularly stressful to turfgrass and it normally recovers from such conditions quickly. However, if these conditions are recurring or if they last for extended periods, the turf will decline and may not survive (Beard and Martin, 1970).

Site inspection and drainage design

Natural drainage is, of course, a function of elevation. However, water not only runs downhill on the surface of the soil, it can infiltrate the soil and subsurface drain. A basin or depression surrounded by higher elevation cannot surface drain, it has to subsurface drain. The water cannot move over the soil, it has to move through the soil. Unless the soil in the basin is deep and sandy, water flowing into the basin during periods of runoff will take some time to either evaporate or subsurface drain. As a turfgrass manager, you should be reluctant to depend on subsurface drainage, even in artificially constructed sand systems. Surface drainage is reliable. Subsurface drainage specifically constructed for the purpose is also reliable if properly constructed, but has a finite lifetime (Murphy *et al.*, 1993b). It may take 20 years or more, but properly constructed subsurface drainage will eventually fail. If you are the turfgrass manager at the end of that finite lifetime, you will not have a high opinion of subsurface drainage. To you, subsurface drainage will not be reliable. Surface drainage, however, is timeless unless construction practices nearby or unlikely natural events destroy it.

Sports fields and golf course greens specially constructed of sand under United States Golf Association (USGA) or similar protocols provide excellent subsurface drainage. However, even these systems can become ineffective if surface drainage is ignored during construction. Consider a golf course bunker or sand trap. If you are a golfer, are interested in turfgrass management, or have ever constructed a bunker, you will realize that in most cases these bunkers are not just holes in the ground. They are highly engineered entities with more than adequate drainage, geotextile mats, special sand and other custom features. Golf course bunkers are specially constructed holes in the ground designed to provide longevity over a typical hole in the ground. Nonetheless, they are still holes in the ground and they still have the attributes of holes in the ground. It may take them longer for drainage to fail compared with the typical hole in the ground but it will, in fact, fail. Bunkers collect all of the debris, silt, clay, and organic and inorganic material from the surrounding watershed that drains into them. A 10-year-old bunker that still drains well is rare and bunkers that collect debris from a watershed of substantial size fail relatively quickly.

A basin or depression in a turfgrass stand is like a bunker. The basin collects debris from the surrounding area that drains into it. At first, the water in a sand system drains down through the sand faster than it can surface drain into the basin. Eventually, however, the healthy turfgrass begins to produce thatch and that thatch begins to break down into mat that exists at various levels of organic degradation. The mat begins to interfere with water infiltration. In addition, dust collects on the sand system. Sediments and salts from rainwater and irrigation enter the system. Eventually, these barriers to infiltration worsen to the point where runoff occurs and the basin begins to accumulate debris on a regular basis. The accumulating debris soon forms a soil layer that holds water and the subsurface drainage, as well as the turfgrass in the basin, begin to deteriorate. Some basins are constructed by design. Other, far less detrimental, basins occur when attempts are made to make a perfectly level surface. It is probably impossible to build a perfectly level surface from sand but even if it is not, the sand would eventually settle inconsistently, forming minor hills and basins. There are some cases, in the game that Americans call soccer, for instance, where the slopes that provide surface drainage are detrimental to the game. In that case, a flat playing field cannot be avoided but the expectations for the life of the field should be reduced. Otherwise, designs for turfgrass systems should, without exception, provide for adequate surface drainage. As a turfgrass manager you should seek out the basins on your area of responsibility and either deal with them or closely observe them. Eventually, declines in turfgrass growth will occur.

Drainage

Providing drainage to your area of responsibility is usually easier than irrigating it, but not always. During construction, natural surface drainage is sometimes destroyed by poor design and it can be difficult to reestablish. Fine soils and compacted soils have high runoff potential and poor infiltration. Low spots in these soils have to be properly drained because infiltration is slow and the soil in a depression will remain saturated for long periods. Whether the site is a home lawn or a playing field, turfgrass systems are generally sites for recreation and standing water or saturated soil is detrimental to use as well as to turfgrass health. It takes more than a prevailing slope or slopes of sufficient fall to drain your site. The water must also have a place to go. As ridiculous as it seems, people seem to ignore drainage in favor of mounds, beds or other features in locations where they want them to be. It is common to see standing water or dead grass near constructed mounds and other sloping features because the natural drainage was destroyed.

The first order of drainage is to pick a place for the water to go. That place should be in the natural drainage pattern. Otherwise, you will be making it particularly difficult on yourself. If you have pockets, depressions, basins, ditches or perhaps an entire large site that has to be drained, follow the natural drainage pattern or your time and labor is likely to be for naught. Basic surface drainage is from the highest point to the lowest point, and if there are ridges between those two points, a path must be constructed so that runoff can flow around the ridge and into the lowest point in the area. If that lowest point is not a lake or a creek then a ditch swale or other feature must be present or be constructed to carry the water into a creek, pond or possibly a sewer. The water has to have someplace to go even if you have to pump it there.

Constructing subsurface drainage

Surface drainage is most often preferable to subsurface drainage, but in some cases subsurface drainage is necessary for cosmetic or functional reasons. Constructing subsurface drains to accommodate water removal from small areas is fairly simple to design and fairly easy to construct. As usual, the design requires a place for the water to go. To get it there requires a ditch of sufficient depth and slope containing a pipe of sufficient size and construction. Corrugated plastic drain pipe is light, flexible and inexpensive, and is available with or without holes and in different sizes. A 4-inch (12-cm) plastic drain pipe is easy to handle and will drain sufficient water for most circumstances. If the pipe is expected to take up water, it has to have holes in it. A pipe without holes carries water from one place to another but it does not take up water from a saturated area. To understand how subsurface drainage works, you need to remember some of your basic soil principles.

If you place a soaker hose on a loam soil on top of sand on top of gravel containing a pipe, what happens (Fig. 8.8)? If you remember, fine particles soak up water from more coarse particles. The most

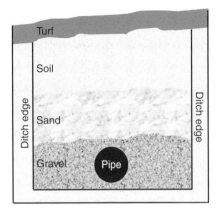

Fig. 8.8. A ditch containing a drain pipe backfilled with gravel around the pipe, sand above the gravel and soil above the sand. Subsurface ditches like this one can be constructed to carry water from intermittently saturated soil into a drain. When the soil on the top of the ditch becomes saturated water moves into the sand. When the sand becomes saturated water moves into the gravel and then into the pipe for drainage. The sand speeds the drainage process by bridging the particle size gap between the soil and the gravel.

difficult part of that principle to understand is the time factor. It is easy to remember that water moves from coarse particles to fine particles until you are asked what happens when water drains from sand into compacted soil, and you get confused. You know from experience that water puddles on compacted soil. You see it happen all the time. The physics tells you that the fine compacted soil should be soaking up the water and actually it is. It just takes a long time (Box 8.2). There is actually pressure on the water to enter the soil, but the compacted soil pores are small and few and fill up quickly. Once the compacted soil fills the physics tells you that the water has no other place to go except into the surrounding soil, which is less compacted and therefore more coarse. So as long as the puddle exists, it fills the compacted soil around it with water, but the transfer of that water through the tiny pores of the compacted soil into the less compacted soil is very slow. Hence, the water remains in the puddle for a long period of time as it drains through the compacted soil. If water drains from sand to highly structured soil, the soil will take it up fairly quickly until it is full, then, if possible, it will slowly drain into another medium.

Because of coarse-to-fine soil physics, the person who applies soil-based sod over sand ends up with dead grass. Whatever water exists in the system remains in the soil that came with the sod. Water will not drain into the sand until the soil in the sod above it is saturated. Consequently, the roots in the saturated soil have no oxygen with which to perform respiration. As you learned earlier, a turfgrass in that situation dies of drought when, in fact, its roots are surrounded by water. For the same reason, a pipe surrounded by gravel under sand under

---

Box 8.2. If you pour a bucket of water onto a layer of sand on top of a layer of gravel on top of a layer of soil containing a pipe, does the water drain into the pipe?

In this system, the water will drain into the pipe if the soil around it is saturated, but depending on soil type, the soil will probably drain too slowly for the pipe to be useful as a drainage water conductor. If the bucket of water is of sufficient volume to saturate the system and cause water to flow into the pipe, the water will have to be poured at a rate slower than the infiltration rate of the soil around the pipe or the system will overflow before water enters the pipe. However, if such a system is used as a drainage ditch it will probably drain saturated soil rapidly enough to be adequate under most conditions without any water ever entering the pipe. The water will drain without entering the pipe because the sand and the gravel are also drainage devices. Water will flow downhill through a ditch filled with sand fairly rapidly and through a ditch filled with gravel even faster. Consequently, a ditch like the one in Fig. 8.8 without the soil on top would drain the water very fast. Not only would the pipe fill as rapidly as water could infiltrate the sand, the sand and gravel would also carry drainage water, but more slowly, downhill toward the drain. The problem with such a system is that if the strip of sand on top of the ditch is wider than 2 or 3 inches (5–8 cm), the ditch will always be visible because the turfgrass growing on the sand will look different from the turfgrass growing on the soil. The sand would leach nutrients much faster than the soil and it would always be dry except for periods immediately following rainfall.

a layer of soil with turfgrass on top drains slowly and does not drain past soil saturation (Fig. 8.8). Nonetheless, it drains rapidly enough and completely enough to eliminate turfgrass decline resulting from saturated soil. Once the puddle effect is removed, the soil drains naturally and it should be rapid enough to prevent turfgrass damage. However, if periods of heavy rain intense enough to keep the soil saturated for 3 or 4 days at a time are common, it would be better to fill the ditch with sand all the way to the surface in the area that tends to puddle. Just a small slit 3 to 4 inches wide is sufficient and such a small slit is rarely noticeable in the turf. The small slit of sand will help the water move into the pipe faster.

Water drains from coarse particles to fine, but it also drains faster from saturated fine soil into sand than it does from saturated fine soil into gravel. In the example (Fig. 8.8), under normal conditions it is rarely necessary to cover the gravel with sand. However, it will speed drainage if you do because the particle size of the sand is more similar to the soil than the particle size of the gravel is to the soil. Therefore, when the soil becomes saturated, water will move faster from the soil to the sand than from the soil to the gravel. Nonetheless, the water will move from saturated soil to gravel, it just takes a little longer. The pipe should always be surrounded by coarse sand or fine gravel to keep the soil from filling up the holes that allow the water to enter the pipe. If you happen to have some old pipe that you are going to use for drainage and you intend to drill your own holes in it, you only need to drill the holes on one side and the holes need to go on the side that is down when the pipe is placed in the ditch. That is correct, the holes go against the sand or gravel that you placed in the bottom of the ditch before you placed the pipe on top of it. Consider air to be the largest of particles. Water will not move into the air in the pipe until it saturates the gravel or sand around the pipe. Therefore, it actually enters the pipe from the bottom, not from the top.

8.5 Chapter Summary

Turfgrass irrigation systems are designed for and are uniquely effective for uniform distribution of precipitation over large or small turfgrass areas. They are normally divided into zones based on the amount of water available, the species requiring irrigation, management intensity, location, type of spray head used and other considerations. Each zone may be operated independently by a designated valve and each head can be adjusted for output and for area covered. The most expensive systems can be adjusted to automatically operate each head independently. Obviously, turfgrass irrigation systems are complicated. However, the principles on which they are based are relatively simple and easy to remember.

It takes pressure to move water through a pipe. The size of the pipe and the pressure at the source determine how much water is available to the system. The highest point in the system will have the lowest pressure and the lowest point in the system will have the highest pressure. Small pipes carry less water and therefore must move water faster to maintain the same flow rate as a larger pipe. The faster the water moves in the pipe, the greater the friction becomes and the harder it is to push the water. Consequently, large pipes are preferable to smaller ones, but they are also more expensive, so over-engineering the system is not always desirable. The water window, the amount of time available to irrigate the site, determines the rate at which water must be supplied to the system in order to accomplish the irrigation task. The amount of water required for a given irrigation task is primarily determined by the plant material to be irrigated, by weather conditions and by consumer expectations.

Evaporation and transpiration combine to determine how quickly water is used or lost from the turfgrass site. The combination of evaporation and transpiration is called evapotranspiration (ET) and can be determined mathematically if the required weather parameters are measured. There are many sources, including the Internet and government weather stations, available to determine the reference ET, which is referred to as ET_o. The ET_o can be adjusted by site-specific factors to determine the irrigation required fairly accurately. However, a more accurate measure uses the ET_o combined with a crop coefficient (K_c). The ET_o multiplied by the K_c for a particular species of turf, for instance, results in ET_c, a fairly accurate measure of water loss from a turfgrass system containing that species.

Although an ET_c may be calculated from weather data accumulated from a station nearby, it is still not completely accurate for estimating irrigation need throughout an entire site. Site-specific management practices can conserve water, save money and improve the uniformity of the turfgrass on the site. Moisture data collection and/or spectral sensing of turf will result in designated areas called site-specific management units that require more or less irrigation than is indicated by the ET_c. Mapping and dividing the site into site-specific management units can result in substantial water conservation and economic savings. In addition, the knowledge and experience of a quality turfgrass manager is necessary to make irrigation adjustments by observation. The use of site-specific irrigation management does not preclude the monitoring of the site by a qualified manager or irrigation technician.

Too little water can seriously damage turf, but turfgrasses can stand longer periods of too little water than they can stand periods of too much water. Standing water or saturated soil will usually damage turf more quickly than drought. Although proper irrigation is rarely ignored, proper drainage is sometimes overlooked. Surface drainage is nearly always more important than subsurface drainage, but both can be used to properly manage water on a turfgrass site. Regardless of which type of drainage is used – surface or subsurface drainage – the water must always have someplace to go. Natural drainage patterns are best. It is much easier to follow the natural drainage than it is to create an artificial drainage pattern.

Suggested Reading

Beard, J.B. (1989) Turfgrass water stress: drought resistance components, physiological mechanisms, and species-genotype diversity. *Proceedings of the 6th International Turfgrass Research Conference*, Tokyo, Japan, pp. 23–28.

Beard, J.B. (1993) The xeriscaping concept: what about turfgrasses. *International Turfgrass Society Research Journal* 7, 87–98.

McIntyre, K. and Jacobsen, B. (2000) *Practical Drainage for Golf, Sportsturf, and Horticulture*. Ann Arbor Press, Chelsea, Michigan.

Pira, E.S. (1997) *A Guide to Golf Course Irrigation System Design and Drainage*. Ann Arbor Press, Chelsea, Michigan.

Puhalla, J., Krans, J. and Goatley, M. (1999) *Sports Fields: A Manual for Design, Construction and Maintenance*. Ann Arbor Press, Chelsea, Michigan.

Suggested Websites

CIT (2009) Center for Irrigation Technology. Available at: http://cit.cati.csufresno.edu/ (accessed 5 August 2009).

Hunter Irrigation Innovators (2009) Available at: http://www.hunterindustries.com (accessed 7 October 2009).

Hunter Irrigation Innovators (2009) Irrigation System Design: Student Workbook. Available at: http://www.hunter-industries.com/Resources/PDFs/Educational/Domestic/ed_004_final.pdf (accessed 7 October 2009).

Hunter Irrigation Innovators (2009) Precipitation Rates and Sprinkler Irrigation: Instructor's Manual. Available at: http://www.hunterindustries.com/Resources/PDFs/Educational/Domestic/precipitation_rates_and_sprinkler_irrigation_workbook_instructor.pdf (accessed 7 October 2009).

Hunter Irrigation Innovators (2009) The Handbook of Technical Irrigation Information: A Complete Reference Source for the Professional. Available at: http://www.hunterindustries.com/Resources/PDFs/Technical/Domestic/LIT194w.pdf (accessed 7 October 2009).

Irrigation Association (2009) Smart Practices. Sustainable Solutions. Available at: http://www.irrigation.org/ (accessed 7 October 2009).

Rain Bird (2009a) Manufacturer and provider of irrigation products and services. Available at: http://www.rainbird.com (accessed 7 October 2009).

Rain Bird (2009b) Landscape Irrigation Design Manual. Available at: http://www.rainbird.com/pdf/turf/IrrigationDesignManual.pdf (accessed 7 October 2009).

Toro (2009a) Available at: http://www.toro.com (accessed 7 October 2009).

Toro (2009b) Do-It-Yourself: Sprinkler Planning and Installation Guide. Available at: http://www.toro.com/sprinklers/pig.pdf (accessed 7 October 2009).

Toro (2009c) Sprinkler Replacement and System Troubleshooting: Toro Irrigation Design Service. Available at:http://www.toro.com/sprinklers/repair.pdf (accessed 7 October 2009).

UC Davis (2009) Estimating Leaching Fraction Requirements. Available at: http://ucce.ucdavis.edu/files/filelibrary/5049/773.pdf (accessed 11 June 2009).

9 Adjusting for Seasonal Conditions and Temperature Stress

Key Terms

For purposes of this text, **seasons** are defined in the traditional sense. In the northern hemisphere, winter occurs from 21 December to 20 March, spring from 21 March to 20 June, summer from 21 June to 20 September and fall from 21 September to 20 December.

Lipid peroxidation occurs when free radicals take electrons from lipids, causing them to destabilize and degrade. Because lipids are important components of cell membranes, lipid peroxidation can be very destructive to the regulated passage of chemical species in and out of cells.

Enzymes and proteins **denature**, meaning that they unfold and become ineffective, at 104 °F (40 °C).

A **heat shock protein** is one of many proteins that may be synthesized in a plant to protect proteins and enzymes from unfolding as a result of excessive heat. A heat shock protein is a form of chaperonin.

Chaperonins are a group of proteins that help to guide the correct folding of polypeptides (long chains of amino acids) into proteins and enzymes without becoming part of the final structure.

Conduction is the transfer of energy from one molecule to another touching molecule. Warm grass leaves can conduct heat to the air surrounding them.

Convection is a mass movement of energy, usually in liquids or gases. The bottom of a pool is always colder than the surface because warm water rises by convection, and the ceiling in a room is always warmer than the floor for the same reason. As warm grass leaves conduct heat to the surrounding air, the warmer air rises.

Equilibrium is a balanced condition that occurs between two entities when they become equal in some respect. When a grass blade is the same temperature as the air around it, the grass blade and the air have reached equilibrium with respect to temperature.

Photorespiration occurs when Rubisco (ribulose bisphosphate carboxylase), the major enzyme in the Calvin cycle of photosynthesis, binds oxygen instead of carbon dioxide. The process only occurs in cool-season (C_3) plants and is the main reason that C_4 photosynthesis is more efficient than C_3 photosynthesis when the weather is warm. It is called respiration because carbon dioxide is lost during the process.

Colligative properties are the physical properties of solutions that are affected by the concentration of solutes contained in them. Water containing a high concentration of solutes, such as the water in a plant cell, boils at a higher temperature and freezes at a lower temperature than pure water.

A **solute** is a substance that is dissolved in water.

Intracellular water is the water inside the cell membrane. Intracellular water is higher in solutes than intercellular water.

Intercellular water is the water contained in and around the cell walls of a plant.

Viscosity is the measure of the resistance of a liquid to flow. Liquids become more viscous as they cool and approach their freezing temperature, and less viscous as they warm and approach their boiling temperature.

Photooxidation is the loss of an electron from a photoexcited chemical species, in this case, chlorophyll. Chlorophyll photooxidation is most likely to occur on clear cold days when sunlight is intense but the temperature is too cool to allow rapid chemical reactions.

Photoinhibition is a light-induced decrease in the activity of the photosynthetic process that can occur when temperatures are either too hot or too cold for the pathway to operate efficiently.

©CAB International 2011. *Turfgrass Physiology and Ecology* (G. Bell)

9.1 Optimal Turfgrass Health is a Result of Flexible Management Under Differing Temperatures

Air and soil temperatures are two of the few environmental conditions that turfgrass managers cannot realistically influence. As you probably know, there are actually athletic fields and golf course greens that are artificially air-conditioned for temperature control during high-stress periods. In these cases, perhaps we have carried plant management a little too far, perhaps not; you can reach your own conclusions. Regardless, it is very unlikely that you will have the opportunity to air-condition your turf, so we can assume for purposes of this chapter that you are going to have to manage turf during normal seasons and also during extreme temperatures that are both normal and above or below normal for your climate. You will need a seasonal plan for managing warm-season (C_4) and/or cool-season (C_3) grasses at your site and you will need to help prepare your plants for temperatures that are significantly above or below normal.

A seasonal plan for managing your grass is one that you develop through knowledge such as that contained in this text and revise according to experience at your site. Each turfgrass species differs in its particular requirements within a set of environmental variables. Consequently, each turfgrass species, and sometimes cultivars within a species, require different management practices or different levels of management to grow their best. Obviously, warm- and cool-season grasses require different seasonal management, and individual species within the C_4 category or the C_3 category also differ in their management needs. Heat stress is rarely of major importance in C_4 grasses and cold stress is rarely of importance in C_3 grasses, but they do occur. Prolonged extremely high temperatures can sometimes damage C_4 grasses and extremely low temperatures or ice cover can damage C_3 grasses. Nonetheless, the management of warm-season grasses is primarily directed toward minimizing cold stress and the management of cool-season grasses is primarily directed toward minimizing heat stress. Consequently, the major emphasis of this chapter will be directed toward the seasonal management of warm- and cool-season grasses, cold stress in warm-season grasses and heat stress in cool-season grasses.

9.2 Seasonal Turfgrass Management

We all know that C_4 grasses do not grow particularly well during periods of cool or cold weather and that C_3 grasses do not grow particularly well during periods of warm or hot weather. We also know that turfgrasses are less green, sometimes yellow or brown, during these stressful periods. Consequently, we can estimate plant health by the plant's level of greenness. As you have learned, grasses can sometimes be excessively green, an unhealthy situation, but usually, green is a good thing. In your experiences, you may have seen C_4 grasses that are yellowish in early spring become quite green in late spring and then less green in late summer, followed by green again in early fall then yellow or brown over the winter. If you live in a temperate region, you have seen similar seasonal changes in cool-season grasses. Although we knew that these seasonal effects occurred, until recently we did not have a means to accurately quantify them. However, the commercial availability of spectral reflectance sensors has now made it relatively easy to document the changes in turfgrass species influenced by season. Figure 9.1 demonstrates the spectral reflectance response of two warm- and two cool-season grasses averaged over at least two growing seasons in Stillwater, Oklahoma. Similar seasonal responses have been documented in scientific journals (Guertal and Shaw, 2004; Xiong *et al.*, 2007). These grasses will respond differently to management depending on season.

Seasonal management of C_3 and C_4 grasses

Warm-season (C_4) grasses can be expected to respond to management inputs best when temperatures are relatively high. We recognize the optimum daytime temperatures for warm-season shoot growth to be from 80°F to 95°F (27°C to 35°C) (Beard, 1973). Consequently, temperatures between 80°F and 95°F should be ideal for efficient use of fertilizer, water and other management inputs for warm-season grasses. Warm-season grasses will use management inputs less efficiently when temperatures are lower than or greater than the ideal range. The climatic temperatures (temperatures averaged by day over the last 30 years) should indicate when management inputs are best utilized and when less management is better in your area. However, let us complicate this approach

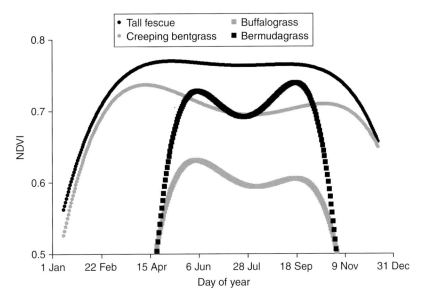

Fig. 9.1. The seasonal spectral reflectance response trend calculated from data measured as the normalized difference vegetation index (NDVI) of tall fescue (*Festuca arundinacea*), creeping bentgrass (*Agrostis stolonifera*), buffalograss (*Buchloe dactyloides*) and bermudagrass (*Cynodon dactylon*) averaged over two growing seasons in Stillwater, Oklahoma.

slightly and attempt to encourage root growth with our inputs instead of shoot growth. The optimum temperatures for root growth in warm-season grasses are between 75 °F and 85 °F (24 °C and 29 °C). The optimum temperature range for root growth in cool-season (C_3) grasses is between 50 °F and 65 °F (10 °C and 18 °C) (Beard, 1973). Soil usually heats and cools more slowly than air. Consequently, the daily maximum soil temperature is nearly always lower than the daily maximum air temperature and the daily minimum soil temperature is almost always higher than the daily minimum air temperature (Fig. 9.2). The daily average soil temperature tends to be slightly higher than the daily average air temperature during the fall and winter but close to the same during the spring and summer in most climates under most soil conditions (Fig. 9.3).

Management practices, fertilizer, water, cultivation, etc. are most effective when your turf has good-to-optimal growing conditions with which to take greatest advantage of your inputs. Warm-season grasses respond best to management during late spring, early summer, late summer and early fall. Cool-season grasses grow best in the spring and fall and will continue to grow roots all winter when the soil is not frozen. It is reasonably easy to adjust management based on temperature forecasts. If we are flexible enough to take advantage of unusually cool or warm periods rather than trying to maintain our turf based only on historical climate we can improve our turf's ability to maintain reasonable health through stressful periods.

Attempting to encourage growth in warm-season grasses in early spring and late fall or cool-season grasses during the summer when temperatures are not conducive for growth can be counterproductive. During these periods, managing for increasing density and greenness will probably harm your turf more than help it.

9.3 High-temperature Stress in Grasses

As you know, chemical reactions increase at an accelerating rate as temperature increases. That means that the metabolic processes in a plant and all of the chemical reactions in its environment speed up with increasing temperature. As temperature increases, both photosynthesis and respiration increase. However, when the leaf temperature reaches about 104 °F (40 °C) photosynthesis begins to decline (Berry and Björkman, 1980). Respiration also begins to decline as temperatures approach approximately

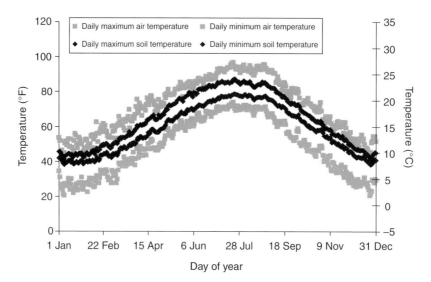

Fig. 9.2. The daily maximum air temperature, daily maximum soil temperature, daily minimum air temperature and daily minimum soil temperature averaged over 13 years (1994–2006) in Stillwater, Oklahoma. Notice how the daily maximum soil temperature is always lower than the daily maximum air temperature and how the daily minimum soil temperature is always higher than the daily minimum air temperature.

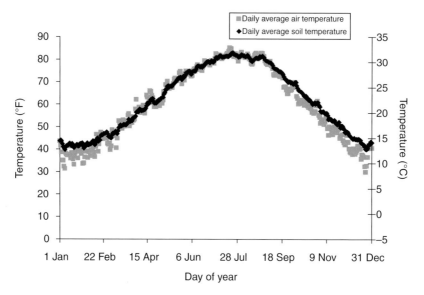

Fig. 9.3. The daily average air temperature (the average of a day's maximum temperature and minimum temperature) and daily average soil temperature in Stillwater, Oklahoma. Notice how the daily average soil temperature tends to be higher than the daily average air temperature during the autumn and winter but about the same during the spring and summer.

104°F (40°C) but the net balance favors respiration (Huang and Gao, 2000). Consequently, photosynthesis slows more rapidly than respiration, resulting in a net loss of carbohydrates. Because of this carbohydrate deficit, plants that have been healthy enough to have adequate stored carbohydrates before stress are more likely to maintain quality during heat stress (Huang and Gao, 2000).

The decline in net photosynthesis during heat stress is believed to be related to a loss of integrity in cell membranes or in membranes contained in chloroplasts and other organelles (DiPaola and Beard, 1992). Cell membrane thermostability is an important component of each plant's ability to tolerate heat (Marcum, 1998). As you will learn later, it is also an important component of cold tolerance in plants. Membranes are not only affected by the loosening of molecular bonds caused by heat, they may also be damaged by a process called lipid peroxidation (Jiang and Huang, 2001). Lipid peroxidation occurs when free radicals, usually oxygen, take electrons from a lipid, in this case a membrane lipid, causing the lipid to destabilize. Lipid peroxidation is more likely to occur in polyunsaturated fatty acids than in saturated fatty acids. For that reason, cool-season grasses, which have more unsaturated fatty acids in their membranes than do warm-season grasses, are more readily affected by lipid peroxidation than warm-season grasses. Polyunsaturated acids also resist cold temperature degradation better than saturated acids, and this is one important reason why cool-season grasses also resist cold temperatures better than warm-season grasses.

Free radicals are produced as an unintentional by-product of many energetic biological reactions. There are free-radical scavenging compounds generally referred to as antioxidants that quench or stabilize free radicals, thus preventing much of the damage that could be caused by them. If you are familiar with human nutrition, you have probably heard the term antioxidant many times. High temperature depresses the synthesis of some of the more important antioxidants in grasses (Xu and Huang, 2004). Consequently, lipid peroxidation and the resulting degradation of membrane integrity is a major cause of high-temperature decline in turfgrasses, especially cool-season grasses (Liu and Huang, 2000). There is some reason to believe that applications of cytokinin(s), a plant hormone, may help to prevent lipid peroxidation in cool-season grasses, but as yet this has not become a technique for practical management of grasses (Liu and Huang, 2002a; Liu *et al.*, 2002).

A turfgrass plant's rate of photosynthesis is also limited by the amount of light, water or carbon dioxide available, but respiration has fewer restrictions. At high temperatures, respiration continues rapidly as long as sufficient carbohydrates, lipids and proteins are available to respire. For that reason, grasses with high carbohydrate reserves and sufficient light, water and carbon dioxide to perform rapid photosynthesis are better prepared to withstand high temperature stress than those without (DiPaola and Beard, 1992).

There is an old saying "starve a cold but feed a fever". Whether or not "starve a cold but feed a fever" is good medical advice is unknown to this author. However, it is important to note that when you have a cold and your system is not functioning well, you are not hungry, but when you have a fever and you are burning a substantial amount of calories, you do get hungry. Your fever is caused by the rapid respiration required for your body to manufacture the compounds needed to fight the pathogen that is causing your illness. Increasing respiration causes high body temperature. When the temperature gets uncomfortable for us we drink a lot of water and perspire to cool ourselves. Plants do the same thing but we don't call it perspiration, we call it transpiration. In the last chapter, you learned that water is the best plant defense against high temperature. However, if the temperature is high enough and lasts long enough, plants will eventually die, just as humans will eventually die from a fever that is not controlled. Both humans and plants use water as a defense against high-temperature stress. The difference between the two is that if water is not enough to relieve high-temperature stress in a human, that human can move to a cooler place, the plant cannot. If the temperature is too high, or the humidity is too high and/or air movement is too low to allow rapid transpiration of available water at the plant's current location, the plant will die.

High-temperature stress in warm-season grasses

Warm-season grasses are more capable of handling higher temperatures than cool-season grasses. The architecture of a warm-season grass, most notably its extensive root system and efficient high-temperature photosynthetic pathway, allows it to thrive in temperatures that stress a cool-season grass. Warm-season grasses do not photorespire, so that particular source of heat, which occurs in cool-season plants during high temperatures, is eliminated. Because they do not photorespire, warm-season plants use water more efficiently during high temperatures than cool-season plants. They are also capable of obtaining more water than cool-season grasses because they tend to have more extensive root systems. Finally, their high-temperature-efficient photosynthetic pathway enables more rapid replacement of the carbohydrates used during rapid respiration caused by high temperature.

Although warm-season plants are more capable of cooling themselves and tolerating high temperature than cool-season grasses, they are not immune to it. When temperatures reach levels greater than 95 °F (35 °C), warm-season grasses begin to stress in a similar manner to cool-season grasses. At that point, temperatures are becoming dangerously close to 104 °F (40 °C), the temperature at which enzymes and other proteins begin to denature (meaning that they unfold and are no longer useful). Both warm- and cool-season plants have physiological defenses that help prevent enzymes and proteins from denaturing (DiPaola and Beard, 1992). These defenses are called heat shock proteins (DiMascio *et al.*, 1994).

Heat shock proteins

Heat shock proteins are in a class of biological compounds called chaperonins (Gething, 1997). In a sense, the heat shock protein surrounds the target enzyme to prevent it from unfolding. Consequently, the enzyme or protein maintains its original characteristics and continues to function in a useful manner. Genes that control heat shock protein synthesis are activated when plant temperature approaches 104°F. The production of heat shock proteins reduces the loss of enzymes to high-temperature denaturation.

Heat shock proteins are most effective during the early stages of heat stress, so they perform best when temperatures increase gradually rather than abruptly (He and Huang, 2007). Slow temperature increases allow gene activation and adequate production of heat shock proteins as the plant's temperature increases. Rapid temperature increases overwhelm the plant's heat shock response. When temperatures rise rapidly, a plant cannot produce heat shock proteins quickly enough to overcome the denaturing of enzymes. In that case, fewer enzymes can be protected and more plant damage occurs.

Conduction and convection

Transpiration is not the only source of plant cooling. Conduction and convection are also processes that help to cool a plant. Transpiration is an active process, meaning that it is instigated or controlled by plant response and that energy is required to maintain it. Conduction and convection, however, are passive processes, and are part of the natural environment. We heat and cool our buildings by conduction and convection. Air-conditioning, for instance, is an active process. We apply air conditioning for the purpose of cooling a building and it works by a process of conduction and convection. However, conduction and convection would occur whether the air-conditioner was present or not. We simply use the air-conditioner to take advantage of conduction and convection processes for our own comfort.

If you heat a cooking pan on a burner, then turn the burner off, the pan eventually cools to the temperature of the air in the kitchen. It does that by conduction and convection, two separate processes that combine to equilibrate the hot pan and the cooler kitchen air. We say that the pan cools to the temperature of the kitchen, but that is not entirely true. The pan and the kitchen air actually reach equilibrium (Box 9.1). The pan not only cools; it also warms the kitchen air. The effect on the air, however, is small compared with the cooling of the hot pan, so we tend to ignore it. However, if we had several of these pans cooling, they would noticeably increase the temperature in the kitchen. All of those pans would be warming the air around them.

Conduction is the transfer of energy from one object or entity to another as the objects or entities contact each other. Objects consisting of molecules at high relative energy transfer energy to other objects or entities with molecules at lower relative energy. In other words, hot objects transfer their heat to cold objects when the objects touch. Except for water, solids, liquids and gases become less dense as they warm and are therefore lighter. Because the air around the pan becomes lighter as it warms, it rises and is replaced by the cooler air around it. As the pan warms the new air this also rises and is replaced. This process continues until the pan is at the same temperature as the air around it. We call that process convection. Once the pan is the same temperature as the air around it, the conduction stops and convection no longer occurs. The temperature of the pan is now in equilibrium with the temperature of the air around it. These conduction and convection processes also occur in the case of warm grass leaves surrounded by cooler air. Conduction and convection are not nearly as effective at cooling grass plants as is transpiration, especially when air temperatures are high. However, the conduction and convection processes provide some high-temperature relief, especially when limited water availability causes grass plants to close their stomates and thereby limit transpiration.

High-temperature stress in cool-season grasses

Although high-temperature stress can occur in warm-season turf, it only occurs at extremely high temperatures and it rarely kills grass. Cool-season grasses can be damaged by heat at much lower temperatures than those that stress warm-season

> **Box. 9.1. A balance scale works by equilibrium.**
>
>
>
> Balance scales were one of the earliest measuring devices and are still in use today. An object of unknown weight is placed in a tray on one side of the scale and small weights are added to a tray on the other side until the weights and the object counterbalance. When the weight of both trays is equal, the balance has reached equilibrium. The amount of weight required to counterbalance the object is exactly what the object weighs.
>
> The equilibrium is considered stable because it does not change unless subjected to an outside influence. If you touch one side of the scale with your finger the equilibrium is disrupted. However, once you remove your finger the scale will quickly return to equilibrium. The equilibrium can be maintained if the same amount of weight is added to both sides of the scale. The weight on the scale will change but it will still be in equilibrium. Equilibriums are maintained in living cells in the same way. When too much of a compound is present on one side of the cell membrane, the membrane allows a portion of it to cross to the other side so that a new equilibrium can be achieved.
>
> Equilibrium in a living cell is considerably more complicated than equilibrium in a simple balance scale. The inside and the outside of the cell may harbor unequal amounts of a particular compound when equilibrium is achieved because many other compounds also influence the equilibrium simultaneously. However, the complicated balance achieved by the cell works in the same manner as the simple balance achieved by this scale.

grasses. Cool-season turfgrass shoots prefer temperatures between 60 °F and 75 °F (16 °F and 24 °C) (Beard, 1973). Temperatures greater than 75 °F begin to cause stress in cool-season plants. High air temperatures and high soil temperatures extending through both days and nights are most detrimental to turfgrass growth. However, it is the high soil temperatures (Xu and Huang, 2000) and high night temperatures (Xu et al., 2003) that cause the most damage. If nights remain cool and/or soil temperatures can be reduced high temperature stress is minimized (Fu and Huang, 2003). High-temperature stress can cause reduced plant density, tiller density, root number and root fresh weight in cool-season grasses (Xu and Huang, 2001). It is likely that shoot growth cessation is actually caused by a net loss of carbohydrates in the roots (Aldous and Kaufmann, 1979). Increases in soil temperatures during the summer season cause declines in the production of new roots in cool-season grasses. By the end of the summer, the ratio of dead roots to live roots increases to an annual high (Huang and Liu, 2003). Root depth is also affected as it tends to become less with increasing soil temperature (Martin and Wehner, 1987). At the end of the summer, cool-season grasses tend to be in their poorest, most vulnerable condition of the year.

Much of the high temperature stress that occurs in cool-season grasses is directly related to their propensity for photorespiration when temperatures are high. You learned about photorespiration in Chapter 3. Photorespiration occurs when Rubisco, the enzyme that binds carbon dioxide in the Calvin cycle, binds oxygen instead. Photorespiration is not a useful process in terms of carbon acquisition. In fact it is called respiration because it results in a loss of carbon. As you know, the occurrence of photorespiration increases as temperature rises because oxygen is more soluble in water than is carbon dioxide at high temperature, and also because, especially when temperature is high, light is intense and days are long, photosynthesis is occurring so rapidly that the air around the turfgrass leaves becomes extremely high in oxygen and low in carbon dioxide. In experiments where oxygen availability was seriously restricted, the photosynthesis of cool-season grasses approached that of bermudagrass (*Cynodon dactylon*) even when temperatures were high (Watschke et al., 1973). In another experiment, low oxygen enabled a nearly two-fold increase in the net photosynthesis of Kentucky bluegrass (*Poa pratensis*) at 95 °F (35 °C) (Watschke et al., 1972). When photorespiration occurs, heat is generated but carbon is lost rather than gained. Photorespiration does not occur in warm-season plants, but it occurs progressively more often in cool-season plants as temperature increases.

In high temperatures, photorespiration results in a substantial decrease in carbohydrate production from that which would occur in a cool-season plant in lower temperatures where C_3 photosynthesis is more efficient. It is this reduction in carbohydrate production that often causes the loss of cool-season plants, either directly or indirectly, when exposed to high temperature for extended periods. Low carbohydrate production is the reason that cool-season plant root systems decrease during warm summers. The reduction in nutrient uptake, and more importantly, in water uptake, by the weakening root system can cause plant death. The loss of energy to photorespiration may also weaken plants sufficiently that they become more sensitive to insect or disease damage.

It should come as no surprise that heat stress and drought stress in cool-season grasses are related (Wang and Huang, 2004). Even warm-season grasses under drought exhibit canopy temperatures much higher than air temperatures (Steinke et al., 2009). Drought stress predisposes cool-season grasses to heat stress (DiPaola and Beard, 1992). Drought-stressed cool-season grasses succumb to heat stress at lower temperatures than plants with adequate water. In contrast, cool-season grasses that have had the opportunity to acclimate to mild drought and higher temperatures, and have been managed properly in the spring and fall preceding the summer can withstand a hot summer in a temperate climate or a mild subtropical climate.

Cool-season grass acclimation to summer seasons

Annual cool-season grass species avoid summer heat by producing abundant seed then dying in late spring or summer. When temperatures cool in the fall, the seeds germinate and the species prospers. Some of these annuals, such as annual bluegrass (*Poa annua*) make nice turfgrasses for certain locations and uses. However, in spite of the ubiquitous nature of annual bluegrass, its general adaptation to most regions as a weed species and its propensity to act as a perennial in some locations under certain management, it still tends to make a poor turfgrass over the summer. We need perennial cool-season turfgrasses that can withstand the rigors of summer and maintain reasonable health during that season. We have several cool-season turfgrasses that can survive summer seasons, but they perform best if they have an opportunity to acclimate to or are forced to acclimate to summer stress before its actual arrival.

Turfgrass response to longer day length

As day length increases, cool-season grasses acclimate by growing more upright. Prostrate grasses, such as creeping bentgrass (*Agrostis stolonifera*) and bermudagrasses (*Cynodon* spp.), tend to tolerate exceptionally low mowing because they expose a large amount of leaf area to sunlight. Consequently, low mowing does not affect a prostrate grass as negatively as it does an upright growing grass. For that reason, cool-season grasses acclimate to long days by growing more upright and reducing the amount of light interception by each leaf. Less light interception results in less heat in the leaf and upright growth exposes more leaf surface area to the air, thereby enhancing both evaporation and conduction. When days are long, rapid photosynthesis is not necessary. As days lengthen, light saturation occurs for longer and longer periods, meaning that photosynthesis is at a maximum rate for long periods during the day (Bell *et al.*, 2000). Because rapid photosynthesis is not necessary, both cool-season grasses and warm-season grasses also reduce leaf chlorophyll levels during summer, as evidenced by reductions in greenness (Xiong *et al.*, 2007). In addition, carotenoid pigment levels increase (Demmig-Adams, 1990). Carotenoids act as receptors of excess light energy that might otherwise damage plants. These adjustments in leaf pigment concentrations cause summer turfgrasses to become less green and more yellow (Fig. 9.1).

As turfgrass leaves grow more upright and less chlorophyll is synthesized, light interception is reduced and the leaves remain cooler. If temperatures warm gradually as summer approaches, cool-season grasses have sufficient time to acclimate to the upcoming stressful season. However, if the temperature warms abruptly and remains high, as sometimes happens, our grasses are unprepared and stress increases. We can help to avoid this situation by subtly forcing our grasses to prepare for summer. It is possible to encourage upright growth by mowing in multiple directions by rotation. Each successive mowing should be performed in a different direction than the last. For instance, mow north and south for this event then east and west for the next one. Upright growth can be further encouraged on intensively managed turf, such as golf course putting greens or bowling greens, by practicing shallow grooming or brushing (Fig. 9.4).

Fig. 9.4. A reel-type mower fitted with a grooved roller and a grooming attachment, devices that improve mowing consistency, encourage plant density and help to encourage upright growth.

Fertilizing cool-season grasses with nitrogen immediately before or during summer is not recommended except for light applications on intensively managed turf such as golf courses or athletic fields where rapid recovery from damage is desired. Nitrogen fertilization encourages chlorophyll synthesis. Encouraging chlorophyll synthesis during a period when your grasses are naturally reducing their chlorophyll content is counterproductive. Withholding nitrogen fertilization in late spring helps the grasses to acclimate to the upcoming summer season.

Managing cool-season grasses to survive summer heat

Being proactive and attempting to help your cool-season grasses through a stressful summer often increases rather than decreases stress. It is the management processes executed in the previous fall and spring that get cool-season grasses through the summer with the least amount of stress. Water is a plant's best defense against summer stress and, as you learned earlier, irrigation management is paramount during the summer. Many other management practices harm the turf rather than help it.

Management practices to avoid during summer stress

It was mentioned earlier that except for light applications to certain types of turf, nitrogen fertilization should be avoided on cool-season grasses during the summer. Conversely, potassium fertilization may help, especially if soil potassium is low. Practices that damage turf, no matter how slight, should be avoided. An exception to that statement is small-tine aerification or spiking when required to penetrate a thatch or soil layer and promote water infiltration and soil oxygenation. Turf on poorly aerated soil is at greater risk of heat-related damage than turf on well-aerated soil (Huang et al., 1998). As increasing soil temperature causes increasing respiration, oxygen is removed from the soil and replaced by carbon dioxide. When this process occurs in poorly aerated soil, cool-season grasses may seriously decline (Rodriquez et al., 2005b). Thatch or layering problems are best addressed during the fall or spring, but sometimes these problems are not noticed until symptoms of stress appear during the summer. Turfgrass roots need an ample supply of water and oxygen even if minor damage has to occur to get it to them.

Cool-season grasses under summer stress are more susceptible to disease and insect damage than the same grasses during other times of the year. Frequent monitoring of turf status is required during the summer. Cool-season grasses can sustain damage quickly during this period and will heal very slowly. Insect or disease pressure must be monitored frequently and dealt with immediately. Damage can spread quickly on stressed turf. Most insecticides and fungicides can be applied safely to cool-season grasses when temperatures are high. However, you must be careful with these applications. There are a few examples of these products, such as sterol inhibitor fungicides, that can cause damage during hot weather. Some products include carriers like emulsifiers or oils that can be harmful in hot weather. Read the label, and check with your supplier before application. It is also important to stay within the range of recommended rates.

Although insecticides and fungicides are usually safe for summer applications, many herbicides can damage grasses during periods of high temperature. Some herbicides that are routinely applied without incident during cooler times of the year can cause damage to desirable grasses during summer. Depending on the herbicide, damage may occur on cool-season grasses, warm-season grasses or both. Sometimes the damage is particular to certain species. Many herbicides that are safe on one cool-season grass in the summer may not be safe on others. The same is often true of warm-season grasses. You should always be careful with chemicals, but summer is a time to be exceptionally careful.

Low-mown turf is more susceptible to heat stress than high-mown turf. High-mown turf has more leaf area above the soil surface where the turfgrass crowns reside. The greater leaf area of the high-mown turf shades and insulates the soil better than does low-mown turf. A close-mown turfgrass canopy can easily exceed the air temperature around it during the heat of an afternoon on a clear summer day (Throssell et al., 1987). Golf course fairways, tees and putting greens are especially vulnerable to high temperature stress because they are closely mown. They also sustain traffic damage, which can be especially problematic during stress periods when recovery is slow.

Efforts should be made during the summer to channel traffic to little used areas, allowing the high-traffic areas to recover. You should make an attempt to route traffic so that the entire turfgrass area is used intermittently, thus giving each small region the greatest possible time to recover before traffic is applied there again. The alternative is to construct mulch, gravel or concrete pathways that can be used to channel traffic away from stressed turf.

Sand topdressing is another practice that should be avoided during high temperature except in amounts small enough that the sand can be watered in or lightly brushed. Sand readily reflects light and heat, and when left in a pile for an hour during lunch on a hot sunny day will almost surely cause the turf around it to wilt and possibly die. Sand also has sharp edges. If topdressing is applied at high enough rates that dragging is required to facilitate canopy infiltration, the sharp edges of the sand will cut the turfgrass leaves as dragging occurs, creating damage that can only heal slowly. Each time the surface receives equipment or foot traffic, the leaves will be pressed against the sharp sand, causing more damage. Topdressing with sand in the summer should be executed with restraint.

A combination of surface fans and light misting of water can effectively reduce canopy temperatures on high-value turf such as golf course putting greens. However, this practice also increases canopy humidity and should be closely monitored for disease activity (Rodriquez et al., 2005a).

Although summer can be a stressful time for turfgrasses, especially cool-season turfgrasses, successful managers can minimize damage through preparation and responsible management. Water is the most important component of a plan for high-temperature stress relief during summer. Proper management during the spring and fall growing seasons preceding summer will minimize stress. Applying tactics to the management program that help the turf acclimate to the coming summer can also help.

9.4 Cold-temperature Stress in Grasses

Although cool-season grasses are generally adapted to very cold temperatures, they are occasionally damaged by subfreezing weather and sometimes by ice cover. A cool-season grass that gradually acclimates to cold weather, however, is rarely damaged by it. Conversely, warm-season grasses are quite sensitive to cold. Temperatures do not have to reach freezing before some warm-season turfgrasses begin to experience stress. This cold temperature stress that may occur when temperatures are below about 50 °F (10 °C) is referred to as chilling stress and only occurs in warm-season grasses (Hale and Orcutt, 1987).

Freezing stress in grasses

Do you remember what colligative properties are? Colligative properties were introduced in Chapter 4 and are part of most introductory chemistry classes. There are four colligative properties of solutions. They are that the higher the concentration of solutes in a liquid the higher the boiling point will be, the lower the freezing point will be, the lower the vapor pressure will be and the greater the osmotic pressure will be. Colligative properties are very important to plants because they keep plant cells from freezing until air or soil temperatures get very low; far below the freezing temperature of pure water. Plant cells contain many different solutes, including sugars, amino acids, nutrient ions and many other elements and compounds.

To understand why freezing and chilling stresses occur in plants, you have to remember some simple plant anatomy. Plants have a cellular structure that is slightly different from animal structure in that each cell has not only a cell membrane, but also a cell wall. The cell walls add support and act in a manner similar to how a skeleton works in vertebrates like us. However, because plants don't move they do not require joints in their support structure, so their anatomy is basically a bunch of semi-rigid cells stacked on top of each other, with each cell wall filled with and surrounded by water. Small imperfections or gaps between cells or between the membrane and the cell wall are filled with water. Because water does not compress, its presence in small amounts around and in the cell walls provides rigidity, but its propensity to flow if space is available also encourages flexibility. The water inside a plant cell is called intracellular water and the water in the cell walls and surrounding them is called intercellular water (Fig. 9.5). If the intracellular water freezes the cell nearly always dies. However, it is more likely that the intercellular water will freeze, damaging membranes and cell walls or, more often, causing cell dehydration.

The intracellular, inside space contains the living parts of the plant. All of the plant's metabolic processes occur in the inside space in water. Consequently, the inside water contains a high concentration of solutes necessary to perform plant growth and maintenance. Activity is always occurring in the inside space. However, activity, but not living

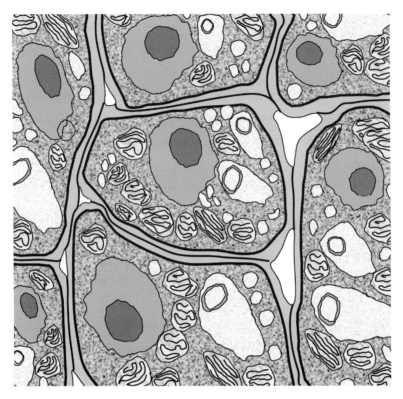

Fig. 9.5. The structure of plant cells. Living protoplasm is surrounded by a membrane which is encased in a cell wall filled with and surrounded by water.

activity, is also occurring in the intercellular, outside space. The outside space provides water and nutrients to the inside space. The outside space is in contact with the xylem, the transpirational plant veins that furnish the water and nutrients for plant metabolism and the water for transpiration. The outside water also contains a high concentration of nutrients and other solutes, but its concentration is not as high as that of the inside space. Consequently, the outside space freezes before the inside space.

Water is the only known compound that expands when it cools. All other compounds shrink as they cool. Consequently, when intercellular water freezes, the expanding water can cause damage to cell walls. Ice crystals are sharp, so the freezing of intercellular water can also pierce cell membranes. However, it is dehydration that most often causes a cell to fail following the freezing of intercellular water.

To understand the cell dehydration process you must remember the colligative property called osmosis. The higher the concentration of solutes in water, the higher its osmotic pressure becomes, meaning that when water of a high solute concentration is separated by a water-permeable membrane from water with an even higher solute concentration, water will move from the high concentration of solute to the higher concentration of solute until the concentrations are equal. Solutes do not dissolve in ice, so for the purposes of solute concentrations, ice is not the same as liquid water. When the intercellular space begins to freeze the amount of water becomes less and less, but the amount of solute remains the same, so the solute concentration increases. When the solute concentration in the intercellular space becomes great enough, water begins to flow from the intracellular space to the intercellular space by osmosis. This flow dehydrates the intracellular space and if the dehydration becomes excessive can cause the cell to die. If water does not flow from the cell to the intercellular space, its solute concentration will not increase and it will probably freeze and die. For that reason, the dehydration process is more likely to save the life of the cell than to kill it. The intercellular and intracellular spaces of a warm-season grass are maintained at lower solute concentrations than those of cool-season grass and the warm-season grass is therefore more likely to freeze.

Chilling stress in warm-season grasses

Chilling stress is a condition that can occur in warm-season plants when temperatures fall below 50 °F (10 °C). It is defined as low-temperature stress in the absence of freezing (Levitt, 1980). Chilling stress results from the loss of selective permeability of cell membranes. It can also occur because of changes in metabolism caused by cool weather, or may occur from a combination of these.

Cell membrane permeability

Cell membranes are primarily comprised of phospholipids. I am sure that you are aware of the effects of cold on lipids. Lard, for instance, is a liquid at high temperature but a solid at low temperature. Vegetable oil does not become a solid until its temperature gets extremely low, but its viscosity (its thickness or resistance to flow) becomes greater as its temperature declines. Plants with cell membranes containing greater amounts of polyunsaturated fatty acids are less likely to experience cold damage than those containing mostly saturated fatty acids (Cyril *et al.*, 2002). Warm-season turfgrass cell membranes contain lipids closer to lard than to vegetable oil and cool-season cell membranes contain more lipids like vegetable oil. Consequently, cool-season cell membranes are not seriously affected by cold temperature until the temperature gets extremely low, but warm-season membranes can be seriously affected when the weather is cool.

In plants, especially warm-season plants, there is a membrane transition from flexible to rigid as the temperature is reduced. You probably remember that cell membranes are like selective sieves. They allow certain molecules to penetrate but not others. If the membrane becomes rigid, however, its selectivity is affected and it begins to permit unwanted compounds to penetrate both into and out of the protoplasm. That, in turn, makes it very difficult for the cell to maintain efficient metabolism, and the cell's controls on metabolic processes may be compromised. Researchers have found extensive damage to inner chloroplast membranes of bermudagrass, but little damage to the more cold-tolerant zoysiagrass (*Zoysia* spp.) inner membranes following chilling stress (Rogers *et al.*, 1977).

If the temperature increases, metabolism may return to normal, but if the chilling temperature persists or becomes lower, cell membranes may become disrupted badly enough that they begin to leak protoplasm noticeably. Water-soaked lesions will form on turfgrass leaves, and the leaves and possibly the plants will probably die (DiPaola and Beard, 1992).

Chlorophyll photooxidation

Photooxidation is a destructive event (Niyogi, 1999). As temperature declines, photosynthesis, respiration and transpiration decline. In this case, the decline in photosynthesis is most detrimental. When temperature declines, photosynthesis slows. The chemical reactions in both the z-scheme and the Calvin cycle are affected. However, light energy interception is not affected. Chlorophyll molecules continue to intercept light energy, but because the photosynthetic system is slow, this excitation energy has nowhere to go. Eventually, some chlorophyll molecules reach an energy level high enough to pass an electron directly to the PSII (photosystem II) receptor. This excess energy passing into the photosynthetic system causes damage that has to be repaired. The chlorophyll that passes the electron to PSII has become oxidized, hence the name photooxidation. The oxidized chlorophyll molecules become free radicals seeking electrons from other compounds, and damaging membranes and other metabolic processes. Chlorophyll photooxidation is most likely to occur on clear cold days when intense light is present, but chemical processes are excessively slow because of cool temperatures.

Dehydration

Motor oil is rated by viscosity because rapidly moving close-fitting metal mechanical systems like gasoline engines require different grades of oil for lubrication according to temperature. A gasoline engine containing an oil of high viscosity and close-fitting parts may be damaged shortly after start-up if its oil is too viscous to squeeze between its close-fitting parts without creating excessive pressure. Once the engine warms, the oil viscosity will be reduced and damage will not occur. However, when the oil is cold and viscosity is high, damage is likely. Therefore, gasoline engines require low-viscosity oils when temperatures are cold.

As temperature cools, oil viscosity increases. Temperature has the same effect on water.

As soil temperatures cool, soil water becomes more viscous. The more viscous the soil water becomes, the more difficult it is for plant roots to absorb it. Consequently cool temperatures slow water uptake, water translocation and transpiration. In temperate and cooler climates, soil water may be frozen for much of the winter. If the soil water freezes when our turfgrasses are still green and active the plants may dehydrate. As photosynthesis, respiration and other metabolic processes occur, water is used. This water, however, cannot be replaced when the soil water is frozen. The same situation can occur when our plants need water but the soil water is too viscous to be absorbed rapidly enough to satisfy plant needs. It is likely that much of the winter turfgrass damage attributed to freezing is actually caused by dehydration.

Turfgrass damage by ice encasement

In many regions, ice encasement of grasses for short periods during the winter is common. Creeping bentgrass has been known to survive under ice cover for up to 90 days (Tompkins *et al.*, 2004). For the most part, cool-season grasses can tolerate this ice encasement for short periods of 2 or 3 weeks. However, longer periods or periods of freezing and thawing of the ice can be very detrimental to the turfgrass below it. Warm-season grasses are more susceptible to ice damage than cool-season grasses.

Ice can cause damage to grasses in three ways. It can crush or pierce crowns and roots as it freezes, it can restrict air exchange for a long enough period that the air under the ice becomes of poor enough quality to damage turf, or it can be pushed into crowns or roots by pressure from traffic or other sources.

Periods of freezing nights and thawing days are likely to cause more damage to turfgrass than constant freezing temperatures. When ice encasement first occurs, it usually does not crush or pierce enough crowns to cause substantial damage. However, if the sun or warm temperature melts the ice daily then it freezes overnight repeatedly, damage is likely to be extensive. As the ice melts, water gathers on the soil surface. Under ice cover, the upper soil is usually frozen. Water from the melting ice helps to melt the very top layer of soil and infiltrates it slowly, but it cannot penetrate far because most of the soil is frozen. Consequently, the very top layer of soil where the turfgrass crowns reside becomes saturated. At night, when this water freezes, it expands and crushes turfgrass crowns.

Slushy, partially frozen conditions such as slowly melting ice or partially frozen wet snow can also cause damage. During such conditions, ice crystals can be pushed into and pierce turfgrass crowns as traffic or another pressure is applied overhead.

Ice encasement basically eliminates air exchange from the living parts of the plants to the atmosphere. Although the turfgrass may be dormant or partially quiescent, its roots and crowns are still performing respiration at reduced rates. Consequently, the respiration process is removing oxygen from the air and replacing it with carbon dioxide. Eventually the plants may die of anoxia, a lack of oxygen. However, low oxygen and high carbon dioxide is a more severe condition than low oxygen alone (Castonguay *et al.*, 2009). If you have areas of high-value turf covered in ice for more than 20–30 days, it would probably be worth your effort to remove the ice. It could make the difference between live and dead turf next spring.

Cold acclimation in turfgrass plants

As temperatures cool in late fall, turfgrasses begin to "harden off" for the winter, which means that they begin to prepare themselves to withstand temperatures that would normally be damaging.

As winter approaches, both warm- and cool-season turfgrasses begin to translocate carbohydrates into their roots and stems (McKell *et al.*, 1969; Dunn and Nelson, 1974). At least in some grasses, these long-chain carbohydrates may be quickly converted to sucrose, which accumulates in the turfgrass crowns and stems when necessary to prevent freezing (Fry *et al.*, 1993; Dionne *et al.*, 2001a). As temperatures cool, shoot growth slows, but photosynthesis continues as long as the plants are green. The carbohydrates produced by photosynthesis and stored in belowground roots and stems for the winter also provide fuel to allow them to continue slow metabolism over the winter after their leaves die. Some grasses of course, depending on species and location, remain green all winter and continue to produce carbohydrates throughout the period. Carbohydrates stored but not used during the winter may provide for rapid new growth in the spring.

During cold acclimation, turfgrasses also accumulate proteins and other nitrogen-rich compounds that decrease in the spring as the plants de-acclimate (Dionne et al., 2001b). This protein accumulation has been implicated in the greater freeze tolerance of some bermudagrasses (Gatschet et al., 1996) and some zoysiagrasses (Patton et al., 2007a,b). So it seems that warm-season grasses that rapidly accumulate carbohydrates and proteins during cold acclimation are likely to be more cold tolerant (Zhang et al., 2006).

During acclimation, grasses also go through a dehydration process. Less water is stored in tissues as winter approaches. This dehydration is perhaps the most important process for plant protection against freezing. Cool-season grasses dehydrate more completely than warm-season grasses and this is believed to be a major reason that cool-season grasses are more resistant to freezing. Dry tissue is less likely to freeze than succulent tissue. Another reason that cool-season grasses are better at resisting freeze damage than warm-season grasses is because fructans, the storage carbohydrate that you know to be used by cool-season grasses, is soluble, and the starch stored by warm-season grasses is not. Consequently, carbohydrate storage in cool-season grasses has colligative properties, but in warm-season grasses it does not. That also helps cool-season grasses to survive cold temperatures that kill warm-season grasses.

In the spring, grasses de-acclimate quickly with warming soil temperatures (Davis and Gilbert, 1970). Sudden cold temperatures following de-acclimation can be stressful. De-acclimated plants, however, are able to partially re-acclimate when soil temperatures drop, thus helping to prevent severe damage (Tompkins et al., 2000).

Managing cold-temperature stress

When we consider temperature, we normally think of air temperature. Although cold air affects our turf, it is soil temperature that is most important. Dead turfgrass leaves in the winter are common to warm-season grasses north of tropical regions and even to cool-season grasses in regions where winters are exceptionally cold. However, it is the crowns (the thick, whitish part of the turfgrass that grows at soil level where grass shoots and roots meet) that determine the life or death, and even the current health of a turfgrass plant, and the crowns are in the soil. Soil temperatures are ultimately more important than air temperatures for the cold survival of turfgrass plants.

A frozen wet soil is more damaging to turfgrass crowns than a frozen dry soil. A wet soil expands more when it freezes than a dry soil, and is more likely to crush crowns. However, wet soils are less likely to freeze and resist rapid temperature declines better than dry ones. Consequently, light fall and winter irrigation can help to prevent freeze damage to turfgrass. The moist soil resists the rapid temperature declines that are most detrimental to turfgrass plants but does not contain enough water to crush crowns as it freezes. Soil moisture is especially important during early cold snaps that occur before the turf has had sufficient time to acclimate to cold temperatures. However, it is important not to apply soaking irrigation. The dryness of the crown is an important component of resistance to freezing. We want the crown to remain relatively dry.

Proper fall fertilization can also help turfgrasses to resist chilling and freezing. It is widely believed that warm-season plants should not be fertilized with nitrogen for a month or more before freezing temperatures are likely to occur. The succulence normally encouraged by nitrogen fertilization could counteract the dehydration of the plant tissue as the grasses harden off for winter. This no- or low-nitrogen protocol is not supported by scientific research (Reeves et al., 1970; Richardson, 2002). However, it is based in logic and seems to make sense, so most managers are reluctant to use nitrogen fertilizer on warm-season grasses as winter approaches. Cool-season grasses are not detrimentally affected by late-season nitrogen fertilization. In fact, as stated in earlier chapters, late fall is an excellent time to promote root growth with nitrogen fertilization of cool-season grasses without encouraging shoot growth.

When cell membranes begin to increase in rigidity with the chilling of warm-season grasses, one of the first membrane-regulated processes affected is potassium exchange. Some of the potassium necessary for cell functions is lost to intercellular spaces when the cell membranes are disrupted. Potassium helps to increase turfgrass resistance to many stresses, including cold. Replacing late-season nitrogen applications to warm-season grasses with late-season potassium applications helps warm-season grasses to resist chilling stress and freezing stress (Juska and Murray, 1974).

Research suggests that high phosphorus levels may interfere with the cold-protection qualities enhanced by potassium (Reeves et al., 1970). In addition, potassium may not be utilized fully unless sufficient nitrogen is also present (Reeves and McBee, 1972). Consequently, light applications of nitrogen should accompany late-season potassium applications for best results. Nitrogen fertilization in the fall has the added advantage of encouraging warm-season grasses to stay green longer in the fall and green up earlier in the spring, thereby extending the growing season for about 20–25 days (Reeves et al., 1970; Richardson, 2002). Greenhouse research found that a fertilizer ratio of 4:1:6 nitrogen:phosphorus:potassium encouraged the best cold tolerance in two bermudagrasses (Gilbert and Davis, 1971). Depending on soil tests, the phosphorus in this formulation may be avoided to reduce potential nutrient runoff that could enhance surface water eutrophication.

Iron applications can also be beneficial to warm-season grasses approaching winter. Iron application encourages chlorophyll synthesis which, in turn, encourages carbohydrate production and storage. In addition, the iron promotes aesthetic color without causing an increase in succulence (Goatley et al., 2005).

Some ice damage and dehydration may be avoided by encouraging grasses to grow deeper roots. In summer, the soil gets cooler as you dig deeper because the finer minerals that accumulate deeper hold more water and because the deeper soil is protected from the sun's radiation. In the winter, the deeper soil stays warmer again because it holds more water and because air temperature has less effect on it. The upper soil may warm from the sun but freeze again at night. Deep soil, however, resists freezing. Builders and construction personnel are always aware of the "frost line" in their area. The frost line is the depth of soil below which the ground never freezes and, although it has nothing to do with turfgrass, is a good example of how the soil reacts to freezing temperatures. The frost line changes with climate. In warm regions it is quite shallow, but in cold regions it can be quite deep. Although the frost line is too deep for turfgrass roots in many areas, a turfgrass plant with a deep root system is more likely to find water in the winter than one with a shallow root system.

Photoinhibition

Photoinhibition, the disruption of the photosynthetic process can detrimentally affect photosynthesis when temperatures are either too hot or too cold (Long et al., 1994). Photoinhibition is not something that we can control, but it is a term that you should be familiar with. It is not unique to warm- or cool-season grasses, and is usually the reason that warm-season grasses yellow when the temperature is excessively high in the summer and that cool-season grasses yellow as the temperature declines in the winter. As in photooxidation, photoinhibition occurs when light energy is absorbed but cannot normally proceed down the photosynthetic pathway. When the temperature is high, photoinhibition occurs because of the disruption to protein structures that can occur at temperatures greater than 104 °F (40 °C). Cold-temperature inhibition occurs when the PSII (photosystem II) reaction center, specifically the D1 protein, is damaged by excess energy. The damaged proteins are constantly repaired and replaced to allow some photosynthesis to proceed even during normal metabolism (Tyystjärvi, 2008). Consequently, photoinhibition usually does not permanently damage photosynthetic pathways and they return to normal when light becomes less intense or temperatures return to levels more conducive for chemical activity.

9.5 Chapter Summary

The best management practice available for maintaining turfgrass under the extreme hot and cold temperatures common to your area is choosing the best species and cultivar for your site(s). Warm-season species and cultivars differ in their tolerance of cold conditions, and cool-season species and cultivars differ in their tolerance of heat (Sifers and Beard, 1993; Dunn et al., 1999; Anderson et al., 2003). In some cases, you never have the opportunity to establish or reestablish the turfgrass that you are working with, but in many cases you do. Be sure to choose wisely. In addition to extension services, local agricultural universities, local growers and local practitioners, selection information is available at the website of the National Turfgrass Evaluation Program (NTEP). If you live in the USA, you can find locations near you that have completed turfgrass quality trials for the NTEP. If not, you can compare your climatic conditions with those in the USA to find the locations that are similar to yours.

There is little that we can do about changes in temperature. Temperature is one of the very few things that affects our turf that we have very little control over. Consequently, a seasonal management plan needs to be formulated to take advantage of the best growing conditions in your region to prepare your turf for the worst conditions. In addition, seasons differ from year to year and we have to be prepared to adjust our management practices so as to best affect our turf under unusual temperature conditions as well as under those that are relatively normal for our region. A flexible management regime is necessary for best management. Knowledge of turfgrass responses to weather and the overall environment is required to make management decisions. Because we cannot seriously affect the temperature at our site, we have to be able to react to it in the most positive manner.

Suggested Reading

DiPaola, J.M. and Beard, J.B. (1992) Physiological effects of temperature stress. In: Waddington, D.V., Carrow, R.N. and Shearman, R.C. (eds) *Turfgrass*. ASA-CSSA-SSSA, Madison, Wisconsin, pp. 231–268.

Suggested Websites

NTEP (National Turfgrass Evaluation Program) (2009) Available at: http://www.ntep.org (accessed 17 December 2009).

10 Growing Grass on Soil, Sand and Salt

Key Terms

Sand, for purposes of this chapter, refers to root-zone mixtures of sand with or without amendments specifically designed as root-zone material for turfgrass systems.

Native soil, or simply **soil**, for the purposes of this chapter, refers to disturbed or undisturbed mixtures of sand, silt, clay and organic material commonly found in natural settings.

Black layer is an anaerobic soil layer where sulfur-reducing bacteria have turned the soil black. Black layer occurs when a soil pocket or layer remains saturated with water most of the time.

Localized dry spot is a condition where the soil has become hydrophobic and restricts water infiltration.

Nutrient leaching occurs when a nutrient, usually nitrogen, moves downward through the soil past the root zone with percolating water before plants can take it up.

Spoon feeding is a series of frequent spray applications containing light rates of nutrients, perhaps in the range of 10 pounds of nitrogen per acre (11 kg ha^{-1}) or less.

A **foliar application** is a spray application of solution at less than 20 gallons per acre (190 l ha^{-1}) in very fine droplets meant to attach to the turfgrass leaves for absorption rather than to the soil or thatch.

A **typical spray application** of nutrients is commonly applied at 20 to 40 gallons per acre (190–380 l ha^{-1}) and contains 20 to 40 pounds of nitrogen per acre (22–44 kg ha^{-1}) and possibly other nutrients.

Saline in the traditional sense refers to a solution high in salts. In this text, it refers to a solution high in salts other than sodium.

Sodic refers to a solution, including a soil water solution, containing a high proportion of sodium ions.

Saline+sodic in this text refers to a solution, including a soil solution, high in both sodium and other salts.

Electrical conductivity (**EC**) is defined as the reciprocal of electrical resistance and refers to a material's ability to carry electricity. It is measured in Siemens. The EC of soil solution is measured on a saturated soil extract and is designated EC_e. The EC of water is designated EC_w.

ESP, the exchangeable sodium percentage, is the proportion of exchange sites on a soil's cation exchange capacity (CEC) occupied by sodium reported as a percentage.

SAR stands for sodium adsorption ratio and is a measure of sodium hazard that is calculated by comparing the amount of sodium in solution with the amount of calcium plus magnesium.

A **leaching fraction** is the proportion of irrigation and rainfall that carries salts with it through the root zone.

10.1 Different Media Require Different Management

In Chapter 7, a review of basic soil properties was provided because of its importance for nutrient retention. In Chapter 8, soil was discussed for its importance in water retention. The perfect soil for plant growth would contain 45% mineral solids, 5% organic solids, 25% water and 25% air. However, that particular soil on a turfgrass site that was required to sustain high pedestrian or vehicle traffic would soon compact and contain too much water and too little air. For that reason, many turfgrass sites, especially those designed specifically to provide characteristics conducive to playing sports, are built on sand. Properly designed and constructed sand surfaces are highly resistant to compaction, but they require different management from most natural soils.

Many native soils have high sand contents, but few have a sand content as high as those designed for turfgrass systems and probably none have sands of such uniform particle distribution as those selected for turfgrass. Turfgrass sand systems are designed to resist compaction, but they are not a particularly good media for growing grass. It takes a knowledgeable individual to grow turfgrass on sand, especially on a sand surface that is under constant stress. However, knowledgeable turfgrass managers tend to prefer working with sand systems because the manager can control nearly all of the inputs.

10.2 Problems With Soil

As we discussed in Chapter 7, natural soil contains both large and small pores called macropores and micropores. Natural soils develop structure that consists of aggregates. The micropores inside the aggregates hold water and the macropores that form between the aggregates hold air. This situation is perfect for plant growth but is easily disrupted if the soil is compressed by traffic. Traffic compression crushes the aggregates and forces them closer together. In a simple sense, traffic compression of the soil turns the macropores into micropores. Consequently, a soil that was once 25% air and 25% water at field capacity is now something like 10% air and 40% water or worse. Following traffic compression, there is still plenty of water but not enough air to properly sustain root respiration. We call traffic compression, or any other type of soil compression, compaction (see Chapter 5) and we deal with it primarily through aerification (aeration), a process that was also introduced in Chapter 5. The term "aerification" is preferred in this text so as not to confuse the cultivation practice with the aeration of water and other media.

Improving soil aeration

Plants, including turfgrasses, are highly adaptable and able to alter their physiology to improve their performance in compacted soils (Agnew and Carrow, 1985). However, the constant traffic and rate of compaction on many turfgrass sites is too much stress for the plant to adapt to and the turf deteriorates (O'Neil and Carrow, 1983). Aerification is a mechanical soil disruption process that is designed to relieve compaction, improve water infiltration and promote oxygen retention on high-traffic soils (Aldous *et al.*, 2001). It is usually accomplished by punching holes in the soil, but slicing the soil, drilling it or disrupting it with high-pressure water or air can also be effective (Wiecko *et al*, 1993; Praemassing *et al.*, 2009). Natural freezing and thawing is an excellent aerifier but one that we have little control over. Natural aerification and mechanical aerification offer the same benefits and usually have the same drawbacks.

Aerification is one of our most beneficial turfgrass management practices (Jaabak, 1993). High-use areas such as golf course tees, greens and fairways, sports fields, grass tennis courts and bowling greens need to be aerified regularly. However, we hesitate to aerify as often as we should because the practice is so disruptive to playing surfaces. Hollow-tine aerification, the process of punching holes and removing cores, is the most effective aerification process that we practice (Baker, 1994). However, it temporarily results in a very poor playing surface, especially for a golf course or bowling green. For that reason, the players who use the turf surface dislike aerification and only allow it because they have been educated to realize its necessity. As managers, we try to balance the agronomic needs of the turf with the needs of our customers. Sometimes the needs of our turf conflict with those of our customers so we have to be creative to satisfy our goals with as little disruption as possible. After all, if it wasn't for the players or the customers, we would not have a job.

Hollow-tine aerification needs to be practiced on most playing surfaces at least twice per year and more often if possible (McCarty, 2001). Agronomically speaking, the best time to aerify is when the turfgrass roots are growing the fastest and need the most oxygen. The fastest root growth occurs in the spring, so from a plant management standpoint, spring is the best time to aerify. From a customer standpoint, however, that may be a poor choice. Winter is often the best time to aerify home lawns, commercial grounds or parks because few people use them at that time of year. If the site is not being used, the soil cores can be left on the surface and allowed to reincorporate naturally. By spring, the cores should be gone but the holes will not necessarily be filled. Macropores will still exist to promote oxygenation and root growth during the spring.

Aerification of sporting sites should be accomplished when the turf can heal most rapidly (Puhalla *et al.*, 1999). For cool-season grasses that is usually in mid-spring and mid-fall. For warm-season grasses the best times to heal occur in mid-to-late spring and mid-to-late summer depending on normal temperatures and conditions. Core aerification timing has to be adjusted using common sense and climate. The needs of your grasses and those of your customers will differ by location and use (Baker *et al.*, 1999a).

Aggressive aerification

Although compaction is a situation that normally occurs on native soil, core aerification should also be practiced on sand, although turfgrass managers who work on sand media need not be as aggressive with their aerification practices as those who work on soil (Praemassing *et al.*, 2009). Some turfgrass managers who have to grow grass on native soil for sporting operations, including golf, practice aggressive aerification. By aggressive I mean that they aerify in multiple directions for a single event, pull relatively large cores, and/or set the spacing between cores as small as possible. In order to provide their customers with the best playing surfaces possible, they aerify less often but they aerify aggressively when they do. Although the turf requires more time to heal following aggressive aerification, it disrupts the playing surface for a shorter period of time than would multiple aerification events during the season. Aggressive aerification is not quite as effective a management tool as multiple aerification events, but it is nearly as good.

Core aerification appears to be a very disruptive process for a playing surface (Fig. 10.1). However, close observation reveals that the process only affects a small portion of the playing surface directly (Box 10.1). For instance, core aerification using ⅜-inch (9.5-mm) diameter tines on 1 by 2 inch (2.5 × 5.1 cm) centers only affects about 5.5% of the playing surface. However, if the surface is aerified in three different directions about 16% of the surface is affected. That calculation is made with the assumption that no two holes touch when, in fact, this is highly unlikely. Nevertheless, the result is a very close estimate. Consequently, an aggressive aerification event such as the one just described can be almost as effective as aerifying three times. A general recommendation for core aerification would be to remove about 20% of the surface area each year (McCarty and Miller, 2002), although the actual amount removed depends on the conditions and use of the site, and should always be subject to adjustment depending on recent events and conditions.

Aerification processes

Core aerification of turfgrass areas is usually practiced at depths of approximately 3 inches (7.6 cm). If you practice a 3-inch aerification two to four times per year, it will not be long before a

Fig. 10.1. The result of an aerification event on a research putting green using ⅜-inch (9.5-mm) tines on 2 by 2 inch (5.1 × 5.1 cm) centers. The cores have been removed from the area in the foreground. This operation only removes about 3% of the putting surface but it looks as if it is more disruptive than that.

> **Box 10.1. A formula for calculating the surface disruption of an aerification event.**
>
> Calculating the surface area removed by a core aerification event is fairly simple when you have a background in basic mathematics. You have to know the diameter of the hole left following core removal so that you can calculate its area. You also need to know the spacing of the tines so that you can calculate the number of holes that will be pierced in a given area. The overall formula is this:
>
> Surface area affected in percent =
> Hole area in inches × no. of holes per sq. ft./144 sq. in. per sq. ft. × 100
>
> or
>
> Hole area in mm² × no. of holes per m²/1,000,000 mm² per m² × 100
>
> The area of the holes is calculated from their radius (r) by the formula for the area of a circle:
>
> $\pi r^2 = 3.14 \times $ (hole diameter/2)²
>
> The number of holes per unit area is calculated by:
>
> (length of area/length of tine spacing) × (width of area/width of tine spacing)
>
> Consequently, if you use ³⁄₈-in. (0.375-in.) tines at 2 × 2 in. spacing:
>
> Hole area = 3.14 × (0.375/2)² = 0.110 sq. in.
>
> The number of holes per sq. ft. = (12 in./2 in.) × (12 in./2 in.) = 36
>
> So the surface disruption is:
>
> 0.110 sq. in. × 36 holes per sq. ft./144 sq. in. per sq. ft. × 100 = 2.75%
>
> If you use 10 mm tines at 50 × 50 mm spacing:
>
> Hole area = 3.14 × (10/2)² = 78.5 mm²
>
> The number of holes per m² = (1000 mm/50 mm) × (1000 mm/50 mm) = 400
>
> So the surface disruption is:
>
> 78.5 mm² × 400 holes per m²/1,000,000 mm² per m² × 100 = 3.14%

compaction layer builds up at the base of the tine penetration (Fig. 10.2). Every 3 to 5 years that compaction layer will have to be disrupted. Deep-tine aerification is the process used to break up shallow compaction layers and improve deep water penetration and soil aerification. It is very effective (Morgan *et al.*, 1965). Deep-tine aerification is performed using deep coring tines or solid spikes, verti-drain units whose tines penetrate on an angle to cause additional soil disruption, or deep drills that can be used alone or with drill-and-fill techniques where an amendment, perhaps sand, is injected into the holes as they are drilled. Regardless of the technique used, deep-tine aerification is usually more disruptive than 3-inch coring. Deep-tine procedures lift and displace more soil, make deeper holes that are harder to fill, require more cleanup and take longer to heal.

Core aerification and deep-tine operations are disruptive, but there are additional procedures that are less severe. Slicing is a process that uses thin, triangular-shaped tines to penetrate the soil, leaving long but very thin slits. These units rarely lift soil and result in minor disruption. Slicing units are usually pull-behind units with a series of disk-type tines that slice the turf as they are pulled across it (Fig. 10.3). Slicing tines may also be mounted on the typical small aerifiers normally used for golf course greens. The slits left in the green barely affect the putting surface.

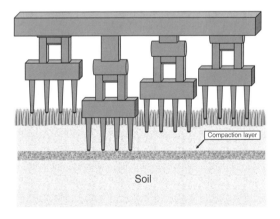

Fig. 10.2. Repeated aerification to a consistent depth is likely to cause a layer of compaction to occur at the base of the tine penetration over time. Deep-tine aerification should be practiced occasionally to break up that layer.

Fig. 10.3. Slicing units are usually pulled behind a tractor or utility vehicle. The blades on this unit are used for slicing golf course fairways and athletic fields. Much finer, less disruptive slicing blades are also available for this unit, as well as shatter core blades, spikers and core aerifiers. Fine slicing blades cause very little disruption of playing surfaces, are fast to use and require almost no cleanup. Although slicing provides very positive aeration, the effects are generally short-lived compared with core aerification.

Water injection is also a method that creates little disruption to a playing surface (Karcher and Rieke, 2005). Extremely small high-pressure nozzles are used to inject small amounts of water through the turf surface, fracturing the soil as they pass over it (Toro, 2010). Once the unit passes, you can hardly tell that it has been there. Air injection units are also available but, at this time of writing, little is known about their use and effectiveness.

It is tempting to develop an aeration program that does not include the disruptive process of core aerification. Slicing or water injection can be accomplished with little or no disruption to playing surfaces. In addition, they require less labor and almost no cleanup. Slicing is considerably faster as well. However, a quality aerification program has to include punching holes and removing cores. Although slicing and water injection can have positive aeration effects, those effects are usually short-lived, and these techniques are best used as supplemental positive practices to a well-designed core aerification program.

10.3 Problems With Sand

In terms of adequate root respiration, sand is superior to most native soils. Sand does not compact to a point that deters root respiration in a turfgrass system unless the sand is extremely fine or it contains fine materials such as silt, clay or organic material. Most turfgrass is managed on soil. However, some of the most highly managed turfgrass systems, such as golf course putting greens, bowling greens, tennis courts and athletic fields are built on sand. Consequently, you may spend considerable time managing relatively small areas of turfgrass to perform up to extreme expectations on sand. It is important to note that managing turf on sand is quite different from managing turf on most soils. Those differences will be discussed later in the chapter. For now, let me assure you that sand is not a perfect system. There are problems that occur on sand that seldom occur on soil.

Black layer

Black layer is not a problem unique to sand, but it is rarely found on soil. Black layer appears as a fairly thin layer of black sand usually close to the surface of a sand root zone (Fig. 10.4). In extreme conditions, such as when soil-bound sod is laid on top of a sand root zone (see Chapter 8), the turf can be pealed back to expose the black layer below. The black layer will smell like rotting eggs, the smell that occurs when sulfur is reduced by bacteria. Black layer smells because it is caused by, encouraged by or occupied by (and we don't really know for sure) sulfur-reducing bacteria such as *Desulfovibrio desulfuricans* (Hodges, 1992a). Black layers are anaerobic, meaning that they contain no oxygen. Consequently, they are believed to be associated with pockets of water-attracting clay, silt or organic material. These fine materials will

Fig. 10.4. A black layer in a sand root zone. This area smells like rotting eggs.

hold water and will not release it to the sand until they are completely saturated. Consequently, a layer of fine material in sand is filled with water most of the time.

Although black layer can be severe enough that turfgrass roots will not penetrate it, this is not always the case. Turfgrass roots are discouraged by black layer, but they can sometimes grow through it, resulting in turfgrass decline rather than death (Cullimore *et al.*, 1990). It is believed that to some extent plant roots can oxygenate the soil as they pass through it. Consequently, at least in some cases, turfgrass plants can break through, or even break up, a black layer. Nevertheless, black layer always causes temporary-to-terminal decline and is always a serious problem. In some cases, black layer or the circumstances that cause it can result in a weakening of the turf to the point that it becomes fatally sensitive to other stresses (Hodges, 1992b).

Black layer is often the result of poor sand construction or poor maintenance practices. Improperly mixed organic amendments added to sand systems during construction can lead to pockets of potentially dangerous anaerobic conditions. Black layer can also be caused by improper topdressing with highly organic materials, soil or soil-containing sand. Sulfur-containing fertilizers or other materials added to a system with an existing black layer can make the black layer worse (Berndt and Vargas, 1992). However, because sulfur-reducing bacteria are not particularly tolerant of low pH soil, additions of sulfur or sulfate materials that lower soil pH can help to reduce the potential for black layer to occur provided that it is not already present (Adams and Smith, 1993).

In most cases, black layer is not a particularly difficult situation to handle. Aggressive core aerification with cores removed followed by topdressing with clean sand that closely matches the original root zone will usually improve the situation. Deep-tine aerification is advisable to make sure that the black layer is broken and that it is infiltrated by columns of clean sand. The more often the aerification is practiced and the more aggressively that it is performed, the sooner the layer disappears.

Localized dry spots

Localized dry spots (LDS) are another detrimental condition usually only experienced in sand systems. Although water-repellent native soils are known to exist, water repellency is usually only a serious problem for turfgrasses grown on sand (York and Baldwin, 1992). As sand became popular as a growing medium for golf course putting greens, it became apparent that certain spots in some greens repelled water. It was soon discovered that the sand particles in these spots were coated with a water-repellent material that was believed to be fulvic acid (Wilkinson and Miller, 1978). Further research and observation tended to confirm that some organic material, possibly fulvic acid, was coating the sand in LDS causing them to repel water (Tucker *et al.*, 1990).

Sand is a drought-prone root-zone medium so it is not unusual for grass grown on sand to exhibit a drought response when conditions are dry. High spots on the site, south-facing slopes, poor irrigation patterns or areas of heavy thatch can look like LDS when irrigation is needed. However, if sufficient irrigation is applied, these areas do not decline. However, LDS are noticeable even under adequate irrigation. They may be identified as puddles during rainfall or irrigation, or as dry turf on an otherwise dew-covered surface. When water is applied to LDS, it does not penetrate the soil, it simply runs away. Consequently, turfgrass root systems are stunted, the soil holds little water and the turf declines rapidly when rainfall is less than substantial.

The problem of LDS is more widespread and harder to handle than black layer. York (1993) conducted a survey of golf course superintendents in the UK to determine how many were experiencing LDS. A total of 112 superintendents were surveyed, and 86% of them indicated that LDS was a problem at their course.

Because the hydrophobic coating on sand particles in LDS is organic, it can be removed with hydrogen peroxide (H_2O_2). However, this is a dangerous procedure for a golf course putting green, bowling green, tennis court or athletic field. Applying hydrogen peroxide is similar to applying light rates of glyphosate (Roundup) to control annual bluegrass (*Poa annua*) in creeping bentgrass (*Agrostis stolonifera*) turf. The light rate of glyphosate may work, but it may kill all of the grass on the surface as well. Hydrogen peroxide has been used successfully many times in recent years to control algae in creeping bentgrass on putting greens with no detrimental affect to the turf. These can be dangerous applications, however. At higher rates hydrogen peroxide can kill almost anything, including turf.

High pH treatments, another potentially dangerous type of application, may also be effective for controlling LDS (Karnok et al., 1993). This is another treatment that is hard to recommend because of its potential detrimental effects on the turf. Karnok et al. (1993) demonstrated that an application of 0.1 molar sodium hydroxide (0.1 mol/l NaOH) sufficient to saturate the upper 2 inches (50 mm) of soil, followed by a water flush of one pore volume, significantly reduced hydrophobicity if repeated three times. The researchers reported that the severity of turfgrass injury depended on the number of consecutive applications and the air temperature. Little or no discoloration was observed when the air temperature was less than 75 °F (24 °C), but injury was severe when the air temperature exceeded 95 °F (35 °C). As a turfgrass manager, you will be faced with making important decisions about which management practices you apply. If you intend to use a potentially dangerous, but potentially effective, tool make sure that you test it thoroughly on turf with little value, such as a nursery or practice green, before you apply it to an area of high importance.

There has been some indication that topdressing with porous ceramic clay can be effective for reducing or eliminating LDS but it has not become a standard practice (Minner et al., 1997). Currently, the application of wetting agents and frequent core aerification have become the management practices most commonly employed for the relief of LDS (Leinauer et al., 2007; Lyons et al., 2009). Wetting agent applications are made early in the season before symptoms appear, and at regular intervals throughout the season, depending on the product.

10.4 Sand System Design and Construction

The United States Golf Association (USGA) Green Section Recommendations for Putting Green Construction (USGA Green Section Staff, 2004) are by far the most commonly used methods for constructing turfgrass sand systems for all kinds of sporting applications. The University of California Sand Putting Green Construction and Management (Davis et al., 1990) method is also popular and a third method, the Airfield System (Airfield Systems, 2009), is lesser known but appears to be gaining in popularity (Xiong et al., 2006). Each of these systems has specific construction recommendations that must be followed or the final product may not meet expectations (Gibbs et al., 1993).

Turfgrass sand systems are constructed not only for compaction avoidance but also to facilitate drainage. Subsurface drainage is extremely slow in native soil compared with pure sand. It is not uncommon for a new pure sand system to drain at 30 inches (76 cm) per hour or more. As a result, the surface can maintain play shortly following a severe storm or after a week of rainy weather. Such conditions provide immense benefits to players and spectators. However, these benefits do not last forever (Baker *et al.*, 1999b).

Turfgrass sand systems have a long but finite lifetime (Murphy *et al.*, 1993b). They gradually become compacted and the drainage eventually slows (Baker *et al.*, 1999b). However, they can provide superior playing conditions for 25 years or more before they need to be replaced. As you have learned, quality turfgrass systems produce substantial thatch. The thatch is then degraded by soil microorganisms to humus. Therefore, healthy turfgrass systems produce their own soil organic materials that will eventually promote compaction and discourage drainage in a sand system. In addition the system is constantly bombarded by silt and clay carried by wind, rainfall and irrigation. Nature is nearly always victorious over man and the sand system eventually becomes more like soil than sand. For that reason, and because these aging properties can encourage surface compaction, surface drainage should not be ignored in sand systems. It is not wise to rely solely on subsurface drainage, even on a USGA, California or Airfield System constructed sand system. It is interesting to note that in spite of superior subsurface drainage, all three of these systems retain more water at the bottom of small slopes than they do at the top (Prettyman and McCoy, 2003; Xiong *et al.*, 2006). Consequently, if a client is specifically interested in a sand system because it drains uniformly, that does not appear to be an advantage. Sand systems drain quickly, but they do not necessarily drain uniformly.

Turfgrass sand systems require sands of a specific particle size distribution in the root zone, and in the USGA system a particular size of gravel is recommended as well. The particle size of the sand in the root zone is selected to match the drainage of the system. A USGA system is constructed to maintain a perched water table – a zone of nearly saturated sand above a gravel layer that resides above the soil base (Fig. 10.5). The California system recommends a root-zone sand that sits directly on the soil base. The USGA system recommends at least 60% coarse-to-medium size sand and no more than 20% fine sand because the system retains water by design, and the finer sands are not required. The California system drains slowly into the subsoil and does not retain water. So the California system recommends that at least 90% of the root-zone material consists of coarse, medium and fine sand combined, and is more tolerant of fine sand because the smaller particles hold more water in the root-zone material. Neither system recommends that more than a minor fraction of the root-zone material be of very coarse or very fine sand, and silt and clay should make up not more than a very small proportion. Sands that have a broad particle size distribution (meaning that they have a substantial percentage (> 5%) of each size category from very fine to very coarse) are easier to compact than sand with a narrow particle size distribution (Davis *et al.*, 1990) (Fig. 10.6). Both the USGA and California systems have proved effective but they require slightly different irrigation management. The Airfield System has not been time tested but early indications are promising (McInnes and Thomas, 2008).

Amendments are often added to sand systems to improve water and nutrient retention (Bigelow *et al.*, 2004; McClellan *et al.*, 2009). The

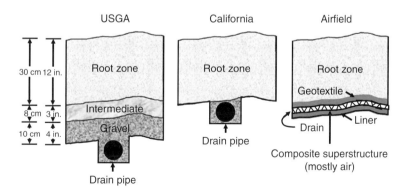

Fig. 10.5. Cross-sectional views demonstrating the general characteristics of the USGA (United States Golf Association), California and Airfield sand systems.

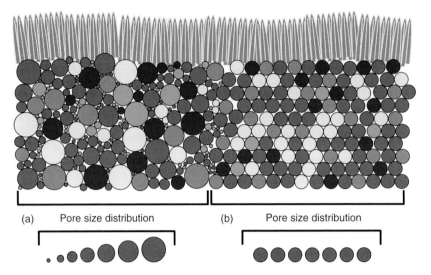

Fig. 10.6. Sands with a broad particle size distribution compact more readily than sands with a narrow distribution. Notice how small some of the pores are in sand A compared with sand B. Sand B is already compacted about as much as possible but sand A can be compacted further as the small particles continue to settle into the gaps left by the larger ones.

amendments could be organic or inorganic. Organic amendments have been used historically but inorganic amendments such as calcined clay, diatomaceous earth, zeolite and others show promise. Organic amendments such as rice hulls and various composts have been used successfully but the most common amendment is peat. The more peat contained in the mix, the tighter the medium will hold water (Li *et al.*, 2005). An 85:15 by volume sand:peat mix is common. The sand:peat mix encourages rapid establishment compared with pure sand (Bigelow *et al.*, 2001; Waltz and McCarty, 2005). The USGA system tends to encourage the use of amendments, but the California system tends to discourage them. Given time, a turfgrass forms its own organic soil layer. Some believe that the addition of organic material into a pure sand root zone can encourage layering problems and inconsistencies in the media. If rapid establishment is required, the best course of action is to add organic material. If time is not an important factor, perhaps it is best to use an unamended root zone. Unless a consensus is reached or research is performed that clearly identifies the best course of action, with a little knowledge and experience you will have the opportunity to form your own opinion.

The important message in this section is this: sand, especially sand in a narrow particle size distribution range, does not compact like soil but it does not hold water and nutrients like soil does either.

10.5 Managing Turfgrass on Soil and Sand

Would you rather manage turfgrass on soil or sand? Without a doubt, managing turfgrass on sand is more difficult than managing turfgrass on soil unless the site is expected to maintain heavy traffic. Where heavy traffic is expected, a turfgrass on sand may be sustainable but a turfgrass on soil very likely is not. If you are a knowledgeable and experienced turfgrass manager you may prefer to manage grass on sand regardless of the level of traffic expected. However, managing grass on sand will require more labor, and therefore more expense, than managing on soil where little traffic is expected. So the expense of growing on sand has to be justified by a noticeable increase in turfgrass quality and your customer has to be convinced that the increase in quality is worth the expense.

Differences in irrigation management

In general, an uncompacted loam soil has a near 50:50 split of micropores and macropores.

The micropores hold water and the macropores hold air. Sand with a narrow particle size distribution such as that normally recommended for turfgrass system construction has very few micropores and lots of macropores. The sand holds very little water but contains lots of air. In most climates, natural rainfall is sufficient to grow reasonably dense, uniform turfgrass on soil but not on sand. In most climates, artificial irrigation will be required for turfgrass grown on sand.

The primary reason that sand is used in turfgrass sand systems is because it does not compact appreciably until finer materials become part of the medium and because it has large pores that drain readily allowing rapid use of the surface following rainfall. Because it drains so well, sand is not a good growing medium unless you have access to water. Turfgrass on sand does not necessarily require more water than turfgrass on soil, but it requires water more frequently. For instance, in most climates it is not necessary to irrigate a home lawn more than once a week during dry periods, and a very general recommendation would be to apply 1 inch (25 mm) of precipitation each week. If that lawn was on a sand system, it would require at least two irrigations a week at 0.5 inch (13 mm) per irrigation. It might even require three applications a week to prevent wilt, depending on the species present, the depth of the root system and the water retention of the medium.

One advantage of a sand system is that it can take more water than necessary with no detrimental effects. Heavy prolonged rainfall can be quite difficult to deal with on soil but not on sand. Turfgrass can not sustain itself in saturated soil for more than a few days before it declines for lack of root oxygen. Sand systems drain readily and are rarely saturated for more than an hour or two. Once saturated, water in a sand system drains through the subsurface irrigation. Consequently, sufficient root oxygen is normally always available. For that reason, managers of sand systems tend to over-water rather than under-water. Although that may not be harmful to the turf in most cases, it quickly becomes expensive if you have to purchase water or purchase energy to operate your pumping stations. So unless you work on one of those extremely rare sites where you can irrigate from a natural aquifer by gravity flow, irrigation is expensive and over-irrigation is a waste. In addition, water and energy are in limited supply and should be conserved. If we don't conserve them we will be one of the first industries rationed. Sand systems require more frequent irrigation than soil systems.

Differences in nutrient management

Nutrient leaching is an event that you are probably familiar with. Nutrients are lost through leaching when their soluble forms percolate downward through the soil in water. Leaching of nitrogen is common in all types of media, especially sand, but most nutrients are held tightly enough by the soil that they resist leaching (Petrovic, 1993). Frequent fertilization in small amounts or the application of nitrogen and potassium in slow-release forms is necessary for turfgrass fertilization in sand systems.

As you know, the cation exchange capacity (CEC) holds nutrients in soil. The higher the CEC, the more nutrients are retained by the soil and the lower the likelihood that nutrients other than nitrogen will be lost to leaching. A typical loam soil has a CEC of about 7 to 16 cmol kg^{-1} of soil (Carrow et al., 2001). A cmol kg^{-1} (centimole per kilogram) is equal to 1 meq. (milliequivalent) per 100 grams of soil, and these are the two units most commonly used to report soil CEC. It is not necessary to understand the units as long as you realize what constitutes a high CEC and what constitutes a low CEC. A fertile clay loam soil may have a CEC from 20 to 50 cmol kg^{-1} and organic matter is likely to have a CEC as high as 140 to 250 cmol kg^{-1}. Sand, in contrast, has a very low CEC. Pure sand is likely to have a CEC of less than 1 cmol kg^{-1} and an 85:15 peat-amended sand will probably have a CEC ranging from 1 to 3 cmol kg^{-1}. Obviously, sand is not a growing medium that retains many nutrients.

Nutrient deficiencies are common in sand systems. For that reason, soil testing of sand systems should be done more frequently (two to three times per year) than it is for native soil. In addition, nutrient management has to be adjusted to accommodate the poor nutrient retention of sand systems. Nutrients should be applied to close-mown sand systems like bowling and putting greens weekly to maintain consistent color and growth. Higher mown turf, athletic fields for instance, may be fertilized once every 2 weeks, but weekly is better. Sand systems generally do not require more fertilizer than soil but they do require smaller amounts per application more often. Although soil rarely requires additional micronutrients as long as the pH is close to normal, sand systems usually respond best to one or two broad-spectrum micronutrient applications annually.

A spray application of nutrients to sand systems weekly is often called "spoon feeding", indicating that only a small amount of nutrients is applied. The term "foliar application" or "foliar feeding" is often used interchangeably with spoon feeding. Spoon feeding, however, implies a light application of nutrients, whereas foliar application implies a light application of water that usually contains only small amounts of nutrients to prevent foliar burn. A typical spray application applies a larger amount of nutrients than a spoon-feeding application and a larger amount of water than a foliar application. My objective here is not to correct the use of these terms but to make certain that you understand what is meant when they are used in this text. There are cases of calcareous sands that have to be fertilized by foliar application because the pH of the medium is so high that many nutrients precipitate from the soil solution and are no longer plant available once they enter the sand. There are some alkaline soils that work the same way. Phosphorus, iron and manganese are the most likely nutrients to be deficient in the case of calcareous sand or alkaline soil (Carrow et al., 2001). In many cases trying to adjust the pH of calcareous sands and some alkaline soils is not possible or not practical. It could take tons of sulfur over 100 years to affect the pH in many of these sands and soils. Consequently, foliar fertilization becomes the only available option for growing good grass. In most cases, alkaline sands and soils are manageable if foliar fertilization is used.

Generally, the nutrients of major concern in sand systems are nitrogen and potassium. Not only are these the two fertilizer nutrients needed in the greatest amount, they are also the two nutrients most likely to be lost to leaching. In a sand system, potassium is not held tightly as it is in soil and it leaches readily just as nitrogen does unless sufficient organic material is present to bind it (Petri and Petrovic, 2001). As a sand system ages, organic material increases in the system and more nutrients are retained (McClellan et al., 2007). So frequent fertilization of an old system (10 years old or more) may not be as critical as that of a new system. Regardless, sand systems always need more frequent fertilization to grow their best turf than do soil systems, provided all other factors are the same. Spoon feeding is the best process to use on sand systems so that nutrients are available as the plants need them but not in an excessive amount which will leach away with the next major precipitation event. An alternative is the use of slow-release fertilizers that release small amounts of nitrogen and potassium over time. As you learned in Chapter 7, slow-release fertilizers are more expensive than quick-release sources but can be applied less often, thereby providing a savings in labor expense.

10.6 Managing Turf on Soils High in Salts

Many ornamental plants, including trees, grow very poorly in salty soils. Turfgrasses in general, are some of the most resistant species to poor soil conditions. Nonetheless, turfgrasses can be difficult to maintain in soils high in salts, especially at the aesthetic and/or functional levels that many of your customers have come to expect. The purpose of this section of the text is to supply you with general information needed to manage turfgrass on soils high in salts. You will learn how salts affect grasses and techniques that can help you manage grasses on salty soil. However, if you are tasked with managing a turfgrass on a soil high in salts, especially sodium, you will need to seek considerably more education than can be supplied here. Additional sources of information are recommended at the end of the chapter. In addition, professional or expert consultation should be considered before attempting to develop a program for the management of turfgrass on a salty site. Each site has unique characteristics, requiring techniques that range from simple leaching solutions to complex programs of drainage reparation, salt remediation, foliar feeding, chemical irrigation management and perpetual leaching practices.

Dealing with salts on sand-based systems

In addition to the differences that we have discussed concerning the management of soil versus sand systems, one more component requires attention. Irrigated sand systems tend to accumulate salts. As was indicated earlier, sand systems require frequent, light irrigation in comparison with soil because the sand systems retain little water. As you irrigate under these light, frequent restrictions the soil is only wet for a short distance into the profile and the improved aeration of a sand system, rapid water use by turfgrass and shallow wetting add to rapid drying of the irrigation applied. An irrigation source that does not contain salts of one type or another is rare. In fact, as will be discussed later in this chapter, irrigation water that is too pure is undesirable. As the water in the soil evaporates the salts in the water are left behind. These salts build up with frequent irrigation and drying until maintaining adequate turf becomes difficult. So they need to be removed periodically.

It is generally not difficult to remove salts from sand systems. It usually only requires a period of heavy rainfall or an irrigation event specifically designed for salt removal. The most common mistake managers make concerning designed-to-leach irrigation events is not applying enough water (Fig. 10.7).

In typical sand used for USGA system construction, up to 30% of the profile consists of macropores. Because the profile is 12 inches (30 cm) deep, the macropores make up almost 4 inches (10 cm) of the profile. If you want to fill all of those pores with water, you have to apply 4 inches (10 cm) of irrigation. Once the pores are filled, you can then begin to apply more irrigation to leach the soluble salts. Consequently, about a 5-inch (13-cm) irrigation event is required to leach the system. On a 12-inch profile that is approximately 3.1 gallons of water per square foot (127 l m^{-2}), a tremendous amount of water. On a USGA system, you will know when leaching begins. The main drain will suddenly fill and water will come pouring out of the entire pipe. Make sure that the drainage pipe is clear before you apply a leaching event and remember that many of your nutrients will be leached along with the other salts.

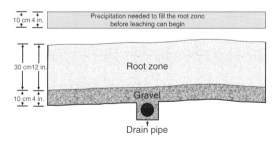

Fig. 10.7. A designed-to-leach irrigation event requires a considerable amount of water. A 12-inch (30 cm) root zone that contains 30% air-filled pore space requires an irrigation event of almost 4 inches (10 cm) just to fill the soil pores before leaching can begin.

Classifications of salt-affected soils

Up to this point, I have avoided the use of the term "saline". Saline traditionally means a solution high in salts. However, in the study of plants and plant management, sodium has a special classification. Consequently, saline as it refers to irrigation and soil water solutions becomes confusing because the reader is not sure whether the author is referring to all salts or salts other than sodium. For that reason, in this text, saline means salts other than sodium. A solution high in sodium will be called sodic and a solution high in both sodium and other salts will be called saline+sodic so that there is no mistake. The United States Salinity Laboratory (USSL) uses these classifications to identify categories of salt-affected soils (USSL Staff, 1954). The USSL classifications are the most widely used in the world but there are countries or regions and at least one continent, Australia for instance, that because of the soils that are most prevalent, may have slightly different classifications.

The soil solution of a saline soil has an electrical conductivity (EC$_e$) > 4 dS m^{-1} and an exchangeable sodium percentage (ESP) of below 15. The unit designation "dS m^{-1}" is an abbreviation for deciSiemens per meter. A Siemen is a measure of electrical conductivity defined as the reciprocal of electrical resistance. Conductivity refers to the ability of a material to carry electricity. There are relatively simple electrical meters used to measure solution conductivity as an Ohm meter or multimeter measures resistance. A conductivity of 4 dS m^{-1} is the conductivity designated by the USSL to distinguish between a saline soil solution and a typical soil solution.

The unit "ESP", the exchangeable sodium percentage, is the proportion of exchange sites on the cation exchange capacity of the soil occupied by sodium reported as a percentage. A measure of 15 ESP means that 15% of the exchange sites in the soil are occupied by sodium. The USSL classifies sodic soils as those soils with an ESP of 15% or greater.

A saline soil is usually easier to manage than a sodic soil. A saline soil will usually, but not always, have a basic pH, but this is normally less than 8.5. Saline soils are often called "white alkali" because they are usually basic and they often have a white crust on them, especially on the low spots where water collects and then dries. Conversely, sodic soils are called "black alkali". Sodic soils have an

$EC_e < 4\,dS\,m^{-1}$ and an ESP > 15. By definition, sodic soils contain considerable sodium but no more than a typical amount of other salts. Sodic soils are nearly always basic, with a pH usually greater than 8.5. These are the most difficult salt-affected soils to manage.

Saline+sodic soils are those with $EC_e > 4\,dS\,m^{-1}$ and ESP > 15. Although this combination sounds devastating, the resulting soil is often easier to manage than a sodic soil. The pH is usually lower than 8.5 and the soil tends to take on the characteristics of a saline soil rather than a sodic soil. In a way, the saline components have a positive buffering effect on the negative sodium component.

The effects of salty soil on turfgrass

You probably remember from chemistry class that salts dissociate easily into cations and anions in water. Salt-affected soils usually contain cations of sodium, potassium, calcium and magnesium and anions of chloride, sulfate, carbonate and bicarbonate that dominate the soil solution. All of these components are found in and, with the possible exception of sodium, are necessary for plant function. It is not the existence but the proportion of these components in the soil solution that determines a soil's plant support characteristics.

Most, but not all, salt-affected soils are found in arid regions. Salts are easily leached by rainfall, but in dry regions rainfall does not occur in a sufficient amount to perform this leaching function. If rainfall equals evapotranspiration, salts usually leach away. Otherwise they remain in the upper soil layers and cause problems for plant growth. Excess salinity affects grasses and other plants by making it more difficult for them to access water and nutrients. Saline or sodic conditions result in poor growth with additional symptoms similar to drought stress (Harivandi et al., 1992). Leaf blades are likely to be narrow and stiff with a blue-green coloring that eventually leads to wilt. As the soil solution becomes more saline its water potential becomes lower, causing it to hold water more tightly. Plants have more difficulty absorbing this tightly held water, which causes drought-like symptoms. The plants are, in fact, experiencing a physiological drought.

True salt-tolerant plants may experience higher rates of shoot growth under minor saline or sodic conditions (Marcum, 1999). However, accelerated root growth is more likely. In fact, even some moderately salt-tolerant grasses will increase root growth when they experience minor saline or sodic conditions (Qian et al., 2001). Bermudagrass (Cynodon spp.) species and cultivars differ in their tolerance to salt (Youngner and Lunt, 1967). However, they all appear to respond to high-salt media by decreasing shoot growth and increasing root growth, which results in increasing root-to-shoot ratios (Dudeck et al., 1983). Most warm-season turfgrasses respond to salt media in the same way (Marcum, 1999), and some cool-season grasses are believed to respond similarly.

As salinity or sodicity increases in a soil solution, its water potential decreases. Consequently, it becomes harder and harder for grasses to take up soil water. Salt-tolerant grasses respond to this decline in soil water potential by extending their root systems. Their roots grow both deeper and with more mass (Marcum, 2008). Most salt-tolerant warm-season turfgrasses also have the ability to lower the water potential within their roots and shoots to encourage the absorption and translocation of water and nutrients into the roots and throughout the plant (Marcum and Murdoch, 1990). The lower the water potential of the media, the lower the plant's water potential becomes (Peacock and Dudeck, 1985; Dudeck et al., 1993).

At least some turfgrass species practice a salt avoidance process called ion exclusion (Volkmar et al., 1998). Ion exclusion occurs when salts are compartmentalized and excreted through salt glands on the leaves. Sodium and chlorine are known to be excreted from leaf salt glands on seashore paspalum (Paspalum vaginatum), manilagrass (Zoysia matrella), mascarenegrass (Zoysia tenuifolia), St. Augustinegrass (Stenotaphrum secundatum), bermudagrasses, Japanese lawn grass (Zoysia japonica), centipedegrass (Eremochloa ophiuroides) and Kentucky bluegrass (Poa pratensis) (Marcum and Murdoch, 1994; Qian et al., 2001). The efficiency of salt gland excretion and the density of salt glands on the leaves are known to influence the salt tolerance of zoysiagrasses (Zoysia spp.). The salt tolerance of bermudagrass cultivars is also determined by the efficiency with which they excrete sodium through their salt glands (Marcum and Pessarakli, 2006). As we learn more about these mechanisms, it may be possible to design management practices to specifically encourage reductions in water potential and efficient ion exclusion.

When the soil medium is high in sodium, tissue concentrations of nutrients such as calcium, potassium, magnesium, manganese and phosphorus are affected (Dudeck and Peacock, 1993). As sodium levels increase in the soil and in the plant these nutrient levels decline. Thankfully, nitrogen and iron do not seem to be affected (Ackerson and Youngner, 1975; Dudeck and Peacock, 1985a). However, high sodium in the soil solution interferes with calcium and potassium uptake, and these are two very important nutrients. Sodium also tends to replace calcium in root membranes, leading to loss of membrane integrity, and it may replace potassium in some reactions, thus disrupting metabolism.

Sodic soils are especially detrimental to plant growth because the high sodium content causes the soil to lose its structure. The resulting soil system is similar to soils that have been highly compacted. Soils are normally very high in calcium and magnesium, and those two elements usually occupy most of the cation CEC. However, in sodic soils, sodium replaces some of the calcium and magnesium in the CEC, causing soil structure to deteriorate. The replacement of calcium with sodium in the soil CEC causes the bonds within soil aggregates to weaken, and the aggregates eventually break up into fine particles of clay and humus, destroying soil structure. Sodic soils not only hold a lot of water and hold it tightly, they become impermeable as the aggregates fall apart and the macropores disappear.

Although sodicity is, in most cases, the most detrimental condition that exists in salt-affected soils, saline soil solutions can sometimes contain toxic amounts of certain salts. While chlorine, bicarbonates or hydroxides can reach toxic levels in saline or sodic soils, boron is the element most likely to be found in amounts toxic to turf (Carrow et al., 2001). Boron toxicity most commonly occurs in arid regions where irrigation water is high in boron. Its symptoms are necrosis of leaf tips, and regular mowing tends to temporarily relieve the stress as the boron that accumulates in the tips is mowed off. According to Carrow et al. (2001), grasses can tolerate soil boron accumulations as high or higher than 10 parts per million (ppm) depending on species and cultivar. However, the authors recommend levels below 1 ppm (1 mg kg^{-1}) for sensitive grasses and other plants.

Developing a management program

The first step in developing a management program for turfgrasses grown on salt-affected soil is proper assessment. Chemical assessments of salt-affected sites can be expensive but are critical for the development of an effective management program (Carrow et al., 2001). The conditions on salt-affected sites can vary widely, often based on elevation, and a sufficient number of samples should be taken to form a truly representative composite. On highly problematic sites, the advice of a professional or expert should be sought. Qualified specialists may be found as commercial consultants, soil and water laboratory analysts, agricultural university personnel and others in your particular field of turfgrass management. These advisors may suggest plans that include foliar application of nutrients, aggressive plans for cultural improvement, major drainage renovations, leaching programs and conversion to more salt-tolerant grasses.

Irrigation concerns on salt-affected sites

In our industry and others, poor-quality irrigation water is a potential source of salts. Clean water is always in high demand. Because lawns, parks, golf courses and even athletic fields are not considered particularly important in comparison with food crop irrigation and other human uses for water, we may have to use poor-quality water or be content with no irrigation at all. Therefore, irrigation and leaching with water containing high salt concentrations is becoming more common for turfgrass managers. The use of this water for irrigation and the amount of leaching that is appropriate following each use has to be considered.

The USSL lists salinity hazards in water as low ($EC_w < 0.25$ dS m^{-1}), medium ($0.25 \leq EC_w < 0.75$ dS m^{-1}), high ($0.75 \leq EC_w < 2.25$ dS m^{-1}) and very high ($EC_w \geq 2.25$ dS m^{-1}). Sodicity is measured in SAR, the sodium adsorption ratio, rather than as the exchangeable sodium percentage (ESP) introduced earlier. The SAR is a comparison of sodium concentration with calcium and magnesium concentration. It is calculated as the number of CEC sites occupied by sodium, divided by the square root of half of the sum of the sites occupied by calcium plus magnesium:

$$SAR = \frac{[Na^+]}{\sqrt{\frac{1}{2}([Ca^{2+}]+[Mg^{2+}])}}$$

The USSL classifies sodicity as low (SAR < 10), medium ($10 \leq SAR < 18$), high ($18 \leq SAR < 26$) and very high (SAR ≥ 26). Furthermore, because salinity has a positive effect on the negative sodicity of a solution, a more powerful analysis results from considering the salinity and sodicity in combination. Even a soil with low sodicity can become more difficult to manage if very pure irrigation water is used. Pure water can cause a high proportion of all the soil salts, including calcium, to leach, thereby causing a disruption in surface stability that affects permeability.

Leaching

The most common method for improving salt-affected sites is to develop a leaching program for the downward removal of salts from the root zone. For this program to work, however, the water and salt must have someplace to go and the soil must be permeable enough to allow adequate infiltration. The portion of irrigation water and natural precipitation that moves through the root zone carrying salts with it is called the leaching fraction, or sometimes, the leaching requirement. The higher the leaching fraction, the more salts removed. In a saline+sodic soil, another salt, usually gypsum, must be applied to help leach the sodium in preference to other less detrimental salts. Gypsum is calcium sulfate and is used to replace sodium ions with calcium ions. Leaching is possible with salty irrigation water, but the leaching fraction must be substantially greater to have a desirable effect.

To permit leaching, irrigation water must be applied to soil saturation and move through the root zone carrying salts with it. Consequently, irrigating past the point where water runs off is not productive. It is best to irrigate for the length of time required to saturate the soil and stop before runoff. Following a short period, perhaps 1 to 3 hours depending on how rapidly the soil drains, irrigation can be resumed again to saturation. This process can be continued, if necessary, until the desired leaching fraction is reached. Leaching fractions are best calculated by a qualified laboratory which will advise you on what information is required and how to gather it.

Salt-tolerant grasses

Compared with most plants, grasses are relatively tolerant of salinity. However, there are some grasses that will grow on salt-affected sites where others won't. More often, turfgrasses will grow on a salt-affected site but will not be able to achieve the aesthetic or functional value expected. In that case, a salt-tolerant species (and cultivar) must be selected to meet site requirements. In some cases, a less salt-tolerant turfgrass may be invaded by more salt-tolerant weeds on a salt-affected site. However, selection of a salt-tolerant turfgrass species and cultivar gives the turfgrass the opportunity to outcompete most weeds.

In the USA, the deicing of highways with sodium chloride has led to the need for and to the selection of salt-resistant turfgrass species and cultivars. For instance, alkaligrass (*Puccinellia* spp.) was first considered for use as a salt-tolerant turfgrass for low-maintenance sites after discovery along salt-affected roadsides in parts of the country where it was not a particularly common species (Butler *et al.*, 1974). Alkaligrass is widely considered the most salt tolerant of the cool-season grasses suitable for low maintenance turfgrass (Marcum, 2008). Inland saltgrass is probably the most salt tolerant of the warm-season grasses suitable for low-maintenance turf (Marcum, 2008). Species, however, differ by cultivar in their adaptability to salt-affected sites. The most popular turfgrass in the northern USA, Kentucky bluegrass, is not particularly salt tolerant (Torello and Symington, 1984), but there are cultivars of the species that demonstrate moderate salt tolerance (Qian *et al.*, 2001). Seashore paspalum is generally considered the most salt tolerant of the turfgrass species suitable for highly maintained turfgrass, but salt-tolerance differs by cultivar within the species (Lee *et al.*, 2004a). There could be cases where a highly salt tolerant cultivar of manilagrass, mascarenegrass, St. Augustinegrass or bermudagrass performed better on a salt-affected site than a poorly salt-tolerant seashore paspalum (Dudeck and Peacock, 1993; Lee *et al.*, 2004b). Take care to select the best cultivars within the species when salt tolerance is a major consideration.

Certain cultivars of creeping bentgrass (*Agrostis stolonifera*) are reasonably salt tolerant (Marcum, 2001). However, in general, this cool-season species can not compete with many warm-season species for salt tolerance. Particular cultivars of tall fescue (*Festuca arundinacea*; synonym: *Schedonorus phoenix*) (Lunt *et al.*, 1961; Greub *et al.*, 1985) creeping red fescue (*Festuca rubra*) (Harivandi *et al.*, 1982; Greub *et al.*, 1985; Torello and Rice, 1986), and perennial ryegrass (*Lolium perenne*)

(Dudeck and Peacock, 1985b; Greub et al., 1985) exhibit moderately good salt tolerance. Salt tolerance of species and cultivars may also be affected by growth stage (Harivandi et al., 1982; Dudeck and Peacock, 1985b; Wu and Lin, 1993) with germination and seedling response considered more sensitive to salt than mature growth (Harivandi et al., 1992). Choosing a salt tolerant species and cultivar is obviously the easiest way to help avoid poor turfgrass performance during construction on a salt-affected site. Renovation to a salt tolerant species and cultivar is also an important consideration on established turf.

10.7 Chapter Summary

Native soil mixtures differ substantially from artificially constructed sand root zones, not only in their constitution but in their management. Although sand is a far better system for many turfgrass surfaces used for sporting purposes than is soil, sand requires a more knowledgeable manager and is not without its problems. Compaction is not a problem on sand systems, but black layer and localized dry spots are fairly common. Sand systems are expensive to construct and have a finite lifetime. A poorly constructed sand system can be very difficult to manage. Nonetheless, a properly constructed sand system is far superior for sites that have to sustain regular traffic than a soil system would be. Not only are sand systems highly compaction resistant, they also drain rapidly, providing firm playing surfaces following rainfall when soil sites could not be used.

Aeration is the main factor that prevents native soils from providing excellent playing surfaces. Most turfgrass areas are required to sustain moderate traffic. For that reason, aerification practices are required to promote enough rooting activity for a turf on a soil site to maintain excellent health and aesthetic value. However, soil has an advantage over sand in nutrient and water retention. For the most part, nutrients are tightly bound and generally available in soil, but are poorly held and easily leached in sand. Soil remains damp following rainfall, providing water for the turf, but sand drains quickly, providing mostly air. For that reason, the systems require different management.

Sand requires frequent applications of nutrients and frequent irrigation compared with soil. Although the total input of water and nutrients are approximately the same, sand requires smaller applications more often. Because sand is a basically inert system, a knowledgeable turfgrass manager has the opportunity to control water and nutrients more precisely and to vary inputs more easily when circumstances change. Soil, however, is more forgiving of mistakes and easier for a novice to handle. Where compaction is a serious problem, sand is advantageous, but if a site is vast or only sustains light-to-moderate use, soil is superior. Root respiration that provides for adequate root growth and plant transpiration determine which medium performs best under the requirements of the site.

Soil salinity and sodicity affect turfgrass by osmotic inhibition of water uptake, by creating ion imbalances in plants and, in some cases, by ion toxicity. Although turfgrasses respond to osmotic stress by increasing their root growth and root and shoot water potential, they can be overcome by physiological drought in salt-affected soils. The ability of turfgrass species to tolerate salt-affected soil ranges from those that are salt sensitive to those that are highly tolerant. Sensitivity to salt-affected soils also differs substantially among cultivars of a particular species. Most of the salt-tolerant turfgrasses are warm-season species, but some cool-season species demonstrate moderate tolerance.

Salt-affected soils are classified as saline, sodic or saline+sodic. In addition to osmotic inhibition, ion imbalance and ion toxicity, sodic soils can destroy soil structure, thereby seriously affecting permeability. For that reason, sodic soils are hardest to manage and require chemical treatment with gypsum or other means to positively affect soil structure.

Leaching is the most common practice for turfgrass improvement on salt-affected soil. A leaching fraction at a problematic site is determined after critical assessment of the soils and irrigation. On sand systems, leaching is practiced regularly, in most cases annually, according to need. Salts tend to build up near the surface of sand systems because the practice of light, frequent irrigation does not allow for leaching. Salt-affected soils and poor irrigation water negatively affect respiration and transpiration.

Suggested Reading

Beard, J.B. and the United States Golf Association (2002) *Turf Management for Golf Courses*, 2nd edn. Ann Arbor Press, Chelsea, Michigan.

Carrow, R.N. and Duncan, R.R. (1998) *Salt Affected Turfgrass Sites – Assessment and Management*. Ann Arbor Press, Chelsea, Michigan.

Carrow, R.N., Waddington, D.V. and Rieke, P.E. (2001) *Turfgrass Soil Fertility and Chemical Problems: Assessment and Management*. Ann Arbor Press, Chelsea, Michigan.

Cockerham, S.T. (2008) Culture of natural turf athletic fields. In: Pessarakli, M. (ed.) *Handbook of Turfgrass Management and Physiology*. CRC Press, Boca Raton, Florida, pp. 151–170.

Harivandi, M.A., Butler, J.D. and Wu, L. (1992) Salinity and turfgrass cuture. In: Waddington, D.V., Carrow, R.N. and Shearman, R.C. (eds) *Turfgrass*. ASA-CSSA-SSSA (American Society of Agronomy-Crop Science Society of America-Soil Science Society of America), Madison, Wisconsin, pp. 207–230.

Marcum, K.B. (2008a) Relative salinity tolerance of turfgrass species and cultivars. In: Pessarakli, M. (ed.) *Handbook of Turfgrass Management and Physiology*. CRC Press, Boca Raton, Florida, pp. 389–406.

Marcum, K.B. (2008b) Physiological adaptations of turfgrasses to salinity stress. In: Pessarakli, M. (ed.) *Handbook of Turfgrass Management and Physiology*. CRC Press, Boca Raton, Florida, pp. 407–418.

McCarty, L.B. (2001) *Best Golf Course Management Practices*. Prentice Hall, Upper Saddle River, New Jersey.

McCarty, L.B. and Miller, G.L. (2002) *Managing Bermudagrass Turf: Selection, Construction, Cultural Practices and Pest Management Strategies*. Ann Arbor Press, Chelsea, Michigan.

Puhalla, J., Krans, J. and Goatley, M. (1999) *Sports Fields: A Manual for Design, Construction and Maintenance of Sports Fields*. Ann Arbor Press, Chelsea, Michigan.

White, C.B. (2000) *Turf Manager's Handbook for Golf Course Construction, Renovation, and Grow-in*. Sleeping Bear Press, Chelsea, Michigan.

Witteveen, G. and Bavier, M. (1998) *Practical Golf Course Maintenance: The Magic of Greenkeeping*. Sleeping Bear Press, Chelsea, Michigan.

Suggested Websites

Airfield Systems (2009) Home page. Available at: http://www.airfieldsystems.com (accessed 19 Jan 2010).

FAO (Food and Agriculture Organization of the United Nations) (1992) The use of saline waters for crop production. *FAO Irrigation and Drainage Paper* No. 48, T0667/E (Rhoades, J.D., Kandiah, A. and Mashali, A.M.). Available at: http://www.fao.org/docrep/T0667E/t0667e01.htm#TopOfPage (accessed 3 Feb 2009).

USGA (United States Golf Association) (2004) USGA Recommendations for a Method of Putting Green Construction: 2004 revision (Green Section Staff). Available at: http://www.usga.org/course_care/articles/construction/greens/Green-Section-Recommendations-For-A-Method-Of-Putting-Green-Construction/ (accessed 6 Jan 2010).

USSL (United States Salinity Laboratory) (2010) Home page. Available at: http://www.ars.usda.gov/Main/site_main.htm?modecode=53-10-20-00 (accessed 1 Feb 2010).

11 The Ecology of Turfgrass Management

Key Terms

Ecology is the interaction of an organism with its own species, with other species and with its environment.

An **environment** is the sum of all natural and artificial conditions under which a particular organism is expected to live and grow.

Applied ecology is the use of theories and models to study and understand human impact on the human environment.

Physiological ecology, also called **ecophysiology**, refers to the response of an organism to environmental factors such as light, temperature, nutrients and others.

A **population** is the sum of individuals of a particular species occupying a particular space at the same time.

The **carrying capacity** of an ecosystem is the number of individuals within a population that can be supported by the current environment and with the resources available. Carrying capacity can also be considered the greatest density possible at the resource level available.

The **law of the minimum** states that population growth is limited by the amount of the most limiting resource during the most limiting period of the year.

Density is the number of individual plants in a given area. In turfgrass, however, we measure density by the number of tillers, also called shoots, instead of by the number of plants.

The **law of limiting factors** states that population growth is not only limited by too little of a resource but is also limited by too much of a resource.

A **community** is a collection of populations – plant, animal, microorganism, etc. that interact with each other either directly or indirectly.

Interspecific competition occurs among different species.

A **niche** is a resource or set of resources and a microenvironment within a community for which a population is uniquely adapted and for which it can successfully compete.

An **ecosystem** is the sum of activities and interactions within a biotic community as influenced by its abiotic environment.

Landscape ecology is a form of applied ecology that recognizes human activity as a factor that combines with natural disturbances to determine the spatial patterns that constitute a landscape.

Intraspecific competition occurs when individuals of a population compete with each other for resources.

A **blend** is a combination of turfgrass cultivars of the same species.

A **mixture** is a combination of turfgrass species.

Biomass is the amount of plant material by weight in a given area. Clipping yield is a measure of biomass but it does not account for the total plant biomass in a given area of land; clipping yield only measures the biomass over a particular height above that land.

The **−3/2 power law of self thinning**, also called the **−1.5 self thinning law**, is a mathematical description of the natural balance that occurs during intraspecific competition between plant biomass and density. In mown turfgrass, the slope of the self-thinning line is different from that in unmown plants and is believed to be $-1/2 = -0.5$.

Turfgrass ecology could be explained as the use of natural means to maintain a managed turfgrass system to a particular level of human expectation with the fewest inputs and the least environmental impact.

11.1 How to Make Nature Work For You

Ecology is a very old science and an even older philosophy. It is deeply grounded in the workings of nature and nature's effects on life, but it does not depend on evolution and the controversies common to evolutionary philosophy. We can see and study ecological activity at any time within the natural environment around us. Because ecological principles have a profound effect on plants, or vice versa, it benefits us to stay informed of advances in ecological science.

Although ecology is a very old endeavor, it did not become a popular science until the 1960s or 1970s, when a majority of people in many countries began to realize that they were having a critical, often negative, effect on their environment(s). At that time, our environmental impact began to become a subject of great interest. Ecology was soon recognized as a means to study and change environmental impact and society began to notice and embrace it.

In very simple terms, ecology is the study of how environments influence the organisms that live in them and how the organisms affect each other. Ecology, however, like nature from whence it is derived, is extremely complex and at least for the present, beyond our realm of total understanding. Nonetheless, many of the principles of ecology that others have studied and developed are sound and can be very useful to us as turfgrass managers.

The human desire to understand our environment and limit our negative impact upon it led to the development of applied ecology, one of the many branches of ecological study. Applied ecology is the use of theories and models to study and understand human impact on the human environment. You could say that applied ecology is in its infancy. Moreover, there are so many major and minor principles that affect human interaction with our environment(s) and our potential for detrimental influence that the study of applied ecology will probably never end. Practitioners of applied ecology endeavor to devise mathematical models that explain human impacts on our environment(s). Such endeavors are ridiculously difficult but nevertheless important. Models can be used to explain what happens for a given human input and how that input affects the local environment. Currently, most of the models that have been developed are not completely accurate but they will improve with time and use as important factors are identified and incorporated.

Applied ecology was not meant to be used by turf-grass managers and other agronomists. As with other sciences, it was meant to be a means of gathering knowledge of the natural world and its many interactions with us and other organisms. Some ecologists, in fact, believe that managed turfgrass should be eliminated. They argue that the inputs required to properly manage turf exceed its environmental benefits. It would not be wise for us to overlook those beliefs. Indeed, these ecologists may be correct. However, turf has many properties that positively influence our environment(s). Instead of criticizing our use of turf, perhaps these ecologists could recognize its benefits and teach us how to manage it using environmentally sound practices. In truth, they have already, but unknowingly, done that for us to some extent. We can use what they have taught us not only to affect our environment(s) more positively but also to manage our turf more effectively. As scientists and practitioners we can build on those lessons to continue learning better techniques.

11.2 Introduction to Ecological Theory

If you are to use ecology to your advantage, it is best to consider it a concept rather than a science or a theory. It is a way of thinking, not a means of addressing a practical problem. If you can imagine always using nature first and synthetic or industrial means of management second, you are thinking ecologically. If you can manage your turf to outcompete weeds to the point that you can physically remove them more cheaply than you can apply a pesticide, you are practicing ecological management. If you can convince your end user or customer that a 1 or 2% weed encroachment is acceptable, you are saving them money and you are also practicing ecological management. Ecological management not only reduces our environmental impact, it nearly always saves money. The major problem is that we have not identified the management practices or management systems necessary to meet our customer expectations using ecological means exclusively. It would benefit us to do so.

Ecological concepts

According to ecology, we live in and are affected by an environment which includes all natural and artificial factors around us. We can think of this environment as huge, incorporating the environment surrounding all humans, or we can think of it as very small – the environment that only affects us personally. Thinking too big is far too complex and too vague for any practical purpose. Therefore, think small, think about the environment that affects you and you can learn how to affect those factors that influence it. Think about the environment that affects a particular species and cultivar of your turf – in a particular location, under a particular mowing height and frequency, with or without irrigation – and at what level it needs to be managed, and you can make progress in ecological management.

We have come a long way in this text, and up to this point everything that we have studied has been a form of ecology. That particular branch of ecology is called physiological ecology. Physiological ecology, also called ecophysiology, refers to the response of an organism to environmental factors such as light, temperature, water, soil and nutrients. We have not only studied the effects of light, temperature, water, soil and nutrients, we have studied the physiological plant activities that occur in response to the factors discussed. We have actually been working within the realm of ecology all along. Now, instead of physiological ecology, we are going to concentrate on applied ecology, a lesser used science for plant management.

An ecological population

A population is a group of individuals of the same species living in a particular location at the same time. The size of the location is determined by the observer. The population of interest could be all of the Kentucky bluegrass (*Poa pratensis*) plants growing in a home lawn or it could be all of the creeping bentgrass (*Agrostis stolonifera*) plants growing on a golf course putting green. The individuals of a population compete with each other and with other organisms. However, if a plant has a minimal amount of everything that it needs it can continue to grow and reproduce throughout its lifetime. Therefore, if the plants in the population have all of the resources that they need, significant competition does not occur. The population continues to grow as offspring are produced more rapidly than aged plants die. Theoretically, the population will continue to grow until it reaches its carrying capacity, which is the number of indivi... supported by the amount... At that point no new plants population nears the carryi... slows and eventually levels of... are replaced in a sort of dyn... (Fig. 11.1).

The idea of carrying capacity began in early ecological study with a proposal called the law of the minimum (Leibig, 1840). The law of the minimum states that a population's growth is limited by the amount of its most limiting resource during its most limiting period of the year. Hence, the law of the minimum not only considers the resources available to the population but also the environment. Population growth may fluctuate with season but the population can only temporarily grow beyond the resources available during its

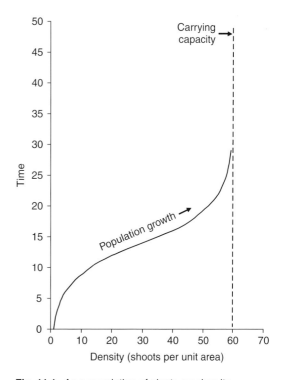

Fig. 11.1. As a population of plants reaches its carrying capacity, population growth slows and eventually levels off in a sort of dynamic equilibrium. When the carrying capacity has been reached, a new plant cannot be added unless an old plant dies. The only way to increase the population is to supply sufficient new resources to increase the carrying capacity.

most limiting season. For example, the density of a cool-season grass lawn may improve in the spring and the fall but its ability to sustain that density will always be limited by the summer season. We already knew that. Ecology did not identify that occurrence for us it simply gave us a principle with which to define it. Now let us use that principle and others to help us make good decisions concerning our turf.

The law of the minimum was replaced by the law of limiting factors (Blackman, 1905). The law of limiting factors states that population growth is limited by too little of a resource but is also limited by too much of a resource. In turfgrass management nothing could be truer; water is good but too much water is not good, nitrogen is good but too much nitrogen is not good. Warm temperatures are nice but hot temperatures are not good; cool is nice but cold is not. With rare exception, in everything that we do, there is a low limit and a high limit between which we must always operate to achieve our greatest success.

The ecological community

An ecological community is a collection of populations that interact with each other. Communities are not only defined by the populations that they contain but often by physical characteristics or, in our case, management levels. A forest and a meadow would normally be considered to be two different communities. A tennis court would be considered as in a different community from the soccer pitch beside it, which would be in a different community from the park surrounding it. Within this complex, there may also be communities of bedding plants that require care too. The idea of a community including grasses, weeds, birds, insects, fungi, bacteria, humans and other organisms gives considerable depth to the concept of interspecific competition. Ecologically speaking, a human playing tennis on a grass court is not using the turf for sporting purposes: the human is competing with the turf for space. If the human travels across the court repeatedly in the same place for a long enough period of time, that human may not win the match with her opponent but she will win the match with the turf for space. She will destroy the turf and someone will have to fix it.

Within a community, there is a struggle among species for dominance. The largest population is not always the dominant one. In our systems we have chosen the dominant species, the turfgrass, but we also have to be able to maintain that dominance. Our management asserts turfgrass dominance, but maintaining that dominance is not always easy. Nature likes diversity and, without our interference, what was once a perfect (in our eyes) turfgrass system will quickly turn into a struggle for dominance among the species present and among the current species and invaders. Ecologists call that succession and it is typical of nature. Our job is to create an environment where succession does not occur. We want our community to remain at a particular stage of succession in perpetuity. However, we want to do that with as little disruption as possible.

A turfgrass that is adapted to the climate where it is grown and mowed on a regular basis is likely to proliferate and may be the dominant species in a community (Box 11.1). Therefore, growing turfgrass where it is naturally adapted and mowing it regularly to remove most of its competitors allows the grass to flourish. In that case, it is reasonably easy to manage in the monoculture that we prefer. If turfgrass is not adapted to the climate where we want to grow it management becomes complex. In that case, managing our monoculture becomes very expensive, both in economic terms and in environmental terms. We have to supply what the turfgrass needs even though it is not naturally available at our location. In most cases, the missing component is water. Because of that, our industry has become undeservedly notorious for luxury water use. If society does not want us to grow turf where it is not naturally adapted, so be it, but if we are asked to do it, we can. All we have to do is to synthetically supply the natural components that are missing. The missing component could be water, or it could be a temperature conducive to growth, a growing medium, a nutrient or a collection of factors. We don't always apply synthetic components; we usually supply natural components synthetically (e.g. irrigation and nitrate fertilizer).

There are usually only a few species that proliferate in a community but many other species that occur in small numbers. For instance, a forest could contain 24 different species of trees but two of those populations (a population is only one species) might make up 44% of the total number of trees (Smith and Smith, 2001). Although the trees

> **Box 11.1. Mowing is a means we use to limit the number of species that can live in a community.**
>
> When we mow, we affect a turfgrass community by preventing the growth of most weeds. Trees and shrubs are automatically eliminated from the site. Most grasses and many broadleaved weeds reproduce by producing stalks that rise above the existing canopy before producing seed. If those plants can only reproduce by seed, they are highly discouraged by mowing. We also eliminate any plant that cannot produce leaves close to the soil surface and any herbaceous plant that, like trees and shrubs, maintains a leading growth structure (an apical meristem) above the soil surface. The lower we mow the more plant species we eliminate. In doing so, we also eliminate those animals and microbes that feed on those plants or have symbiotic relationships with them. We call this community perfect; a naturalist would probably call it practically sterile. We would both be right.
>
> If we were to allow a turfgrass site to naturalize, the turfgrass, if it was chosen properly for the site, would probably be one of the most abundant species providing that the natural environment was best suited for grassland. However, it might not dominate a grassland community and most certainly would not dominate if the area was one where trees proliferate. In contrast, if we were to allow the site to naturalize but continued mowing on a regular basis, the turfgrass would probably be able to maintain its dominance. Turfgrasses are extremely competitive plants. If the turfgrass could not dominate in a natural but mowed environment, then it is not adapted to the site and we are probably the only thing keeping it alive.

make the forest there are hundreds of other species also living in that community. Each of those species has found a niche, a resource(s) in a conducive microenvironment for which they are uniquely adapted. A niche allows a population to live and successfully compete within a community. Each population within the community has a niche that allows it to be a part of the community. If a species cannot find a niche within a community it cannot compete with other populations in the community and therefore cannot survive. It would be nice if we could eliminate the niche that allows turfgrass insect and disease pathogens to live within a turfgrass community. However, the turfgrass is one of the resources that enable the existence of a niche for turfgrass pathogens. In most cases, we would have to eliminate the turfgrass in order to destroy the niche for its pathogens.

An ecosystem

A system is a complex in which the parts interact to produce the behavior of the whole. Ecosystems can be very large or very small. Ecosystems tend to take on the size of their environment. An ecosystem is a combination of a biotic community of species and its abiotic environment. Ecologists tend to study large ecosystems and use energy-dependent models to define the most efficient way to conserve those systems in their natural state or to determine how humans can exist within those ecosystems with the least environmental impact. Our goals are similar yet different. We hope to manage turfgrass systems to meet the expectations and demands of our customers in the most natural way and with the least environmental impact possible. Depending on our customers' expectations, we may have to use synthetic means to achieve satisfactory results. However, our objectives are always to use the least synthetic inputs possible and to affect our surroundings little if at all. In order to do that, we must study the ecology of our ecosystem and for that purpose we should be thinking small. Our ecosystem is the turfgrass community and its environment. Our management should affect that ecosystem and no other. That may seem like an impossible task and it probably is, but that is the ideal that we should strive for.

Ecosystem stability

A natural ecosystem is in constant flux. It is in a constant mode of succession. Succession can be cyclic, occurring over periods that we can study and predict, but it is normally a very slow process involving changes in population dominance within the ecosystem's community over extended periods. Over such extended periods an ecosystem's environment also changes, causing flux in the ecosystem.

Cyclic changes in predator/prey relationships can also cause flux. This flux is similar to changes caused by season and is called oscillation. As a population builds, its predator numbers also build as the predator's resource (the population) increases. Eventually, predator numbers increase enough to cause a decline in the population. As the population declines, predator numbers also decline, thus completing the cycle which, assuming that the population has sufficient resources, begins again. Disturbances also cause ecosystem flux.

Natural disturbances that cause changes to an ecosystem could include fire, flooding, wind damage, ice damage or drought. A relatively stable ecosystem will return to its pre-stress level following a disturbance. If the disturbance is severe, however, recovery may take considerable time. In the turfgrass world, we would renovate following a severe natural disturbance. Some man-made disturbances can also be severe. Mining, logging, cultivation and the ever-present expanding presence of the human population cause serious disturbances.

For turf-management purposes, disturbances are minor but always present, often at varying levels of severity. We usually have traffic damage that disrupts the system. Cultivation practices regularly affect the quality of the turfgrass population and the ecosystem in general. There are seasonal factors to consider, and damage caused by invading weeds, pathogens and pests. Management mistakes are particularly common causes of flux in the turfgrass population. Because of all these potential fluctuations, it is difficult to manage turfgrass at the density of its carrying capacity with any consistency. The population always fluctuates around carrying capacity, sometimes even exceeding it temporarily (Fig. 11.2). Part of your job is to manage carrying capacity and minimize disturbances. We will discuss that later.

Sustainability

Sustainability has become a favorite word in political and social circles. Consequently, it tends to take on multiple meanings which depend on the person using the term. The concept of sustainability is vague, but the simple definition of the term is not vague and it is most certainly something that we should strive for. A sustainable system is one that perpetuates under proper management. In crop production, a sustainable system is one whose yield per unit of time equals its expected production per

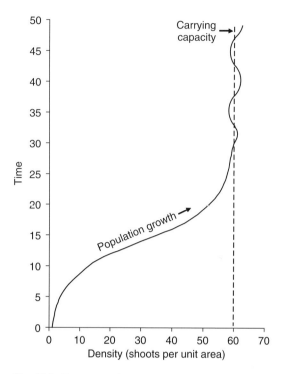

Fig. 11.2. Because turfgrass populations fluctuate in response to many factors, a turfgrass population never achieves its carrying capacity for more than a very short time. Instead, turfgrass numbers fluctuate around carrying capacity.

unit of time. In other words, it is a system that can be expected to yield a certain average amount of product over a given period of time, usually a year, and maintain that yield perpetually. The concept of sustainability in crop production is based on harvest. In a turfgrass system, sustainability is based on aesthetic value, functional use or both.

A more socially oriented definition of sustainability may include a clause that the sustainable system should not interfere with other sustainable systems and should not have a detrimental effect on its ecosystem. Those are also terms that we should strive to meet effectively. In doing so, we can maintain better turf, or at least acceptable turf with less social and political pressure.

Landscape ecology

Landscape ecology is a relatively new branch of ecology. It can be described as natural ecology that recognizes human activity as a factor which combines with natural disturbances to determine the spatial patterns that constitute a landscape. Obviously, managed turfgrass surfaces are part of landscape ecology. Because of our ability to manage turfgrass ecosystems to near perfection, our industry – lawns and golf courses especially – has come under scrutiny. In some instances, poor alternatives to lawns have been suggested as landscapes that use water and nutrients more conservatively. That may be true in climates where turfgrass is marginally adapted, but it is usually not true where turfgrass is adapted (Erickson *et al.*, 2008). Turfgrass has a positive influence in residential applications (Beard, 1973). As the study of landscape ecology matures, the advantages of turfgrass in low-input landscapes will probably be recognized. Whether or not that is true, landscape ecology may provide us with additional information that helps us to manage turf more conservatively without loss of quality.

Intraspecific competition

Populations not only have to compete with other populations for space and resources, the individuals of a population have to compete with each other. As the population reaches carrying capacity, individuals within it compete with each other for resources. We call that intraspecific competition. Intraspecific competition occurs among individuals of a species and cultivar and among cultivars in blends. A blend is a combination of cultivars of the same species. A mixture is a combination of turfgrass species. Competition among cultivars of the same species is intraspecific and competition among species is interspecific. We will consider interspecific competition in Chapters 12 and 13. For now, let us study the intraspecific competition that occurs within a population.

Intraspecific competition explains many things that we know to be true of turfgrass. For instance, mowing a turfgrass lower without exceeding the limit of its adaptation causes turf density to increase. Density can not increase beyond carrying capacity; by definition, carrying capacity is the maximum density that can be supported by the ecosystem. So by reducing the mowing height of the turf, we were able to increase the carrying capacity of the ecosystem. How can that be?

When mowing height is lowered biomass is reduced. Biomass is the weight of plant material in a given area. Biomass and density have an inverse relationship within available resources (Fig. 11.3). Consequently, if we lower mowing height, thereby reducing the biomass of a mature turfgrass population, fewer resources are needed to maintain biomass and the surplus resources can be used to increase density. As a mature turf is mowed lower, reducing the biomass to a point below the full use of the resources available, the plants will allocate the excess resources to producing more tillers and/or daughter plants, causing an increase in density. Eventually, enough tillers and/or daughter plants are produced that resources again become limiting. At that point, intraspecific competition occurs. The new plants have to compete with the other new plants and with their parents for space and resources. Some new plants will not survive and a small number of less competitive parents may die. Soon the population will become fairly stable around the new carrying capacity.

Intraspecific competition is a means of competitive selection. In all likelihood, the strongest plants for the environment will survive. Intraspecific

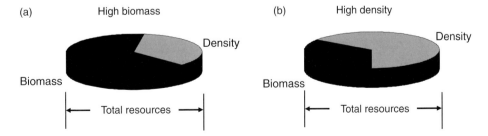

Fig. 11.3. As the plant biomass and density of a mature turfgrass population have a relationship within the available resources, when biomass is reduced, a surplus of resources is available to fuel an increase in density. The pie charts represent the resources available to the population, and the segments within them represent the portion of those resources allocated to biomass and to density. (a) When biomass is high, density is low. (b) When biomass is low, density is high.

competition occurs among plants of the same species. The plants share a common niche. Consequently, competition is normally severe. In a seeded population, where minor genetic variation exists, selection will occur. However, in a vegetatively propagated cultivar, the plants are genetically identical so elimination happens at random and may occur as a result of interference or damage from an outside source. Because turfgrass plants are so highly competitive with each other, increases in density usually result in stands of immature plants struggling with each other for available resources. Immature plants are difficult to manage because they are more susceptible to unfavorable changes in environment, they are more easily damaged and they are more susceptible to pest problems.

Resource allocations between plant biomass and density are ecologically described by the law of self thinning. The density of a population will continue to increase until it reaches the carrying capacity of its ecosystem. At carrying capacity, no new individuals can be added unless an individual dies. In a natural situation, once the population reached the carrying capacity, intraspecific competition would result in some individuals being eliminated as others increased in biomass. Selection would occur as the plants best adapted to the environment were able to out-compete the lesser adapted ones. Overall growth in biomass would continue, for the most part, until the growth of older plants slowed and more resources were available for the birth of new plants. As a population matures in nature, biomass is added at a rate of approximately three parts biomass at the expense of two parts density, i.e. increasing biomass has a negative effect on density (Fig. 11.4). The relationship has come to be known as the −3/2 power law of self thinning or the −1.5 self thinning law (Yoda *et al.*, 1963; White and Harper, 1970). Mown turfgrass, however, does not respond as a natural plant and its self-thinning line is believed to have a slope of −0.5 (Lush, 1990).

11.3 Turfgrass Ecology

Turfgrass ecology has never really been defined, but it could be explained as the use of natural means to maintain a managed turfgrass system to a particular level of human expectation with the fewest inputs and the least environmental impact. The idea behind turfgrass ecology is to allow nature to work for you. The only way to do that is to understand the natural boundaries of your ecosystem and to work within them. The law of limiting factors tells us that more than a minimum of a

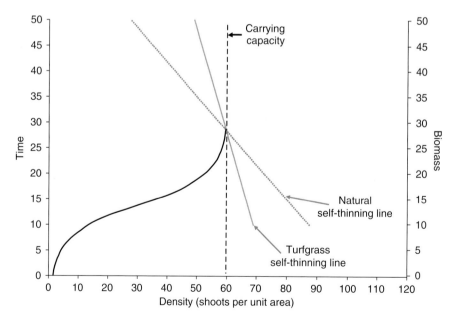

Fig. 11.4. Natural self thinning occurs along a balance between biomass and density that results in a line with a slope equal to −1.5. Mown turf however, self thins along a line with a slope of −0.5.

resource is good but too much of a resource is not good. We need to know our boundaries and we need to know how those boundaries are likely to change with changing environments and sudden disturbances.

Sexual reproduction results in genetic differentiation among offspring of the same parents. It follows then that any seeded cultivar, with the possible exception of Kentucky bluegrass, an apomictic (asexually reproducing) species, results in a stable but wide selection of genetically different individuals. So a synthetic cultivar, as a seeded variety is called, is a group of very similar but slightly different plants. It would seem that, under a given set of circumstances, some of these individuals would be more competitive than others. This appears to be true of turfgrass, because seeded populations, especially in golf course greens where competition is extremely fierce, become patchy, and what appear to be vegetative offspring of the same parent form identical patches that are easy to differentiate. If individual selection occurs within a population, then cultivar selection must occur within cultivar blends (Lickfeldt et al., 2002a). In a cultivar blend, competition is occurring on two levels, one among individuals and a second among cultivars. Fortunately, however, cultivars or biotypes are so similar that unless they differ by a particularly important characteristic, such as mowing height, or particularly widely by drought tolerance, they can be managed in the same way (McGuan et al., 2004).

The competitive advantages of a cultivar are probably more important than management practices for determining cultivar dominance (Lickfeldt et al., 2002b). Cultivar dominance or species dominance in a seed mixture may be determined during establishment. A cultivar that germinates more quickly than the others, or grows more vigorously as a seedling, has a distinct advantage. That particular cultivar may not be the best suited to the environment but it establishes more rapidly and has a competitive advantage over the smaller seedlings of the other cultivars. If it reaches maturity first, it shades its competitors and easily wins the battle for space. Once the blend is established the best performing cultivar for that particular environment would be expected to dominate. However, that may not be the case. Again, the most aggressive cultivar has the advantage. Spreading grasses, those that grow by stems, differ in their ability to spread rapidly. Most warm-season (C_4) grasses and many cool-season (C_3) grasses spread by stolons, rhizomes or both. Their propensity to form daughter plants is a characteristic of their biotype (Karcher et al., 2005a,b). Consequently, the most aggressive of these biotypes has a propensity to dominate the stand. Although a dominant cultivar may increase its share of the stand with time, it is highly unlikely that it will eliminate the other cultivars (Brede, 2004). Nor is it likely that a dominant individual plant would claim more that a small portion of a mature turfgrass stand. That is to our advantage.

Turfgrasses are blended or mixed for a reason. Blends and mixes confer genetic diversity. Genetic diversity makes a stand stronger by increasing stability. A genetically diverse stand of forage grasses, legumes or turfgrass is potentially better able to resist disease, insect damage and even weed encroachment (Chen et al., 1997; Picasso et al., 2008). That said, however, weak cultivars or species do little to improve blends and mixtures (Dernoeden et al., 1998). Each cultivar or species in the blend or mixture must be reasonably well adapted to the site or the genetic diversity that it adds to the stand has little effect. If we were to quit managing a turfgrass site it would quickly convert to a natural stand and within 2 to 3 years it would be a mix of many plant species. That ecosystem would be quite stable, meaning that it would resist change. Each of many plant species would have found a niche in the community and established a population. Other organisms, including microorganisms and insects, would also have diversified and would be in competition for resources and space within the community. Many pathogens would exist, but in all likelihood each would be specific to a different plant species. Consequently, if one plant species contracted disease, the proximity of like plants, separated by non-susceptible plant species, would slow the spread of disease to the point that little damage would be likely to occur and it would hardly be noticed in the community. Although we cannot reach that extent of diversity in a turfgrass monoculture, we can take a lesson from nature and diversify as much as possible. Hence, blends and mixes are usually advantageous.

Ecological study reminds us that stability is a natural characteristic that a turfgrass monoculture is lacking. Therefore our systems are prone to rapid fluctuation. Rapid fluctuations are detrimental to the health of a community and to the sustainability of an ecosystem. Nature teaches us that stability is preferred in a biotic community. Consequently, it is beneficial for us to maintain stability in all things turf related and we have the tools to do so. We have integrated pest management plans that will be presented in the next two chapters, and we plan and schedule management such as mowing and fertilization to reduce rapid changes in growth and promote stability. Changes in season can have a profound effect on plant communities (Lush, 1988a). We need to pay more attention to what is happening around us and be flexible to changes in weather and other environmental conditions. We can predict how plants will respond to changes in season (Koski *et al.*, 1988; Xiong *et al.*, 2007). We can formulate management plans that are easily adjusted to current conditions. We have discussed how to do that with fertilizer, irrigation and cultural practices, and we have determined what inputs are required to positively affect photosynthesis, respiration and transpiration. There is no prescription for turfgrass management. You have to accept what nature requires from you and remain alert to what nature offers you.

Although it is difficult to manage a turfgrass system that nature is happy to destroy, we do have some positive things working for us. For one thing, we can adjust carrying capacity and we can also affect density and biomass. Anything that we can do to encourage photosynthesis, respiration and transpiration increases carrying capacity. Consequently, we have already discussed many methods that we can use to affect carrying capacity. We need light, water and carbon dioxide to promote photosynthesis. Soil oxygen promotes respiration. Sufficient irrigation is used to satisfy transpiration needs and we can add nutrients that supply the elements needed for vegetative and reproductive growth. In order to increase carrying capacity we simply have to determine which of the physiological processes is limiting further growth of tillers (vegetative growth) and daughter plants (reproductive growth), and which factors are needed to encourage it. We can do that through knowledge, experience, or trial and error. Of course, there comes a time when inputs become excessive. The law of limiting factors states that population growth is limited not only by too little of a resource but also by too much. Therefore, we must always be looking for that upper limit of nitrogen, water and other resources that causes a decline in carrying capacity when an increase is expected.

When you add a limiting resource to a turfgrass system you increase carrying capacity. The law of self thinning, however, does not change, although the self-thinning line now begins at a higher maximum density (Fig. 11.5). Turfgrass management objectives nearly always include achieving the maximum turf density that the ecosystem can sustain. Once you have achieved the maximum carrying capacity attainable, you need to determine the minimum acceptable biomass.

Ecological doctrine tells us that the greatest biomass is achieved at relatively low density (Kandel *et al.*, 2004). Relatively low density also encourages seed production. Plants are more likely to reach full maturity if intraspecific competition is minimal (Bednarz *et al.*, 2006). The relationship between turfgrass biomass and density is governed by self thinning (Lush, 1990). However, we have a great deal of control over biomass through mowing (Lush and Rogers, 1992). Therefore, we have a great deal of control over density as well. Within a turfgrass species' range of adaptation we can choose its mowing height, in this way controlling its biomass and determining its density. It would seem advantageous then to lower mowing height to the lowest acceptable height, thereby increasing density to the maximum. However, mowing low is rarely a wise choice because as biomass decreases and density increases, maturity declines and the stand becomes more easily damaged or disrupted (Spak *et al.*, 1993) (Fig. 11.6). Turfgrass mown at low mowing heights is always in a juvenile state.

Mature plants have a competitive advantage over immature plants (Kendrick and Danneberger, 2002). Immature plants are easily damaged. Therefore, populations of immature plants are unstable. Rapid fluctuations occur in the turfgrass population and the overall health of the system is difficult to sustain when the population consists of immature plants. A particularly knowledgeable turfgrass manager is required to grow grass at low mowing heights. In addition, the system requires an exceptionally high amount of labor, energy and resource inputs. Low-mown systems such as putting and bowling greens are sometimes required. Otherwise low mowing should be avoided. Low mowing is not economically or environmentally sound. If you want to make your job easier and your turf system more sustainable, mow as high as your end-user expectations will allow.

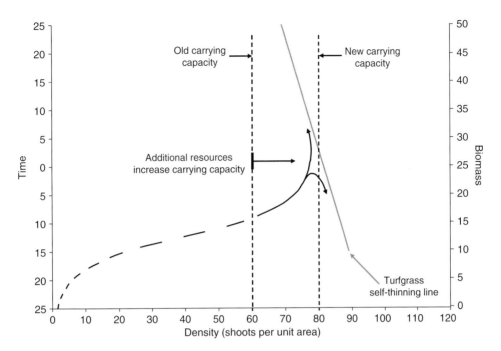

Fig. 11.5. Carrying capacity increases and density increases when a limiting resource is added to the turfgrass system. With an increase in carrying capacity, the self-thinning line moves to a new maximum density but the relationship between biomass and density does not change.

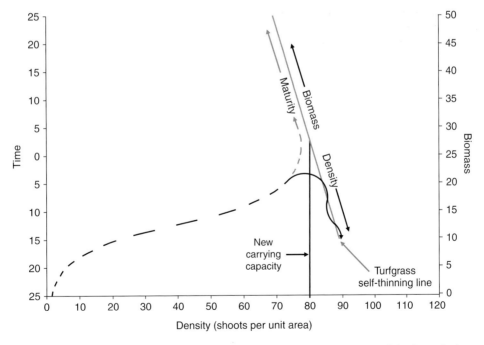

Fig. 11.6. Mowing low to improve density reduces biomass and results in a turfgrass stand that is easily damaged or disrupted. Compare with Fig. 11.5, which shows a stand with the same increased carrying capacity (resulting from the addition of a limiting resource).

11.4 Chapter Summary

Ecology is the science of natural systems. It can instruct us on what to expect from our turfgrass under differing environments and management. By following ecological principles as closely as possible, we can take advantage of natural phenomena and reduce our economic and environmental impact.

According to ecology, a turfgrass of the same species living in a stand at a particular site constitutes a population. If it is a mixed stand of turfgrass species, each species constitutes a population and the populations compete for space and resources. The turfgrass and other organisms at the site and those that may visit the site make a community. Each population is a part of the community because it is able to occupy a particular space and can compete for a particular set of resources. The combination of resources and microenvironment that allows a species to compete within the community is called a niche. The combination of the biotic community and abiotic environment is called an ecosystem. We manage the turfgrass and its competitors in such a way as to make the turfgrass(es) the dominant population(s) in the ecosystem. In many respects, we can also manage the environment.

To promote the turf as the dominant species, we provide resources that help it compete with other populations. We provide light, water and carbon dioxide to promote photosynthesis, soil oxygen for respiration and irrigation for transpiration. We provide nutrients for consistent growth and cultural management that affects the turf positively and/or its competitors negatively.

Rapid fluctuations caused by disturbances are detrimental to the turfgrass because they affect its competitive ability. We try to maintain stability and consistency so that the turfgrass can flourish and maintain its dominance in the community. One practice that encourages stability is to maintain the turfgrass at the oldest maturity possible. We can affect turfgrass maturity by adjusting mowing height higher to favor biomass over density. Although density is normally a condition that we encourage, we can manage more economically with the least environmental impact and the highest stability when biomass is favored. Therefore, ecology encourages us to maintain at the highest mowing height that is acceptable to our end user and our purpose. By taking advantage of natural phenomena, we can promote sustainability in our turfgrass system, thus reducing the cost of our labor and minimizing our effect on surrounding ecosystems.

Suggested Reading

Barbour, M.G., Burk, J.H., Pitts, W.D., Gilliam, F.S. and Schwartz, M.W. (1999) *Terrestrial Plant Ecology*, 3rd edn. Benjamin Cummings, Menlo Park, California.

Danneberger, T.K. (1993) *Turfgrass Ecology and Management*. Franzak & Foster, Cleveland, Ohio.

Park, D.M., Cisar, J.L., Erickson, J.E. and Snyder, G.H. (2008) Influence of landscape and percolation on P and K losses over four years. In: Nett, M.T., Carroll, M.J., Horgan, B.P. and Petrovic, A.M. (eds) *The Fate of Nutrients and Pesticides in the Urban Environment*. ACS Symposium Series 997, American Chemical Society, Washington, D.C., pp. 107–132.

Sachs, P.D. and Luff, R.T. (2002) *Ecological Golf Course Management*. Ann Arbor Press, Chelsea, Michigan.

Schulze, E.D., Beck, E. and Müller-Hohenstein, K. (2002) *Plant Ecology*. Springer-Berlin, Heidelberg.

Smith, R.L. and Smith, T.M. (2001) *Ecology and Field Biology*, 6th edn. Benjamin Cummings, New York.

12 Managing Competition Among Plant Species

> **Key Terms**
>
> A **habitat** is the place where an organism lives. In this case, it refers to the place occupied by a turfgrass community. The habitat is defined by its location and by its environment.
>
> **Habitat management** is the use of management practices that alter the environment of a turfgrass ecosystem to favor turfgrass persistence and discourage its competitors and antagonists.
>
> **Spot spraying** is the use of herbicide by locating and spraying individual weeds in the turfgrass canopy. This is only feasible when the weed population is low.
>
> **Broadcast spraying** is the typical practice of spraying a pesticide over an entire area in an attempt to remove all of the weeds from the site without having to locate them.
>
> **Allelopathy** is the production by plants of toxic chemicals that selectively inhibit the growth or reproduction of other plants.
>
> A competitive plant **strategy** is the sum of genetically fixed physiological and morphological adaptations required to conquer a habitat and to persist with optimal use of resources.

12.1 Manage Your Sites to Favor Turfgrass and Discourage Other Plant Species

In many parts of the world, the use of herbicide applications for managing turfgrass nearly weed free is discouraged. Although herbicide use for producing food crops is tolerated, the use of herbicides for managing urban crops such as turf that are primarily produced for aesthetic value is discouraged socially and sometimes governmentally. Although the pesticides that we use are very much like and sometimes exactly the same as the medicines that we take, they are usually deemed more dangerous. Their use is also considered, rightfully so, to be less important and more likely to have a detrimental effect on natural environments. In many, perhaps most, instances pesticides applied to turfgrass are considered undesirable. We have to accept that and limit or eliminate pesticide use. Under pesticide restrictions, perfect or near-perfect turfgrass is not possible at our current level of knowledge. However, we are good at what we do and we can achieve near-perfect turf with limited pesticide use and, in most cases, acceptable turf with no pesticide use. If we manage our systems to encourage turfgrasses and discourage other plant species we can potentially learn to manage turf with minimal weed encroachment with no herbicide use. For the present, we can most certainly use our knowledge of ecology to develop better plans for permanent weed reductions.

If we were to choose exactly the right species or combination of species for a particular turfgrass site and use all of the techniques at our disposal to reduce weed pressure and encourage turfgrass growth, we could probably meet the expectations for a home lawn, athletic field, or perhaps even a golf course fairway, by using mechanical weed removal. However, we would also want to balance the economic cost of herbicide application with the economic cost of the hand labor needed for mechanical removal. That would be more difficult on a large property like a park or a golf course, but might be possible on a home lawn or athletic field. Ecological management is all about using natural factors to manage turfgrass at a particular level of human expectations. As you learned in Chapter 11, natural communities and human expectations are often not compatible. However, we can learn to use particular natural factors to encourage turf dominance within a natural community. For reasons that you will learn in Chapter 13, especially concerning predator/prey relationships, we will probably not be able to manage today's turfgrass monocultures on a sustainable basis without the use of fungicides and insecticides. For that reason, we may have to lower our expectations. However, it is quite likely that we can learn to manage desirable turf at high expectations with little or no use of pesticide.

12.2 Competition Among Turfgrass Species

Competition among different turfgrass species within a community is by definition a form of interspecific competition. In many instances, however, it does not matter to us which species survives or which dominates the stand as long as the stand meets our expectations for color, density, uniformity and the other factors that our customer deems important. In fact, species mixtures improve the biological diversity in a community. Diversity was introduced in Chapter 11 as a positive factor for the stability and sustainability of an ecosystem. Consequently, in the ecological management of a turfgrass system, cultivar blends and species mixtures are preferred. The blends that were discussed in Chapter 11 and the mixtures to be discussed in this chapter are more naturally resistant to disturbances than single-cultivar or single-species monocultures.

Interspecific turfgrass mixtures

Turfgrass mixtures and blends are ecologically favored over single species and cultivar monocultures. However, if the mixed community is to have an ecological advantage compared with other plant species all component grasses must be well adapted to the site. Each turfgrass must be able to persist and sustain itself in competition with other plant species. In a turfgrass system, the most important plant competitor of each turfgrass should be another turfgrass. In such an ecosystem weed encroachment is discouraged.

Research has indicated that cultivars of Kentucky bluegrass (*Poa pratensis*), creeping red fescue (*Festuca rubra*), tall fescue (*Festuca arundinacea*) and perennial ryegrass (*Lolium perenne*) were differentially adapted to various sites along the Italian peninsula and Italy's major islands (Annicchiarico *et al.*, 2006). In a study in Canada, the proportion of each species in mixtures of the same four species (Kentucky bluegrass, red fescue, tall fescue and perennial ryegrass), including multiple cultivars of each, changed in proportion between that seeded in 1988 and that in the mature stands 4 years later (Hsiang *et al.*, 1997). In this study, perennial ryegrass was the dominant species in mature stands, possibly because of its rapid germination and superior seedling vigor compared with the other species. Kentucky bluegrass and creeping red fescue were competitive with the perennial ryegrass, but the proportion of tall fescue in stands was much smaller than that which had been seeded. Tall fescue has better heat tolerance but poor cold tolerance compared with the other three species and might not have been able to compete well for that reason. Three of the four species and many of their cultivars appeared to be relatively well adapted to the site and able to survive and compete for resources. Such a mixed stand would normally be able to reduce the competitive ability of weed species better than a single stand of one individual turfgrass species. It is also unlikely that any of these turfgrass species would be completely eradicated by natural competition in any of these mixtures. They would always be present to some extent. Although perennial ryegrass was dominant at the Canadian site, Kentucky bluegrass dominated perennial ryegrass in a study in Pennsylvania (Brede and Duich, 1986). In warmer regions of temperate zones, tall fescue is the most competitive of cool-season (C_3) turfgrass species and tends to dominate mixed stands (Bremer *et al.*, 2006). In species mixes in northern Missouri, tall fescue dominated Kentucky bluegrass but not perennial ryegrass (Dunn *et al.*, 2002). In a slightly warmer climate, such as that of southern Missouri, tall fescue would probably dominate perennial ryegrass as well, but both species would persist creating a stronger stand than either would alone if they were both reasonably well adapted to the environmental conditions.

Cool-season grasses are often overseeded into warm-season (C_4) grasses in the USA to provide a green cover in winter. The two species compete with each other during spring and fall but they dominate during the seasons, summer and winter, for which they are best adapted. The most difficult portion of the overseeding process is the transition from winter cover (cool-season) to summer cover (warm-season). In many instances, portions of the cool-season species will persist in the warm-season species through much of the summer, demonstrating how difficult it is to eradicate even a poorly adapted species from a community. In cooler regions, where the cool-season species is poorly adapted to the summer heat but also somewhat tolerant, it may be extremely difficult to remove with cultural practices alone during summers when temperature and humidity are lower than normal (Horgan and Yelverton, 2001). Persistent mixtures of cool- and warm-season turfgrasses have also been attempted where both species are reasonably well adapted.

In Utah, research was attempted to mix fine fescues (*Festuca* spp.) with buffalograss (*Buchloe dactyloides*), a native species, for sustained, year-long green color in low-maintenance situations (Johnson, 2003). Both types of grass are well adapted to low maintenance and performed reasonably well in monoculture. The mixed stands also performed reasonably well, but the fine fescues dominated the buffalograss to an extent that summer greenness was marginal and the experiment was not deemed successful. In Missouri, a similar project was attempted to promote year-long greenness, primarily in athletic fields (Dunn *et al.*, 1994). The use of common bermudagrass (*Cynodon dactylon*) was tested in separate combination with Kentucky bluegrass, perennial ryegrass and two fine fescues. Because athletic field use was to be tested, wear was applied to the plots regularly, which reduced the effectiveness of the fine fescues in the fescue/bermudagrass mix. However, the Kentucky bluegrass/bermudagrass mix and the perennial ryegrass/bermudagrass mix persisted. Bermudagrass would be expected to be the most wear tolerant and to recover most rapidly from injury compared with the other two species. Nonetheless, the bermudagrass was the weakest component of the mixtures because it was the most poorly adapted species for the region. The bermudagrass had poor quality in monoculture and was dominated by both Kentucky bluegrass and perennial ryegrass in mixtures. In most cases, the species best adapted to the natural environment dominates the community, but human activity also constitutes a portion of the turfgrass environment and human use can also determine species dominance.

Where two species are more or less equally adapted to the natural environment, it is human activity and management that determine species dominance. In the UK, in mixtures with perennial ryegrass, creeping red fescue and colonial bentgrass (*Agrostis capillaris*) had poor cover compared with perennial ryegrass when subjected to wear (Gore *et al.*, 1979). However, both species did very well when wear was not present. A species' ability to resist wear or some other human influence affects the dominance of a turfgrass community (Sorochan *et al.*, 2001; Samaranayake *et al.*, 2008). Mowing height and nutrient input can also affect species dominance in a mixed stand (Carrow and Troll, 1977). If we use management practices to assert turfgrass dominance we will probably not eliminate weed encroachment but we can reduce it substantially.

12.3 Competition between turfgrass and other plants

In a turfgrass system, plants other than turfgrasses are considered weeds. In fact, some types of turfgrass in another turfgrass may be considered weeds in some locations under certain circumstances. Our propensity for uniform single-species turf makes management, especially weed management, difficult. Because all turfgrasses share a similar niche, interspecific competition among them is fierce. However, for that same reason, they tend to exclude other plants from that particular niche. Nonetheless, weeds are persistent and opportunistic colonizers that make the most of available resources or disturbances in the turfgrass canopy (Fig. 12.1). Some weeds, dandelion (*Taraxacum officinale*) for instance, are so good at utilizing resources and exist in a niche different enough from turfgrass that they can invade even the most dense turfgrass canopies and be highly competitive. Most weeds, however, are excluded from a dense turfgrass canopy and we can use management practices to discourage others.

It is not possible to completely eradicate weeds in turfgrass. We can come very close to eradicating a particular weed species using herbicides, but we are unlikely to ever remove it completely. Pesticides are a part of most turfgrass management plans although they should not be the first alternative employed; they should be the last. We can come close, in some cases close enough, to customer expectations without pesticide use. If we strive to minimize weed populations through habitat management we may be able to meet customer expectations without pesticides or to meet their expectations by spot

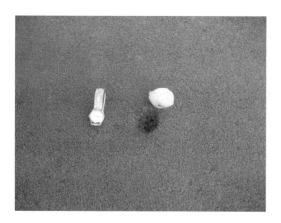

Fig. 12.1. Hail damage is an unlikely but potential source of disturbance. This damage opens the turfgrass canopy and makes a good seed bed for an opportunistic weed. Photo by Craig Evans, Stillwater, Oklahoma.

spraying herbicides rather than broadcast spraying them. Spot spraying is environmentally friendly in that it requires very little use of pesticide compared with a broadcast spray that is applied over an entire area. Spot spraying is the act of locating weeds and spraying them individually wherever they are found. Habitat management is the use of management practices that alter the environment in a turfgrass habitat to favor the turf and exclude weeds and other competitors or antagonists. Environmental factors both alone and in combination can affect the population density of desirable and non-desirable species (Gaussoin and Branham, 1989).

Turfgrasses and trees are an example of plant types that do not exist well together. They negatively affect each other wherever they are found, usually to the eventual exclusion of the turf wherever the competing trees are reasonably well adapted. Earlier in the text, you learned that carbon, hydrogen and oxygen are the most important basic nutrients for plants and that no life exists without photosynthesis. For photosynthesis to occur, light and two of the basic nutrients, water and carbon dioxide, must be present. Turfgrass eventually loses the ecological competition with trees because the trees capture most of the light. Turfgrasses are very competitive with trees in most other respects though, and there is much that we can learn from that competition.

In addition to light, trees may be more competitive for water than turfgrass (Fales and Wakefield, 1981). The biomass of a tree is considerably greater than the collective biomass of the turfgrass plants that can fit onto the surface of its root zone. The tree requires more water and has a deep root system to retrieve water when necessary. Within the turfgrass root zone, the grass may be competitive with the tree for water, but deeper in the soil, the tree has the advantage. Consequently, as the soil surface dries between rainfall events, turf must tolerate the dry conditions while the tree can reach deeper into the soil to find water. It seems that not only are the shoots of the tree more competitive than the shoots of the turf for light, but the roots of the tree are also better competitors for water. However, there is more to be considered.

Tall fescue and Kentucky bluegrass provide serious competition for commercially produced eastern redbud (*Cercis canadensis*) and pecan (*Carya illinoinensis*) trees (Griffin *et al.*, 2007). Two other tree species, silver maple (*Acer saccharinum*) and honeylocust (*Gleditsia triacanthos*), reduced the quality of creeping red fescue (*Festuca rubra*), roughstalk bluegrass (*Poa trivialis*) and Kentucky bluegrass in spite of sufficient water and nutrients (Whitcomb, 1972). Research indicates that turfgrasses are able to compete with or outcompete trees for nitrogen and potassium uptake (Nielsen and Wakefield, 1978; Tworkoski and Glenn, 2001). However, in the production management of peach (*Prunus persica*) trees, some turfgrasses may provide desirable competition because they provide erosion protection while outcompeting other damaging weeds (Tworkoski and Glenn, 2001). Some weeds damage trees and some trees damage turf, and vice versa, by producing chemicals that are toxic to other species.

Many species are known to produce chemicals that are selectively toxic to other species (Barbour *et al.*, 1999). These chemicals are called allochemicals, and they may be produced as root exudates or may be present throughout the plant. The production of chemicals by one plant that selectively inhibit the growth or reproduction of another plant is called allelopathy. Some cool-season turfgrasses, creeping red fescue, perennial ryegrass and Kentucky bluegrass have been shown to produce root exudates that inhibit the growth of the flowering dogwood (*Cornus florida*) tree and forsythia (*Forsythia intermedia*) (Fales and Wakefield, 1981). Bermudagrass and tall fescue are known to produce root exudates that affect the growth of young pecan trees (Smith *et al.*, 2001). Conversely, pine (*Pinus halepensis*) needles have an allelopathic effect on bermudagrass and tall fescue (Nektarios et al., 2005). Allelopathy can sometimes be useful for the suppression of weeds in turfgrass.

Researchers in Michigan found that mulched maple (*Acer* spp.) and oak (*Quercus* spp.) leaves reduced dandelion counts in Kentucky bluegrass by at least 53% (Kowalewski et al., 2009). Researchers in Iowa found that corn (*Zea mays*) gluten meal inhibits root growth of many monocotyledonous and dicotyledonous weed species (McDade and Christians, 2001). Corn gluten meal is commercially available for pre-emergence treatment of some weeds, such as crabgrass (*Digitaria* spp.). Future research is likely to produce more naturally occurring allochemicals for use as alternatives to synthetic herbicides.

The ecology of weed competition

Ecologists have attempted to define plant competition by hypothesizing a number of different but similar competitive plant strategies. They define the term "strategy" as the sum of genetically fixed physiological and morphological adaptations required to conquer a habitat and to persist with optimal use of resources (Schulze et al., 2002). In this case, a strategy is not a plan but a means of adapting to the conditions of a habitat in such a manner that a plant can tolerate the environment and exist and reproduce using the resources available. The first published competitive plant strategy was that of Initial Floristic Competition based on a model by Egler (1954). This early model assumes that a large number of species are present at germination but that as the long-lived slowly developing species mature the short-lived species gradually die off. We can control the plants that live longer than turf by mowing, so we could only hope that succession was that simple.

A more recent theory is that of K- and r-strategies (MacArthur and Wilson, 1967). The K classification refers to plants that are long lived, grow slowly and accumulate large amounts of biomass. An r classification refers to plants that are short lived and reproduce rapidly. A large amount of an r-strategy plant's energy goes into seed production rather than biomass. Presumably, as the K-strategy plants mature, they will eventually out-compete the r-strategy plants, but that is not the case in turfgrass management. In fact, the theory is a good one but nature is considerably more complicated than these two simple categories would imply.

The K- and r-strategy theory has been criticized because it does not take disturbances and stress into account. For that reason, Grime (1974, 1977) proposed a three-strategy model called the C-S-R model. The "C" stands for competitor and constitutes long-lived perennial plants that populate sites with adequate resources and have to endure little stress. The "S" stands for stress-tolerant plants and refers to C-type plants but those that can tolerate sites with few resources and severe environmental stresses. The "R" plants are basically the same short-lived rapidly reproducing plants identified as r in the K- and r-strategy. The C-S-R model also proposes that many plants are not pure C, S or R strategists, but employ combinations of two or of all three strategies.

A third model, the Resource Ratio Model, suggested that all plant species are limited in their distribution by resources (Tilman, 1988). Some of these resources are scarce and the plants that are successful in competing for these limiting resources will prevail. These limiting resources are most often light and nutrients, specifically nitrogen. There is also a fourth model, the Facilitation–Tolerance–Inhibition Model, proposed by Connell and Slatyer (1977). Facilitation, tolerance and inhibition are not plant strategies but pathways of succession. Each requires that changes or disturbances occur in the ecosystem before succession can proceed.

All of these models have merit and are probably far more right than wrong. However, none of them can fully predict the consequences of plant competition. Environments are constantly changing or evolving and disturbances occur without warning. Disturbances are rarely positive occurrences for a turfgrass manager, but they often are positive or neutral in a natural environment. Fires, for instance, rejuvenate grasslands or high winds topple trees in forests making way for new growth and creating diverse habitat that is particularly attractive to many species. For us, however, change is something to be avoided unless it makes the grass greener or denser or the surface more consistent.

Managing weed competition

Some turfgrass weeds are K-strategists and some are r-strategists. For basic identification, the ones we refer to as perennials are K-strategists and the ones we call annuals are r-strategists. Turfgrasses cannot completely replace either of these weed types, so both weed types are consistently antagonistic. Some of these weeds are "S" or stress-resistant strategists. No matter where in the world you are located, there are weeds that take advantage of compacted soil, poor fertility, shade, saturated soil or some other environmental or human-induced stress to colonize a niche in a turfgrass community. Within that particular habitat, they are more competitive than turfgrass. Consequently, we have to change that microenvironment at that specific location to make it more conducive to turf and less conducive to the stress-tolerant weed. Spraying the weed with a herbicide is not the answer. That is only a short-term fix. Changing the environment to favor turf at the expense of the weed is a long-term solution.

The Resource Ratio Model suggests that some resources are likely to limit the population growth of our turf and that those resources are likely to be sunlight and nitrogen. That is nothing new to us. We know that tree removal is often necessary for adequate light with which to grow turfgrass and we are well aware that nitrogen is our most important nutrient. Our goal is to facilitate the growth of the turfgrass population by providing all of the ingredients most conducive for the turfgrass and least conducive for its competitors. By doing that, we create an ecosystem that is intolerant of most invaders and we grow strong turfgrass plants that inhibit the growth of invading populations. We attempt to predict disturbances and avoid them if possible. When they occur, we should rapidly move to repair damaged turf before invasion occurs. Invasion of an open space is inevitable (Fig. 12.2). It is unlikely that spreading turfgrass plants will cover the damage before interloping weed seeds germinate and grow. Such are the basics of controlling weeds in turf.

For many years, we have divided weeds into three categories to facilitate study and control. We call them summer annuals, winter annuals and perennials. So for turfgrass purposes we subdivide the r-strategy weeds into two categories based on their season of strongest vegetative growth. We recognize one category of K-strategy weeds, but we could consider two. It might be a good idea to divide the perennial weeds into categories of cool- and warm-season plants to define their strongest seasons. However, based on what we have learned about ecology, our predecessors seemed to know what they were doing. So let us continue using the traditional groupings for study.

Fig. 12.2. This turfgrass damage was caused by drought. If the turf is not renovated quickly, weeds will soon take over the damaged areas and, once established, they will be difficult to remove.

In order to do battle with a weed you must know what it is. You must at least know whether it is a summer annual, a winter annual or a perennial. The more you know about the weed, its environmental preferences, limiting resources and life cycle, the better you will be able to minimize its encroachment. The first thing you need to do is find a good weed identification site for your area on the Internet, locate a local weed expert and buy a good weed identification and management book or books that cover the weeds most common in your region. Be prepared to identify the weed and study it so that you know how best your turf can compete with it.

Annual weeds live less than a year. They germinate, grow rapidly, reproduce (usually by producing abundant seed) and then die before or during their most stressful season. Biennial weeds are similar to annual plants but they grow vegetatively during one year, then produce seed or tubers and die the next year. Annual weeds are very opportunistic. Dormant seeds of annual plants usually have to experience a period of cold temperature (summer annuals) or warm temperature (winter annuals) before they can break dormancy and germinate. For that reason, summer annuals germinate in the spring and winter annuals germinate in the fall (Fig. 12.3). In addition to the temperature change, there are normally other factors typical for each species that must happen, or factors that must be present, before dormancy is broken and germination occurs. If all other factors are satisfied, all seeds must imbibe (take up) water before they can germinate.

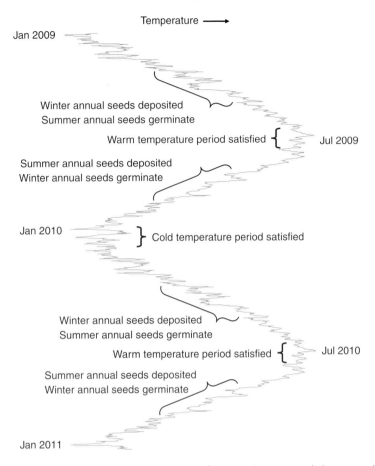

Fig. 12.3. Winter annual weeds flower and seed in the spring. Once the dormant seeds have experienced the extended warmth of summer, they are ready to germinate in the fall. Summer annuals flower and seed in the fall, break dormancy by experiencing the extended cold temperatures of winter and are ready to germinate in the spring.

Dormant seeds of an annual species can remain in the soil for a single year or can be there for several years, depending on the species. Consequently, seeds of an annual weed may accumulate in the soil for quite some time waiting for the right conditions to germinate (Fig. 12.4). We call that accumulation a soil bank. Annual species deposit great numbers of seeds in the soil bank each year. Perennial species also add seeds to the soil bank, but most perennial species produce small numbers of seeds compared with annual species so they have fewer seeds in the bank. When conditions are right, the seeds germinate.

If you use a cultivator of some kind to turn over or disrupt the soil at a particular location in the spring or fall then add water regularly for a few days, seeds begin to germinate. Although perennial species are present, the most pervasive annual species will dominate the initial stand. If you prepared this site as a turfgrass seedbed and applied sufficient seed, the turfgrass should dominate, but weeds, mostly annual weeds, will also be present. Annual weeds are very opportunistic and tend to invade any open area in a turfgrass stand (Lush, 1988b). If you prepare a seedbed and treat the area like a seedbed, weeds will germinate and, in fertile soil, will reach full cover rapidly. Weed seed does not have to be applied; weed seed is already present in the soil bank. Consequently, if you intend to grow turf, the turf must out-compete the weeds for space and for resources. Weed competition can be highly detrimental to the establishment of a turfgrass stand (Maddox et al., 2007). Nonetheless, some weed encroachment is inevitable. The encroachment can

Fig. 12.4. These winter annual weeds (henbit, *Lamium amplexicaule*) germinated in this warm-season turfgrass (bermudagrass, *Cynodon dactylon*) as the grass was nearing dormancy in the previous fall. The weeds could not compete with the turf if the grass was growing vigorously.

be reduced by applying more turfgrass seed at high rates (Busey, 2003). However, as you learned in Chapter 11, intraspecific competition will occur as the stand matures. The more seedlings in the stand, the longer it will take to reach maturity, and the more juvenile the plants, the easier they are to damage or infect. By increasing the seeding rate beyond the normal requirement you are trading one risk for another. The decision should depend on the amount of weed encroachment expected, and that is most affected by time of year.

When you plan to establish or renovate a site, choosing a grass that germinates quickly and grows and spreads rapidly is desirable. In fact, overseeding regularly with a rapidly germinating and vigorously growing species such as perennial ryegrass can be an effective method for reducing the opportunistic weed encroachment that often occurs on athletic fields (Elford *et al.*, 2008). However, rapid germination and establishment does not necessarily indicate the best adapted plants (Gardner and Taylor, 2002). The best adapted species are those that can dominate the community at maturity. Creeping bentgrass (*Agrostis stolonifera*), the most widely used species for golf course putting greens in the USA, is often invaded by annual bluegrass (*Poa annua*). As both creeping bentgrass and annual bluegrass are cool-season grasses that can grow at low mowing heights they are very competitive with each other. The best periods for the establishment of creeping bentgrass are also the best for annual bluegrass. Consequently, unless the site is completely sterilized, annual bluegrass often occurs in seedbeds meant for creeping bentgrass. Even though creeping bentgrass is competitive with annual bluegrass, some cultivars are more competitive than others (Beard *et al.*, 2001; Henry *et al.*, 2005). For that reason, if annual bluegrass competition is expected, a highly competitive cultivar of creeping bentgrass should be seeded, which must also be adapted to the environment and resistant to the stresses common to the site. Choosing the right species and cultivar or the right mix or blend can save considerable maintenance following establishment, regardless of the site or the grass that you are working with.

A dense, mature turfgrass stand is weed resistant. The full canopy cover of dense turf shades the soil, thus preventing some weed species from germinating (Arrieta *et al.*, 2009). Management practices such as irrigation and aeration may also restrict some weed species (Henry *et al.*, 2009). It is possible that some turfgrasses may produce allochemicals that affect certain weeds. Allelopathy is suspected between perennial ryegrass and zoysiagrass (Japanese lawngrass, *Zoysia japonica*) but as yet has not been found between turfgrasses and weeds (Lickfeldt *et al.*, 2001; Zuk and Fry, 2006). Although many management practices are effective for reducing weed competition, the most effective practice is mowing, specifically mowing height.

In a study in the state of Washington, weed encroachment was reduced in perennial ryegrass when mowing height was increased from 1.5 to 2.5 inches (3.8–5.1 cm) (Miltner *et al.*, 2005). In Maryland, a tall fescue mowing height of 3.5 inches (8.8 cm) had less smooth crabgrass (*Digitaria ischaemum*) encroachment than tall fescue mowed at 1.3 or 2.2 inches (3.2 or 5.5 cm) (Dernoeden *et al.*, 1993). In many instances, fertilizer also affects weed encroachment into turfgrass. A combination of fertilizer and high (4 inches versus 2 inches) mowing height reduced dandelion and white clover (*Trifolium repens*) encroachment into a mixture of Kentucky bluegrass, perennial ryegrass and creeping red fescue by 75% (Calhoun *et al.*, 2005). After removing the dandelion and white clover with herbicide, the fertilized plots also had a slower re-infestation than the non-fertilized plots. As you learned in Chapter 11, a mature plant nearly always has a competitive advantage over a seedling. You also learned that low mowing does not allow turfgrasses to mature to their greatest potential. Consequently, high mowing still removes weeds that cannot sustain mowing stress, increases the maturity of the turf so that it is more competitive and creates canopy cover and shade that further discourage weed growth and, in some cases, weed germination.

It is generally believed that nitrogen fertilization favors grasses, especially turfgrasses, over most other plants. However, because turfgrass does not require high amounts of phosphorus fertilizer, high phosphorus fertilization is likely to favor weeds rather than turfgrass. Phosphorus is also the most likely element to encourage eutrophication of surface water, and over-fertilization with phosphorus should be avoided (Soldat and Petrovic, 2008). Many plants, including grassy weeds such as large crabgrass (*Digitaria sanguinalis*), crowfootgrass (*Dactyloctenium aegyptium*) and annual bluegrass respond well to high soil phosphorus levels (Hoveland *et al.*, 1975). Other plants, including large crabgrass and crowfootgrass, respond positively to high soil potassium levels (Hoveland *et al.*, 1975). Low potassium has also been used to reduce dandelion populations (Tilman *et al.*, 1999). Turfgrasses, in general, respond well to high potassium levels, but care should be taken and adjustments should be made if weeds appear to respond as well as your turf to a particular potassium fertilization plan. There are so many combinations of mowing height and fertilizer that could be practiced for weed reduction that it is difficult to choose a course of action. You should begin with the program most likely to encourage your turfgrass species' growth and density, and make small adjustments by observation to discourage weed encroachment. Every situation is likely to require constant monitoring. There are so many different weeds and so many possible habitats that on-site observation and trial and error decision-making is the best course of action. Weed reduction without herbicides takes considerable time and knowledge. That is why most people are reluctant to work with it when so many reliable herbicides are available. An ecological plan, however, is a permanent improvement. A herbicide is only a temporary fix. Constant monitoring also gives you the opportunity to spot spray or to mechanically remove weeds when they are young and much easier to control (Fig. 12.5).

The timing of cultural practices can be very important for weed reduction. We know the life cycles of most weeds and, in some cases, models for

Fig. 12.5. These perennial weeds (dallisgrass, *Paspalum dilatatum*) in a warm-season turfgrass (bermudagrass, *Cynodon dactylon*) have been allowed to mature and spread. They are considerably more difficult to remove now than they were when they were young.

weed emergence have been defined (Danneberger and Vargas, 1984; Fidanza *et al.*, 1996; Kaminski and Dernoeden, 2007). All too often disruptive cultivation is scheduled to coincide with the best periods for weed germination and development or the worst periods for turfgrass recovery. Either situation is detrimental. Choosing the best possible procedures to enhance turfgrass activity is the primary focus of an effective long-term weed control plan. However, all plans have alternatives. It is not necessary to conduct cultural procedures that open a turfgrass canopy during times when weed germination is most likely. If the turfgrass is a spreading type and it has maturity, it can fill in rapidly during periods that are not quite as conducive to weed germination as others. Such periods may include early spring for both warm- and cool-season grasses, before the soil is warm enough for most summer annual weeds to germinate, or late fall for cool-season grasses, after the primary germination period for winter annuals has passed. These periods are also better for seeding (Box 12.1). The high percentage of turfgrass seeds present in the seedbed have a better competitive advantage with weed seeds either before or after the period in which nature has determined that most weed seeds should germinate.

12.4 Biological Herbicides

Biological herbicides could fit into the predator/prey relationships that will be presented in Chapter 13. However, they are not entirely natural.

> **Box 12.1. When are conditions best for seeding a cool-season turfgrass?**
>
> Let us assume that you want to establish a home lawn with a cool-season (C_3) grass and that you can choose the time most conducive to do it. We normally recommend seeding a cool-season grass during the fall rather than the spring because the competition with summer annual weeds in the spring can be sufficient to slow development of the cool-season turfgrass until summer. During the summer, annual weeds have a particularly strong advantage over cool-season turf. In the fall, there is no competition with summer annuals, only with winter annuals and perennials, with which the turfgrass is highly competitive. Consequently, you would probably suppose that the best time to seed would be as soon as summer temperatures began to cool into the range conducive for cool-season turfgrass growth, so that the grass would have as much time as possible to mature before winter. Although that plan is intuitively correct, there is actually a better time.
>
> Let us assume that your major weed competition in the fall will be from annual bluegrass (*Poa annua*).
>
> Research demonstrates that annual bluegrass can germinate at nearly any time during the fall and spring, but that its germination is greatest in the early fall (Kaminski and Dernoeden, 2007). Therefore, if we seed in the early fall and we have a seed bank relatively high in annual bluegrass our grass will experience severe competition from the weed. We need an alternative method.
>
> Instead of seeding the turfgrass in the early fall, we can prepare the seedbed exactly as we normally would, keep it moist just as we would to germinate turfgrass and wait for the annual bluegrass to germinate. Because you have created and maintained a perfect seedbed, most of the weed seed present will germinate. After 3 to 4 weeks you kill the weeds with a nonselective herbicide and seed your turf. By that time, you have removed most of the weed competition and there is still plenty of time for the turfgrass to grow and develop before the winter.
>
> If using a nonselective herbicide is not possible, hand removal of weeds is great, or tilling the weeds under will work reasonably well.

Biological herbicides are formulations of living organisms that prey on weed species in a turfgrass community. Biological weed control is a synthetic application of these organisms to the community with the intention of causing harm to the plants they selectively use for food or, we could say, the plants that they selectively infect. Biological herbicides are usually disease-causing pathogens.

Natural toxins and biological herbicides should not be confused with one another. The allochemicals already presented in Chapter 11 are toxins. However, we are more familiar with toxins that potentially affect us. Plant products such as caffeine and nicotine are toxins that kill insects and, in great enough concentrations, also kill us. Although caffeine is considered to be a safe food product, a concentrated dose could be fatal. Many fruit seeds contain cyanogenic glycosides that can be fatal in high enough doses. Some people believe that naturally occurring toxins make harmless pesticides when used as directed. Others believe that natural toxins are tolerable but synthetic toxins are not. Some think that we should ban them all. This is a politically charged issue, as is the use of disease-causing agents. However, pathogens that attack plants rarely attack humans and those that did would not be used as biological control agents. Consequently, the use of a biological control agent is considered by most to be preferable to the use of a synthetic herbicide. For that reason, recent research has begun to target the use of biological herbicides as an environmentally sound means of weed control.

The organisms in biological herbicides may be present in the turfgrass community before an application is made. However, they are rarely present in sufficient numbers to infect and kill weeds until environmental conditions are exactly right for infection. Actually, if the pathogen was not present before application, it is likely that the organisms applied will not survive long enough to cause any real damage to the target weeds. The greatest advantage of a biological herbicide is that it is a naturally occurring organism. That advantage is also a detriment. If the environment was conducive to support pathogen numbers great enough to cause sufficient weed damage, the organisms would already be present in high numbers. Consequently, the herbicide has to be applied when the environment is sufficient to sustain it for a period long enough for it to infect and kill the target weeds to which it was applied. For that reason, biological herbicides often fail. Synthetic herbicides rarely fail. Turfgrass managers are reluctant to spend money on a biological herbicide that could fail when they have a synthetic herbicide in which they are confident. They make that decision, not because they are not concerned for our natural environment but because they have little choice from an economic viewpoint. Reliable biological herbicides are needed so that we have economically as well as environmentally sound alternatives to synthetic herbicides. Applications of biological herbicides when conditions are conducive, and of synthetic herbicides if they are not, is another possible management strategy (Bewick, 1996).

The fact that the organisms used as weed pathogens in biological herbicides can only exist in damaging populations when environmental conditions in the habitat are particularly conducive for disease is not only a restriction, it is also an advantage. Many of the pathogens that are used as biological herbicides are classified in the same genus as the pathogens that cause disease in turf. *Sclerotinia sclerotiorum*, for instance, has been successfully tested in controlled environments as a biological herbicide for dandelion (Riddle *et al.*, 1991). *Sclerotinia homeocarpa* causes dollar spot, a disease of many turfgrasses, with particular severity on creeping bentgrass and Kentucky bluegrass. *Puccinia* spp., the same genus of organisms that causes rust on perennial ryegrass, Japanese lawngrass and other turfgrasses, has demonstrated potential for the control of nutsedge species (*Cyperus* spp.) (McCarty and Tucker, 2005). *Bipolaris* spp., from the genus of pathogens that causes leaf spot on turfgrasses, and *Pyricularia* spp., from the genus that causes gray leaf spot on turfgrasses, both have potential for use in control of goosegrass (*Eleusine indica*) (Figliola *et al.*, 1988). Finally, a biotype of *Xanthomonas campestris*, the bacterial pathogen that succeeded in eliminating 'Toronto' creeping bentgrass as an economically viable cultivar, has been suggested for use in reducing populations of annual bluegrass in bermudagrass (Johnson, 1994). With the exception of *X. campestris*, these pathogens are all fungi. However, fungi, and especially bacteria, are well known for their ability to develop resistance to fungicides and antibiotics, and for their ability to rapidly evolve to infect alternative hosts. Therefore, if these pathogens were able to exist in the ecosystem in great numbers for a long period, there is a chance that they could adapt to infect turf. For that reason, we are happy to see them die after they kill the weeds.

12.5 Chapter Summary

Ecological weed control is not easy. It is not something that you think about, develop a plan for and proceed with successfully. Instead, it is something that you think about, develop a plan for, and constantly monitor, customize and adjust. It is necessary to identify the weed species at your site and to study them. They should be classified into winter annuals, summer annuals and perennials. You should learn how those weeds differ from your turf. Which nutrients do they need in the greatest amounts compared with the turf? How much water do they require? Can they stand low mowing? Do they need light to germinate? Is there some factor such as compaction or low nutrition for which they are better suited than the turf? These questions and others need to be answered before you can provide the situations that favor the turfgrass and discourage the weed. Obviously, using herbicides is easier, and is necessary when near-perfect turfgrass surfaces are required. However, habitat management is far more permanent than herbicide applications.

Choosing the right turfgrasses for the environmental conditions and customer requirements is necessary for successful habitat management. Blends and mixtures are normally more competitive than monocultures, but each species and cultivar must be adapted to the environment for the combination to be successful. Rapidly germinating and vigorously spreading species are desirable, but the species that are most competitive at maturity are more important for long-lasting weed suppression. Certain weeds can often survive conditions that are not conducive to vigorous turfgrass growth. Such conditions may be compaction, poor drainage, low fertility or soil salts. If the turfgrass is expected to compete, these conditions will have to be permanently repaired. Otherwise, the weed will always have the upper hand.

Most weeds cannot persist under regular mowing. However, there are enough that can to make our job very difficult. Weeds are opportunistic and tenacious. They are considerably easier to remove when they are young than after they mature and spread. Higher mowing heights tend to discourage most weeds and provide a competitive advantage for the turf. A dense turfgrass canopy shades the soil and makes it difficult for some weeds to germinate and difficult for other weeds to survive turfgrass competition. Nitrogen fertilization usually favors the turfgrass, but phosphorus fertilizer often favors weeds. Potassium is highly desirable for turfgrass growth, but too much potassium may lead to weed encroachment. Finally, turfgrass damage needs to be addressed quickly before weeds have an opportunity to get established in damaged areas.

For the most part, ecological weed control is a constant process of addressing problems and devising solutions. It requires perpetual study, observation and problem solving but the long-lasting benefits are particularly rewarding.

Suggested Reading

Busey, P. (2003) Cultural management of weeds in turfgrass: a review. *Crop Science* 43, 1899–1911.

McCarty, L.B. and Tudker, B.J. (2005) Prospects for managing turf weeds without protective chemicals. *International Turfgrass Society Research Journal* 10, 34–41.

Soldat, D.J. and Petrovic, A.M. (2008) The fate and transport of phosphorus in turfgrass ecosystems. *Crop Science* 48, 2051–2065.

Suggested Websites

Georgia Turf (2010) Available at: http://commodities.caes.uga.edu/turfgrass/georgiaturf/index/ (accessed 1 March 2010).

University of California (2010) UC IPM Online: Statewide Integrated Pest Management Program. 'How to Manage Pests.' Identification: Weed Photo Gallery. Available at: http://www.ipm.ucdavis.edu/PMG/weeds_intro.html (accessed 10 August 2010).

Virginia Cooperative Extension (2010) Virginia Tech Weed Identification Guide. Available at: http://www.ppws.vt.edu/scott/weed_id/rightsid.htm (accessed 1 March 2010).

13 Managing Competition Between Turf and its Pests

> **Key Terms**
>
> **Integrated pest management** (IPM) is a plan that combines different pest-management strategies to manage pest damage below a threshold level.
>
> A **threshold level** is the level of turfgrass damage that your customer is willing to accept. It is a turfgrass manager's job to maintain the turfgrass stand below the threshold level determined by the customer.
>
> A **single tactic system** is the use of a single strategy to control a turfgrass predator. Single tactic systems can be very reliable in the short term but they are always temporary. The most common single tactic systems for predator control are breeding for genetic resistance and pesticide applications.
>
> **Predation** is the consumption of one organism by another and in its broadest sense includes parasitism.

13.1 How Can We Affect Relationships Between Predator and Prey?

For many organisms, turfgrass is food. The grasses from which turfgrasses were derived were, and in many cases still are, forages for cattle, sheep and other mammals. In fact, cattle, sheep, goats, horses and a few other animals served as the first mowing devices. That is not the case today. We rarely combine turfgrass and forage grass for aesthetic or functional turf. However, turfgrass is still food for insects and microorganisms. Therefore, turfgrass is a resource for those organisms and, as you learned earlier, an increase in resources usually means an increase in the organisms that use them. When turfgrass is lush, green and dense, as we like it, turfgrass pests become numerous, sometimes lethal. For that and other reasons that you should already understand, as turfgrass becomes more perfect, its maintenance becomes increasingly difficult. As it reaches its carrying capacity, especially a carrying capacity that has been artificially induced by our management, it becomes quite vulnerable to damage. Management inputs beyond those that occur naturally increase the maintenance required of us severalfold. Turfgrass management and its maintenance do not sustain a 1:1 linear relationship. Every positive management input requires severalfold more maintenance. As turfgrass density increases, more intraspecific competition occurs. The severe competition among grass plants causes each plant to become weaker. The weaker plants are more easily damaged. In addition, the presence of a substantial source of food, the nice dense turf, attracts more predators. For that reason, the most perfect turfgrass stand is the stand most likely to crash.

13.2 Predator/Prey Relationships

The natural environment is one huge predator/prey relationship. All organisms, including humans, compete within their species for space and resources. The individuals best adapted to the local environment survive and multiply. However, because we, as humans, are at the top of the food chain, we tend to forget that everything else is prey. Although humans may be attacked and eaten by large animals, such a case is rare and generally only occurs because the animal has lost its fear of us or was starving. Most large animals are hunted, and as long as that continues they will fear us and avoid us instead of considering us as food. Every other plant, animal or microorganism, though, is considered food for something. Every organism has to either photosynthesize or eat, and there are a lot of insects and microorganisms that eat turf.

The typical predator/prey relationship is based on random encounters. As a prey population grows, the likelihood of its predators encountering prey increases. As predators encounter more prey the predator populations become healthier, fewer die and more reproduction occurs. The predator populations grow because there is an abundance of prey. However, as the predator populations increase and more predators encounter prey, the prey population becomes unhealthy, many die and reproduction slows. The decline in prey is then followed by a decline in predators and the cycle begins again (Fig. 13.1).

Of course we know that the predator/prey relationship is not random. Predators seek prey and are only found in habitats where prey exist. However, the random model is a good illustration of how the relationship works. In addition, predators may prey on more than one species and concentrate their efforts on an alternative species when a primary species declines. In that case, although the population of primary prey has declined, predator numbers may remain the same, but the predators will be more concentrated in habitats conducive to the alternative prey species than in habitats conducive to the primary prey. Such a response is common in fall army worms (*Spodoptera frugiperda*) in regions where wheat (*Triticum aestivum*) is plentiful. Once the wheat crop is harvested, the army worms may use turfgrass as an alternative food source. Regardless, the random predator/prey model generally fits managed turfgrass sites very well, mostly because sites managed for turf are nearly all turfgrass. Therefore, alternative prey or predators of other species are only present in the turfgrass habitat in very small numbers.

Human impacts on predator/prey relationships

Humans have a tendency to focus their crop production efforts on a very narrow scale instead of considering the crop, in our case turf, as part of a much larger ecosystem. We forget that there is a multitude of activity occurring on our site. Even a somewhat inert pure sand system has a huge microbiological community (Bigelow *et al.*, 2002; Elliott *et al.*, 2004). These microbial populations are affected by local environments in the same way as turfgrass (Giesler *et al.*, 2000). They live, reproduce and die in a turfgrass community. They compete with each other, prey on each other, and sometimes prey on turf. There are a number of animals and birds, including humans, which include turfgrass sites as part of their habitat. A turfgrass site is part of a much larger ecosystem and is affected by that system. Turfgrass is part of the human community. Turfgrass, in fact, does not exist in the wild. Forages exist in the wild but turfgrass, by definition, only exists because of human intervention.

Humans have a huge impact on turf. Human activities on managed turf determine at least a portion of the predator/prey relationship in the community. The more highly managed the turf, hence the most desirable turf, the more impact that humans have on it. In a natural environment, the predator/prey relationship exists in relative balance. However, we can use knowledge, experience and creativity to influence the predator/prey relationship on managed turf, and that is what we are about to consider in the rest of this chapter.

13.3 Managing Turfgrass Predators

There are many organisms that we consider to be turfgrass pests. Some organisms are competitors, like weeds, and others are predators, like insects and disease pathogens. In Chapter 12, we discussed interspecific competition primarily from other plants such as dandelions (*Taraxacum officinale*)

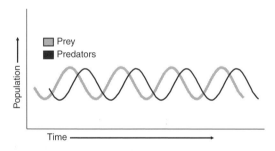

Fig. 13.1. In a random predator/prey relationship, predators increase after a sufficient lag time following a prey increase. However, as the predators feed on the prey, the prey population begins to decline. As the prey resource declines, the predator populations decline and the cycle begins again. The random model of predator/prey encounters fits turfgrass populations fairly well because where management is practiced, turf is nearly all turfgrass. Consequently, the turfgrass is the only prey present in large numbers and most of the predators present are predators of turf or predators of turfgrass predators.

and crabgrass (*Digitaria* spp.). In this section of Chapter 13, I will briefly introduce integrated pest management (IPM), a system of pest control strategies that you are probably familiar with. Turfgrass IPM is designed to guide the planning of pest control programs for both competitors and predators. Its use is desirable for designing weed control plans as well as plans for predator control. The ecological management of interspecific competition is a basic part of IPM. Both management systems take similar approaches to IPM, but IPM has a much broader scope. It would be valuable for you to consider how the ecological management of interspecific competition presented in Chapter 12 fits into a complete IPM program. However, in this chapter we are primarily concerned about how predator/turf relationships work. In the next section, you will see how the ecological principles of turfgrass/predator relationships can affect management plans that enhance an IPM program.

In the USA and many other countries, IPM, or something similar, is the accepted method for managing turfgrass pests. Most turfgrass managers practice IPM, whether they realize it or not. The processes of IPM are multifaceted and cover practically every technique for pest control. Our study will be more specific to natural factors that influence turfgrass pests and how to use them effectively. However, the techniques presented would probably be the most important part of an IPM plan.

When developing or revising an IPM plan it is not only the technology that makes a difference, it is also the philosophy. The IPM process is a combination of strategies used to maintain a particular turfgrass site under a desired threshold level of pest damage. The threshold level is determined by the end user, your customer. Those thresholds are less likely to be based on nature conservation than they are on economics. Sadly, most people are 100% in favor of nature conservation as long as it does not affect what they want to do or cost them money. Fortunately, managing turfgrass to near perfection is expensive in both time and money. Therefore, most customers are happy to endure some weed encroachment and a little insect or disease damage rather than pay for perfection. The exceptions are bowling greens, putting greens and tennis courts, where the condition of the turf can drastically influence play. On those areas, the damage threshold is 0%, meaning that no more than 0% damage will be tolerated. In other words, the turf is expected to be perfect. We can't do perfect but we can come awfully close if we use every strategy available to us, including pesticides.

Turfgrass IPM

As mentioned earlier, IPM is a collection of strategies incorporated into a pest management plan. One of the philosophies of IPM is that all alternative strategies should be employed before pesticides are applied. At our current state of knowledge, maintaining turfgrass at a 0% damage threshold or close to it is not possible without pesticides. Therefore, pesticides are always part of the plan where perfection is desired. If our customers demand perfection then we have no choice but to employ pesticides. We can encourage our customers to accept something less than perfection but we are not the ones who determine the threshold, they are. So if we are to maintain our profession indefinitely we need to strive to manage perfection without pesticides.

According to most pest management experts, there are six basic strategies always considered when developing an IPM plan. They are:

1. Regulatory strategies: government intervention to prevent damaging pests from being introduced into an area, for the removal of noxious weeds, or to restrict products or procedures that encourage pest population growth.
2. Genetic strategies: the development of cultivars or species with genetic resistance or other adaptations that make the target organism, our grass, more competitive with its pests.
3. Cultural strategies: these include mowing, irrigation, fertilization and soil cultivation to help our plants be more competitive or to remove conditions that encourage pests.
4. Physical strategies: the actual capture of insects or animals, hand removal of weeds, dew removal, equipment cleaning and other labor-intensive practices to discourage the spread of disease.
5. Biological strategies: the use of one organism to affect another, such as a natural predator or a competitor. Some would include naturally occurring toxins in this category, but I will not.
6. Chemical strategies: the use of naturally occurring or manufactured chemicals to kill pests.

Many of these strategies can be employed effectively to discourage a pest. By combining strategies, the pest control program is more continuous, more reliable and more effective. A good IPM program is a long-term solution, not a temporary treatment. The opposite of an IPM program is a single-tactic system.

Single-tactic systems

A single-tactic system is the use of a single strategy to control a turfgrass predator. Single-tactic systems can be very reliable in the short term, but they are always temporary. The most common single-tactic systems for predator control are breeding for genetic resistance and pesticide application. Both of these strategies are included in IPM planning because they can provide excellent control. However, neither strategy is reliable long term.

Turfgrass species are attacked by many different organisms, depending on climate and use. However, a turfgrass species in a given region usually only has major problems with a small number of particular insects or pathogens. Bermudagrass (*Cynodon dactylon*), for instance, is somewhat resistant to most diseases but is highly susceptible to spring dead spot (*Ophiosphaerella herpotrica* and others) disease in regions where winters are fairly cold. Tall fescue (*Festuca arundinacea*) is highly susceptible to brown patch (*Rhizoctonia solani*) disease and St. Augustinegrass (*Stenotaphrum secundatum*) and fine fescues (*Festuca* spp.) tend to attract chinch bugs (*Blissus insularis*). Kentucky bluegrass (*Poa pratensis*) is often damaged by bluegrass billbug (*Sphenophorus parvulus*) and perennial ryegrass (*Lolium perenne*) by gray leaf spot disease (*Pyricularia grisea*). Breeders have or may be able to develop cultivars resistant to these highly important diseases and insects but, over time, the insects and diseases may adapt to feed on the resistant cultivars, or the old diseases or insects may be traded for new ones. Therefore, the breeding of resistant cultivars requires continuous attention, and breeding alone is not completely effective for predator control.

When Kentucky bluegrasses, for instance, were developed to resist melting out disease (*Helminthosporium* spp.), the most important disease of the species at the time, other diseases, summer patch (*Magnaporthe poae*) and necrotic ring spot (*Leptosphaeria korrae*), went from being minor predators (diseases) to major ones. For some unknown reason, St. Augustinegrasses that are resistant to chinch bug damage are more susceptible to St. Augustinegrass decline, one of the very few viral diseases of turfgrass. St. Augustinegrass decline is generally considered to be the only economically important viral disease of turf. The St. Augustinegrass decline virus also attacks centipedegrass (*Eremochloa ophiuroides*), resulting in the disease called centipedegrass mosaic. In the USA, gray leaf spot was a disease of St. Augustinegrass and tall fescue, but it was not a major disease on perennial ryegrass until recently (in the last 15 years or so). The gray leaf spot pathogen had to adapt to prey on ryegrass. Breeding and genetics may be our strongest line of defense against predators, but because predators readily adapt to alternative hosts, it is not always effective as a single-tactic system.

Insects and microorganisms are highly adaptable. For that reason, predator resistance to a single-tactic system is likely to occur over time. Chemical strategies are extremely effective for predator control but, as was mentioned in Chapter 12, their effects are quite temporary. One application must be followed by another and another. If, after a period of intense chemical control, it appears that you have eliminated the problem, just wait a year or two and it will probably reappear. It is highly unlikely that you will ever be successful in completely eliminating a turfgrass predator. They will nearly always come back, and when they do, they may have developed resistance to your previous practices. That is not an uncommon occurrence.

Multiple chemical applications encourage predator resistance. Because every individual of a species usually has a unique genetic composition, the likelihood that a few individuals will be resistant to a pesticide application is high. With insects and especially with microorganisms that breed rapidly, a resistant population can occur. A resistant predator population is most likely to occur following multiple applications of a single pesticide or different pesticides that use the same mode of action.

Following the first pesticide application a small portion of predators survive. Survival includes a portion of the susceptible individuals and all of the resistant ones. The remaining predators reproduce, resulting in more resistant individuals. As more susceptible individuals are eliminated by recurring pesticide applications, the population comes to comprise mostly resistant individuals (Fig. 13.2). By using multiple applications of a chemical with the same mode of action you have almost eliminated intraspecific competition among resistant and susceptible predators. Because of your actions, only resistant predators exist. Continuing applications

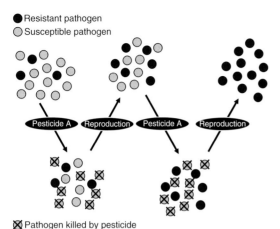

Fig. 13.2. Predator resistance. With the first application of a pesticide, much of the predator population will be killed, but many are likely to survive. A few of those that survive may be resistant to the mode of action of the pesticide. These resistant individuals grow and breed and the resistant population builds. With repeated applications of pesticide with the same mode of action, most of the predators remaining will be resistant and a pesticide with a different mode of action will be required to kill them.

will not only reduce intraspecific competition, they will also kill beneficial organisms in the same class as the predators, thus inhibiting interspecific competition as well. The result will be a very high population of resistant predators. Continuing applications of the same products will further increase the resistant predator population and make the damage worse. If you are in a situation where you have to use pesticides to meet customer expectations, you must use them wisely and carefully. Do not encourage a resistant predator population. Make sure that you kill what you intended to kill and do not kill unrelated organisms, your turf for instance, or your neighbor's garden. Remember that insecticide and fungicide applications kill not only predators, but related beneficial organisms. These products are broad-based to kill multiple insect predators and fungi. When you kill the pathogen, you also kill some of its competing beneficial organisms. Be careful what you apply and when you apply it.

Knowledge is a weapon

In order to effectively manage a pest, information and knowledge are required. Knowledge is a valuable weapon and, in this case, is even more important than experience. If you intend to do your best to manage a turfgrass pest, predator or competitor, you must have an intimate knowledge of that organism. Books, libraries, conferences, newsletters, magazines, Internet sources and personal contacts are all sources of information and knowledge. You may like to do things yourself, make your own decisions and avoid large gatherings, but if you intend to do the best possible job with your turf, you have to have sources of information, including conferences and personal contacts (Box 13.1).

Every turfgrass predator has a history. More importantly, it has a life cycle, conditions that it likes and those that it dislikes, susceptibility to certain other predators, and symptoms and signs that identify its presence. You must be familiar with those characteristics in order to control it. Disease and insect damage do not occur randomly,

---

Box 13.1 Building a "network" of information sources.

You can't know everything, and conversations with people experiencing the same problems and those who have solved them or done research on them is invaluable. Business people call an individual's list of personal contacts a network and they consider networks to be extremely valuable sources of information. Those with the best networks can gather information about any number of subjects from all over a country, perhaps the world, with a few phone calls. They can effectively screen potential employees, learn about new laws, find a new job, keep track of industry developments and investigate many other situations using their network. A network is an extremely valuable resource. A good network takes years to develop, but it grows rapidly following your first few contacts as others refer contacts to you. Developing a network is not the least bit difficult, you simply have to be willing to meet new people and to listen to new ideas. Most importantly, you have to belong to organizations and attend the meetings where you are most likely to meet other people in your profession.

they occur for a reason, and they are, in most cases, predictable. We do not know everything that we need to know about turfgrass predators, partially because they are so easy to kill using pesticides and partially because they are biotic organisms that easily adapt to multiple environments. However, we are always learning more about turfgrass pests and if you want to or are forced to use fewer pesticides, you have to learn everything that you can about a particular pest to manage it effectively.

Keep watch on your site

You may be a parent or an older brother or sister. You may have been a baby-sitter. During your life you probably have had at least some small responsibility for looking after young children. When you are responsible for a young child, you have to keep a close watch on the child and the environment around that child. Potentially dangerous situations happen fast and without warning.

Managing turfgrass is similar but easier than having responsibility for a small child. There is usually some warning before a predator attack, but you must be paying attention to see it. When watching a child you try to place dangerous objects out of reach and provide a harmless environment. You do the same with turf, but as with a child, you realize that there is no such thing as a completely harmless environment. You have to keep watch. Where there is turfgrass there will always be turfgrass predators.

As a manager, you must develop a system for monitoring your entire site. The site, large or small, requires daily inspection by someone knowledgeable in the symptoms and signs of invading predators. As with any human disease, early detection is the best possible means of correction. Every large turfgrass area has predator "hot spots". A hot spot is an area that is most vulnerable, often for reasons unknown, to a particular predator or pest. Specific diseases or insect damage will usually occur in a hot spot before they occur on the rest of the site. Therefore, the hot spots are the places you monitor most closely. They act as early warning devices once you identify them.

Part of identifying hot spots for certain diseases and insects is record keeping. Records can be journals, maps or photographs. Photographs are under-utilized in turfgrass management by all but instructors. If a turfgrass manager is to train assistants and laborers to be on the lookout for certain symptoms and signs of impending predator problems, photographs can be a very valuable source of instruction. More importantly, a wide-angle photograph is an instant record of location and the photograph can be labeled easily for date, either automatically by the camera or digital file or by hand. Close-up photographs can be included to show the intensity of the damage and to help with identification. A digital photograph is an instant source of information that can pinpoint the location where an assistant found a potential problem that you need to observe, and it is a record that may help you with early detection of similar occurrences in the future. With photograph in hand, you do not need the assistant to guide you to the site.

You will probably want to keep a map of locations where problems have occurred so that you know where to look for similar problems in the future. You may also want to keep a journal of specific techniques that you used to attempt a cure. Record those attempts that worked and those that didn't for future reference.

Watch for the environmental conditions that precede predator problems. Insect problems generally occur by life cycle. Usually there are only one or two life stages of insects that are damaging to turfgrass and they only occur during certain periods of the year. Therefore, monitoring and some control techniques are only needed for a few weeks or less each season. You have to know the life cycle of the insect to determine when these periods occur.

A good turfgrass manager is always aware of changing environmental conditions and what those conditions mean for possibilities of predator damage. Turfgrass pathogens are always present in some form or another. However, turfgrass diseases only occur when the environment is more conducive for the pathogen population growth than it is for turfgrass growth. There are usually multiple environmental factors that affect disease pressure (Vargas et al., 1993). The period of likely disease occurrence is extended when the turfgrass is weakened by poor management or other stress (Vargas, 1994). Consequently, turfgrass health is an important part of disease management as it is of weed management. That is usually not the case where insects are concerned. Some damaging insects may seek out the healthiest turf for laying their eggs and causing problems. Because insects attack according to stage of life cycle and pathogens attack during certain environmental conditions, there is some warning as to when to expect insect and disease problems. That is one reason why record keeping, especially recording the dates of occurrence, is so important. A good plan for predator control hinges on a good monitoring program and good record keeping.

13.4 Affecting Relationships Between Predators and Turf

The ecological concept of predator control is to use the predator's natural enemies against it. With that statement, the first thing that probably comes to mind is biological control. We could use nematodes to control white grubs (*Phyllophaga* spp., *Cyclocephala* spp., *Popilia japonica* and other beetles of the order Coleoptera) or bacteria to control fungi but, as you learned in Chapter 12, biological control is not particularly reliable. Although biological control may be part of your pest control program, it is the abiotic enemies of a predator that are most reliable in facilitating its demise. Just as we can influence the local environment to favor photosynthesis, respiration and transpiration, we can influence the local environment to reduce predation. However, throughout any discussion of predator management, it must be remembered that an increase in turfgrass is an increase in its predators' resources. More turf attracts more turfgrass predators and healthier turf feeds more predators. Therefore, it might seem that from an ecological standpoint we cannot have healthy dense turf without high predator numbers. Remember, though, that turfgrass and its predators do not share the same niche. Turfgrass may be a source of nutrition for the predators, but there are other biotic and abiotic factors that determine whether or not a turfgrass predator can live in a turfgrass community.

Cultural practices that affect predators

Mowing height alone, irrigation alone, nutrition alone, thatch reduction alone or other cultural practices alone will probably not have a significant impact on turfgrass predators but, in combination, management of cultural practices for specifically discouraging targeted pests can provide adequate protection in many cases. Spring dead spot disease pathogens, for instance, are a major predator of bermudagrass in temperate regions (Fig. 13.3). The disease is difficult to control with fungicides, although low mowing, 0.5 inches (13 mm), seems

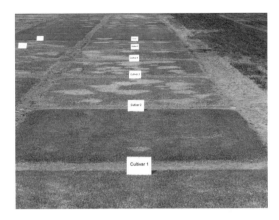

Fig. 13.3. Spring dead spot is a common disease of bermudagrass (*Cynodon dactylon*) in temperate climates where winters are cold. The damage that you see on cultivars 3–6 is severe. The turf in the patches is dead and will not recover. Cultivars 1 and 2 are resistant to the disease, but if conditions are exactly right they will also be infected. However, with proper management, the occurrence of the disease in cultivars 1 and 2 can be maintained below a low threshold level with no pesticide applied.

to help relieve spring dead spot pressure compared with higher heights (1.5 inches, 38 mm) of cut (Martin *et al.*, 2001). The spring dead spot pathogen seems to prefer a relatively basic soil pH, so N fertilization with an acidifying fertilizer such as ammonium sulfate tends to discourage the disease. However, N applications late in the growing season are likely to encourage the disease (Dernoeden *et al.*, 1991). Vertical mowing and aeration for thatch reduction also reduces the susceptibility of bermudagrass to spring dead spot (Tisserat and Fry, 1997). None of these practices alone is sufficient to provide protection, but together and in combination with the proper irrigation to help mask symptoms, bermudagrass turf can often be managed below damage thresholds with a combination of cultural practices and no pesticide.

In rare cases, a single cultural practice can substantially reduce predator damage. Bermudagrass may contract dollar spot (*Sclerotinia homeocarpa*) in late spring, but an application of N fertilizer is often enough to relieve disease pressure (Carrow *et al.*, 1987). Nitrogen fertilizer also helps to reduce dollar spot in Kentucky bluegrass, hybrid bluegrass (*Poa pratensis* × *Poa arachnifera*) and perennial ryegrass (Treadway *et al.*, 2001; Teuton *et al.*, 2007). It has also been known to reduce anthracnose on annual bluegrass (*Poa annua*) and is recommended as a control measure for rust disease (*Puccinia* spp.) on multiple grasses (Vargas, 1994; Inguagiato *et al.*, 2008). Nitrogen fertilization affects many turfgrass diseases. Sometimes, the source (organic or inorganic) of N fertilizer has an effect on disease (Landschoot and McNitt, 1997). In most cases, N fertilization discourages disease, but in the case of insects, the opposite may be true. Fertilized St. Augustinegrass, for instance, is more susceptible to chinch bug than non-fertilized St. Augustinegrass (Busey and Snyder, 1993). Although N fertilization is often a positive means for reducing disease pressure, it has also been known to encourage gray leaf spot disease on perennial ryegrass (Williams *et al.*, 2001).

Some insects, such as black turfgrass aetenius (*Aetenius spretulus*), prefer irrigated turf over dry turf (Jo and Smitely, 2006). In some cases, a micronutrient has an effect on a predator, as in Mn suppression of take-all patch disease (*Gaeumannomyces graminis*) in creeping bentgrass (*Agrostis stolonifera*), but that is rarely the case (Heckman *et al.*, 2003). Then again, soil pH often affects diseases such as creeping bentgrass dead spot (*Ophiosphaerella agrostis*), spring dead spot and take-all patch (Couch, 1995; Kaminski and Dernoeden, 2005). The point of all of this is that random assumptions are not a part of predator management. Predators and their prey must be managed specifically. You also must be familiar with how your management of one predator affects another. Irrigation management, for instance, that has a positive effect on the control of one predator may have a negative effect on the control of another (Jiang *et al.*, 1998). Also, you must monitor and record your efforts and their effects. For many years, chinch bug damage was considered to be mostly a problem in drought-stressed St. Augustinegrass but recently that has been questioned (Vazquez and Buss, 2006). Researchers who found no difference in chinch bug damage on irrigated and non-irrigated St. Augustinegrass suggest that chinch bug damage may simply look like drought damage or that the chinch bug populations breed more rapidly in the warmth of drought stressed turf. Acidifying fertilizers help to prevent earthworms and microarthropods in some instances (Potter *et al.*, 1985), but not others (Backman *et al.*, 2001). Scientific or professional advice is a great source of good information, but you are responsible for developing the programs that work best on your site. Make sure to monitor your programs, especially when changes in activities are made. There may be conditions on your site that require different management from that which is normally recommended.

Genetic resistance to predators

One means of enhancing turfgrass resistance to predators is through the use of mixes and blends. Many predators, especially disease pathogens, are quite specific to the cultivars and species of turfgrass that they prefer (Rose-Fricker *et al.*, 1997). In addition, cultivars of turfgrasses and turfgrass species differ in their level of susceptibility to predator attack. By mixing or blending we can provide variable predator preferences and plant resistance to help slow the spread of disease or insect damage (Fig. 13.4). Many diseases spread plant to plant through extended hyphae. If all of the surrounding plants are resistant, the disease has nowhere to go. In addition, if only a portion of the stand is affected, the visual quality and playability of the turf will be better than that of turf in a stand of uniformly susceptible plants (Abernathy *et al.*, 2001; Brede, 2005). Presumably, predation by common insects or pathogens would remove the susceptible species or cultivars from the mixed stand, but that is not always the case (Golembiewski *et al.*, 2001). Often, the predator damage is not severe enough to kill the plants affected. If the damage is slowed sufficiently by the mixed grasses, many susceptible plants may

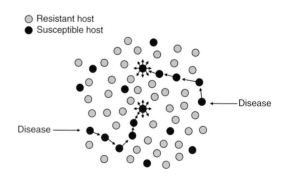

Fig. 13.4. Turfgrass mixes and blends are likely to contain species or cultivars with variable resistance to predators. Because susceptible plants are interspersed with resistant ones, damage spreads more slowly, fewer plants are damaged and visual quality and playability is less affected than it would be in a single species, single cultivar stand.

escape and reproduce, thereby maintaining the mixed genetics in the stand.

Turfgrass breeding programs are constantly striving to develop new varieties that are resistant to a species' most common predators. Much progress has been made toward both disease and insect resistance in grasses, although many predator problems remain a significant challenge.

Biological control of predators

Biological control of predators is deemed preferable to applications of synthetic pesticides by most people. Whether or not a synthetic (that is, an artificial) application of a biological organism is really any different from a synthetic application of a manufactured pesticide is a subject for debate. However, society appears to believe that biological control mechanisms are less dangerous to the human environment than manufactured pesticides, and public perception is important. Consequently, development of biological mechanisms for turfgrass management has become a subject of interest. The greatest success of biological control methods in turfgrass has not been in the application of biological agents but in the use of seed-borne fungi called endophytes.

Endophytes

Endophytic fungi are a group of microorganisms that populate a turfgrass plant and can be borne by a turfgrass seed. Consequently, the fungi in the seed grow with the plant. These fungi form a symbiotic relationship with turfgrass and other plants much like the relationship between plants and mycorrhizae that you probably studied in biology. In the case of endophytes and turfgrass, the fungi live in the plant and facilitate a resistance to some insect predators. Although white grubs, the most common insect predators of turfgrasses, do not appear to be affected by endophytes, other insects are (Potter *et al.*, 1992). Chinch bug damage in perennial ryegrass is directly related to the number of plants infected with endophytes (Richmond and Shetlar, 2000). Endophytes are also known to discourage feeding by billbugs, most caterpillars and various types of webworms. In mixtures of Kentucky bluegrass and perennial ryegrass, bluegrass billbug (*Sphenophorus parvulus*) damage was directly related to the proportion of endophyte-infected perennial ryegrass in the stand (Richmond *et al.*, 2000). As little as 40% of endophyte-infected perennial ryegrass overseeded into a stand of Kentucky bluegrass can significantly reduce damage by bluegrass webworm (*Parapediasia teterrella*) (Richmond and Shetlar, 1999). Endophytes are generally capable of surviving the same conditions as their turfgrass hosts, but some may be affected by certain conditions, such as severe cold (Rochefort *et al.*, 2007). In some cases, it may be found that a particular endophyte confers other positive characteristics in addition to insect resistance. For instance, research has demonstrated that at least one endophyte-infected tall fescue cultivar maintains higher leaf turgor pressure under drought stress than its uninfected counterpart (Richardson *et al.*, 1993), while other cultivars may not (White *et al.*, 1992). Researchers in New Jersey found that endophyte-infected perennial ryegrass was not only resistant to insect damage but also to red thread disease (*Laetisaria fuciformis*) (Bonos *et al.*, 2005). To this date, seed-borne endophyte infected turfgrasses are our greatest success for biological pest management in grasses. Unfortunately, endophyte infection has only been successful in perennial ryegrass and the tall and fine fescues; it is currently not commercially possible in most turfgrass species.

Other potential methods for biocontrol of turfgrass predators

There have been favorable reports concerning composting and the application of compost for the suppression of turfgrass fungal pathogens. The application of organic composts derived from some materials is believed to alter the turfgrass community enough to suppress pathogens. The bacteria in some types of compost are believed to compete directly with parasitic fungi that attack turf, and many bacteria that exist in compost have been identified as antagonistic toward turfgrass fungal pathogens (Boulter *et al.*, 2002). However, as yet, actual successes in disease reduction in applications of compost in the turfgrass industry have not been reported. This idea, though, is a good example of using ecological tactics to benefit turfgrass management and is likely to produce positive commercial products with further study.

Certain waste products or extracts derived from the processing or tilling of crop plants have been tested as sources of biocontrol agents. A crop of oilseed rape (*Brassica napus*), for instance, tilled into a sod farm infected with sting nematodes

(*Belonolaimus longicaudatus*) was effective in reducing nematode populations compared with sites under continuous bermudagrass (Walker *et al.*, 2005). Extracts of poinsettia (*Euphorbia pulcherrima*) and spotted spurge (*Chamaesyce maculata*) have also demonstrated promise in the control of sting nematodes (Cox *et al.*, 2006). Other uses of crop plants, their extracts or post-process waste products have also been tested for antagonistic activity on turfgrass predators. None have yet reached a stage of widespread use. Another possibility for biological control of predators is the application of nematodes or fungi that prey on insect pests, or bacteria that feed on pathogenic fungi.

Many of these biocides have been tested and a few have been found effective enough to reach commercial production in limited amounts. None has reached the general industry acceptability of BT (*Bacillus thuringiensis*), a bio-insecticide commonly used in ornamental horticulture. A bio-fungicide or bio-insecticide must not only be effective for pest control, it must also be able to live in a turfgrass environment under differing environmental conditions. Strains of these organisms differ in their ability to infect the target pest and also in their viability and persistence in a turfgrass community (Thompson *et al.*, 2005). Because they are living organisms, they require special care during manufacture and are expensive to produce. However, many turfgrass managers would pay a high price for them if they knew that they would be effective. Some of these products may normally be effective but may fail in a particular season for unknown reasons (Tomaso-Peterson and Perry, 2007). Such is the fate of biological organisms. They require certain resources that we may not realize are missing in any particular situation. Biocides and other biological predator controls will probably be commonplace for turfgrass use in the future. At present, though, turfgrass managers rely on pesticides, where possible, not because they do not care about environmental management but because pesticides hardly ever fail (Wang *et al.*, 2001).

13.5 Making a Management Plan

A pest management plan begins with the criteria that you will use to divide your site(s) into pest management categories. Categories will differ by turfgrass species, mowing height, customer expectations and any other factors that you deem important. These discriminating factors, with special preference for customer expectations, will be used to determine pest damage thresholds for each category. An athletic field complex might have three or four pest management categories. The primary field might have a very low pest management threshold such as 2 or 3% damage. Practice fields might have a higher damage threshold, 10%, and the surrounds may be acceptable at 20%. Home lawns and commercial property would be managed to thresholds primarily determined by the home or business owner. A home lawn would have one threshold value but a commercial property might have two or three. The idea is to manage without allowing damage to exceed the threshold levels but to set the levels as high as possible. A golf course provides an excellent example.

A golf course, depending on manager and membership, would probably be divided into at least three pest management categories. The highest priority units, the putting greens, would be Category 1. Category 1 might have a zero tolerance or 0% pest damage threshold. The tees and fairways, Category 2, might have a 10% damage threshold and the rough, Category 3, a 20% damage threshold. A 20% damage threshold is high. A 2 or 3% weed encroachment, for instance, is quite noticeable. Insect and disease damage, however, are not particularly noticeable until the grass actually turns brown. If 20% of your turf is brown, that is a lot of brown turf. Obviously, you also have to consider what you call damage. Does the turf have to be brown or just yellow to be considered damaged? For our purposes, let us go with brown as damage and let us call this golf course Any Valley Country Club.

Any Valley CC is located in a temperate region and has creeping bentgrass putting greens, Kentucky bluegrass fairways and tall fescue rough. The putting greens are susceptible to a number of fungal diseases, primarily dollar spot, and insects, primarily black cutworm (*Agrotis ipsilon*). The major problems on the fairways are also fungal diseases, mostly summer patch, and the primary insect problem is bluegrass billbug. Problems in the rough are brown patch and white grubs.

Let us consider how you are going to manage the greens at Any Valley CC to resist pest damage. Because the Category 1 threshold is 0%, you are going to have to include pesticides on the putting greens but you are going to use them wisely. If we do everything that we can ecologically with our fairways (Category 2) and rough (Category 3) our pesticide use will be extremely limited. We will rarely have to apply pesticides to maintain fairway and rough at threshold damage levels, but we will use them if necessary.

Before you develop a pest management plan for the putting greens, you have to do some research on dollar spot disease and black cutworm. You find that, like most fungal diseases, dollar spot prefers damp conditions. Consequently, sun is one of its natural enemies, as well as dew removal, air movement and temperatures warm enough to facilitate drying but not high enough to stress the turf. We can use these natural enemies against the pathogen. In the absence of rainfall, the primary time for dampness is following morning dew. Dew contains guttation fluid that you learned about earlier. Guttation fluid contains sugars, proteins and nutrients that fungi feed on, and increases their resources. Consequently, dew removal is an important process for dollar spot reduction (Williams *et al.*, 1996). Dew can be removed immediately after sunrise by mowing or poling (also called whipping) but mowing works best (Ellram *et al.*, 2007). You will also find that the dollar spot pathogen can exist as dormant hyphae (sclerotia) in the thatch and reproduces by spores, both of which can be carried by mowing equipment and other tools that touch the putting surface. So hosing off mowers after mowing one putting surface and before mowing another can help to delay the spread of the pathogen, and cultural practices to properly manage thatch can help to reduce the disease.

Delaying the spread of a pathogen is usually synonymous with managing it. As was mentioned earlier, diseases occur when the environment favors the growth of the pathogen over the growth of the turf. For that reason, if you can delay the spread of disease until the environmental conditions change, you are exercising some control over it. Therefore, healthy turf is an enemy of disease. Dollar spot is more likely to occur on dry turf than it is on turf that is irrigated properly, so irrigation is a weapon. Nitrogen fertilization is also an enemy of dollar spot. Dollar spot spreads from one plant to another most rapidly by colonizing wounds. The rapid growth and healing encouraged by adequate N fertilization helps to delay the spread of dollar spot. Finally, there are cultivars of creeping bentgrass that exhibit low levels of dollar spot resistance (Fig. 13.5). Although you will not be renovating sections of your putting greens if you maintain near 0% damage, there may be an opportunity for a large-scale

Fig. 13.5. Although this putting green looks like all one cultivar, it is not. In fact there are several cultivars in this photograph, all in areas of the same size as the rectangular areas severely infected with dollar spot (*Sclerotinia homeocarpa*). All cultivars of creeping bentgrass (*Agrostis stolonifera*) can be infected with dollar spot, and all of the cultivars in the photograph show symptoms. However, some cultivars are severely infected while others have minor infection.

renovation for which you can choose a cultivar or blend with some resistance to dollar spot. Unfortunately, as far as we know at present, there are no cultivars of creeping bentgrass with resistance to cutworm damage (Williamson and Potter, 2001). Although renovating patches of putting greens is not a practice to be considered, it would be a practice that you would want to consider in damaged sections of fairway and rough.

According to Couch (1995) the dollar spot pathogen grows slowly when daytime temperatures reach 60 °F (16 °C) and rapidly when daytime temperatures are consistently between 70 and 80 °F (21–27 °C). The nighttime humidity must also reach 85% before dollar spot becomes a significant problem. It is interesting to note that even reasonably dry climates can have a nighttime humidity that reaches 85%. Cold air holds less water than warm air, so as warm air cools overnight the relative humidity increases and is normally highest when the night's temperature is lowest. The night temperature is usually lowest about an hour before sunrise. Early-morning humidity should be monitored closely. We can't adjust the temperature or humidity but we can use them as indicators for pesticide applications.

Pesticides should be applied only when necessary. If you develop a good ecological management plan for the common predators on your Category 2 and Category 3 turf, few pesticides will be needed and you will only apply them when the environmental conditions are exactly right for disease infection or when disease or insect damage is present and approaching the damage threshold. On your Category 1 turf, however, you will probably be more effective, more efficient and use less pesticide by applying disease applications preventatively (before you see symptoms) than you will be by applying curatively (after you see symptoms) (Box 13.2). Most fungicides require applications threefold or fourfold greater to cure a disease than to prevent one. Consequently, using preventative applications when disease pressure is known to be high is a conservative measure for areas where near-zero infection is required.

You will begin those preventative applications when the environmental conditions are conducive for disease occurrence and end them when the danger has past. You should have a good reason for applying preventative pesticide measures on your Category 1 turf instead of just applying general blanket protections. You will probably need insecticide to prevent cutworm damage as well, but those applications can be made at the first sign of insects or symptoms. Constant monitoring during the most likely damage periods is required. Visual inspections should occur daily and a simple soap flush, 1 ounce of lemon-scented dishwashing liquid in 2 gallons of water sprinkled over 1 square yard of turf (30 ml of liquid in 7.6 l of water over 1 m^2), should be performed weekly (Niemczyk and Shetlar, 2000). At the first sign of insects or damage, an insecticide application is warranted. Monitoring should continue, as multiple generations usually occur. You should also study the

---

Box 13.2. Prevent or cure?

Because the surfaces have to be so precise, if you are managing areas such as golf course greens or bowling greens you will have to set your threshold damage level very low, usually near 0%. That kind of intense management is not possible without the use of pesticides. In most disease cases (not insects), it is usually safer and cheaper to apply preventative (before symptoms appear) applications instead of curative (after symptoms are seen) applications. That is because the fungicide rate required to cure a disease outbreak is much greater than the rate required to prevent one. Consider this example:

We set our threshold level for disease damage on the putting greens at Any Valley Country Club at 0%. The greens are creeping bentgrass (*Agrostis stolonifera*) and the most common disease that occurs is dollar spot (*Sclerotinia homeocarpa*). Chlorothalonil is a very popular contact fungicide used to control dollar spot on creeping bentgrass putting greens. It is sold under the trade name of Daconil Ultrex (and others). A preventative application of Daconil Ultrex is 1.8 ounces per 1000 square feet (0.55 g m^{-2}) at 7–21 day intervals, depending on the environmental conditions. A curative rate is from 3.7 to 5.0 ounces per 1000 square feet on a 14-day schedule. So the curative rate is two to three times greater than the preventative rate.

Let us assume that we can usually manage dollar spot preventatively with applications at 14-day intervals and that we have 6 weeks in the late spring and 6 weeks in the early fall when conditions are conducive for disease. Consequently, to be sure to prevent disease, we would make three applications in the late spring 14 days apart and three in the early fall on the same interval. That would require a total of 10.8 ounces of fungicide per 1000 square feet (3.4 g m^{-2}) of greens. Suppose that we wait until we observe disease to apply fungicide. The best case scenario is that no disease would occur or that it would only occur once in the spring or once in the fall. That is unlikely, however. The most likely best case scenario would be disease observation followed by two applications of fungicide at 3.7 ounces per 1000 square feet 14 days apart in late spring and two applications in early fall to cure the disease. That would require a total commitment of 14.8 ounces per 1000 square feet (4.5 g m^{-2}), or about 1.3 times more pesticide than the preventative program. A worse case curative scenario would be three applications in the spring and three in the fall at either 3.7 ounces per 1000 square feet, or a worse case still, 5.0 ounces per 1000 square feet, with those applications totaling 22.2 or 30.0 ounces of pesticide per 1000 square feet (6.8 or 9.2 g m^{-2}), respectively. In the two worse cases, the curative program would require two to three times more pesticide than the preventative program. So although a preventative fungicide program appears to be a poor economic and environmental choice, in cases where a very low threshold level is required, it is often a better alternative, both economically and environmentally, than a curative program.

insect to develop a program like the one you developed for dollar spot. Cutworms, for instance, like to live in burrows in the thatch so thatch control is a preventative measure. They also like to live in aerification holes, so if aerification absolutely has to be performed during the season that is conducive to cutworm damage, an insecticide application afterward will probably be required. If possible, postpone the aerification and save an application.

The best time to apply an insecticide for cutworm control is in the late afternoon as the insects hide during the day and feed at night. The fresh insecticide on the leaves when the cutworms begin to feed is very effective. The insects also tend to live in surrounding areas and crawl onto the greens at night. Therefore, if the green is treated, the green surrounds should also be treated. As I stated earlier, if you must use a pesticide, use it wisely.

As the superintendent of Any Valley CC, you will also have to learn about summer patch and billbug, and brown patch and white grubs. There will also be other predators of primary concern in your area that you will have to deal with on Category 1, 2 or 3 turf. You will have to learn about them and develop economic and environmentally sound programs to discourage them. Learning and devising procedures to reduce predator damage takes time and effort. However, once the protocols have been determined, their execution becomes normal and reasonably easy. Good turfgrass attracts high numbers of predators. You must be knowledgeable and you must be vigilant.

13.6 Chapter Summary

In order to survive, organisms either have to perform photosynthesis or obtain nourishment directly or indirectly from an organism that does. Because they perform photosynthesis, turfgrass plants provide energy to a variety of different species, including insects and diseases. As a turfgrass stand gets more dense and healthy, it provides more resources for its predators. Insects sometimes seek dense, green turfgrass in which to lay their eggs so that their offspring can obtain the best nourishment. Healthy turf is usually more resistant to disease than unhealthy turf, but that does not necessarily mean that a high pathogen population is not present. In most cases, it means that the turfgrass can resist disease in spite of a fairly high pathogen population. If environmental conditions become highly conducive to the growth and reproduction of the pathogen, disease will occur and spread rapidly.

Single-tactic systems are not very effective for controlling turfgrass predators. If single-tactic systems are effective, it is only on a temporary basis. For that reason, pesticides only control the predator until the next outbreak, and genetic resistance only controls the predator until the predator adapts. Fortunately, in the case of genetic resistance to a predator, it may take a considerable time for the predator to adapt. In other cases, the predator may adapt quickly or other predators may become dominant and more damaging. A combination of effective management strategies is the best course of action for reducing predator damage. Adjusting environmental conditions to favor the turfgrass over the predator is a permanent improvement. Mixing control strategies makes it substantially less likely that a predator will become resistant to a management plan.

Devising an integrated management plan that permanently deters predator damage requires in depth knowledge of both predator and turfgrass species. It also requires constant observation and good record keeping. Turfgrass managers can improve a turfgrass site for characteristics such as light and air movement. The manager can use resistant mixtures or blends of resistant turfgrass species and cultivars. He/she can adjust irrigation, fertilizer, mowing height, soil pH and many other factors to manage the situation best for turfgrass growth and least for predator growth. In order to do so, he/she must be willing to learn about the grass and the pest, to be especially vigilant and, in some cases, very creative.

Suggested Reading

Boulter, J.I., Boland, G.J. and Trevors, J.T. (2000) Compost: a study of the development process and end-product potential for suppression of turfgrass disease. *World Journal of Microbiology and Biotechnology* 16, 115–134.

Burpee, L.L. (1993) Integrated control of turfgrass diseases: research and reality. *International Turfgrass Society Research Journal* 7, 80–86.

Couch, H.B. (1995) *Diseases of Turfgrasses*, 3rd edn. Krieger Publishing, Malabar, Florida.

Fidanza, M.A. and Dernoeden, P.H. (1997) A review of brown patch forecasting, pathogen detection, and management strategies for turfgrasses. *International Turfgrass Society Research Journal* 8, 863–874.

Mann, R.L. and Newell, A.J. (2005) A survey to determine the incidence and severity of pests and diseases on golf course putting greens in England, Ireland, Scotland, and Wales. *International Turfgrass Society Research Journal* 10, 224–229.

Nelson, E.B. (1997) Microbial inoculants for the control of turfgrass diseases. *International Turfgrass Society Research Journal* 8, 791–811.

Niemczyk, H.D. and Shetlar, D.J. (2000) *Destructive Turf Insects*. H.D.N. Books, Wooster, Ohio.

Potter, D.A. (1991) Ecology and management of turfgrass insects. *Annual Review of Entomology* 36, 383–406.

Potter, D.A. (1993) Integrated insect management in turfgrasses: prospects and problems. *International Turfgrass Society Research Journal* 7, 69–79.

Potter, D.A. (1998) *Destructive Turfgrass Insects: Biology, Diagnosis, and Control*. Ann Arbor Press, Chelsea, Michigan.

Potter, D.A. (2005) Prospects for managing destructive turfgrass insects without protective chemicals. *International Turfgrass Society Research Journal* 10, 42–54.

Raikes, C., Lepp, N.W. and Canaway, P.M. (1996) An integrated disease management (IDM) strategy for winter sports turf. *Journal of the Sports Turf Research Institute* 72, 72–82.

Sachs, P.D. and Luff, R.T. (2002) *Ecological Golf Course Management*. Ann Arbor Press, Chelsea, Michigan.

Vargas, J.M. Jr. (1994) *Management of Turfgrass Diseases*, 2nd edn. CRC Press, Boca Raton, Florida.

Watschke, T.L., Dernoeden, P.H. and Shetlar, D.J. (1994) *Managing Turfgrass Pests*. CRC Press, Boca Raton, Florida.

Suggested Websites

CENTERE (2010) TurfFiles. Center for Turfgrass Environmental Research and Education (CENTERE), North Carolina State University Turfgrass Management Group. Available at: http://www.turffiles.ncsu.edu/Default.aspx (accessed: 2 April 2010).

EDIS (2010) Turf Diseases. EDIS (Electronic Data Information Source), UF (University of Florida) IFAS (Institute of Food and Agricultural Sciences) Extension. Available at: http://cdis.ifas.ufl.edu/topic_turf_diseases (accessed 2 April 2010).

Iowa Waste Reduction Center (2010) Golf Course Pollution Prevention Guide. Available as a pdf at: http://www.iwrc.org/resources/index.cfm (accessed 12 August 2010).

Landscape America (2010) Lawn and Landscape Problems. Available at: http://www.landscape-america.com/problems/problems_index.html (accessed: 2 April 2010).

University of Georgia Cooperative Extension (2010) Turfgrass Diseases in Georgia: Identification and Control. Available at: http://pubs.caes.uga.edu/caespubs/pubcd/B1233/B1233.html (accessed: 2 April 2010)

14 Making the Right Decisions

Key Terms

A **patio** is an outdoor sitting or dining area that is usually paved.
A **pergola** is an open-roof structure of rafters and cross members or latticework supported by columns.
The **Turfgrass Transition Zone** describes a geographic band in the USA where cool-season (C_3) grass adaptation meets warm-season (C_4) grass adaptation. The zone is characterized as a climate suited to heat tolerant cool-season grasses and cold tolerant warm-season grasses.

14.1 Learn to Use Your Knowledge to Your Advantage

What if you were a lawn or landscape professional and you were asked to design a small turfgrass area purely for ornamental purposes? You would begin by asking questions of the customer to find out exactly what their preferences were. You would consider your options based on those customer expectations and formulate a plan for customer approval. As questions arose, you would answer them. As problems were presented, you would solve them. When the site was finished, you would either assume the maintenance or provide the customer with a management plan to maintain the turfgrass at the expected level of aesthetic value. If the job was done well, you would receive praise and referrals from your customer that could amount to a significant amount of business. So let us see how you might do that using an actual example of a homeowner's back lawn site. A partner and I seeded the lawn for this project in the fall of 2006.

The assignment was to build a patio constructed of pavers covered by a pergola connected to an existing wooden deck attached to the house by a walkway also constructed of pavers. The patio was to be used as an entry to the small shed on the lawn and as a covered but open sitting area for use by the homeowners and visitors during nice weather. The existing small planting bed was to be extended the full width of the backyard and a simple border constructed to define it from the lawn. The existing lawn was a cold-tolerant common bermudagrass (*Cynodon dactylon*), which is used for most home lawns in the area. The home was in Stillwater, Oklahoma, which is in the United States Turfgrass Transition Zone. The Turfgrass Transition Zone is a geographic band that extends across the mid-southern USA at the border of the temperate climate zone and the subtropical climate zone. Both warm-season (C_4) grasses and cool-season (C_3) grasses may survive the transition zone climate. However, the cool-season grasses should be fairly well adapted to warm summers and the warm-season species have to be fairly well adapted to cold winters. In addition to cold tolerant bermudagrass, good choices for home lawns in Stillwater and in the transition zone are zoysiagrass (Japanese lawngrass, *Zoysia japonica*) and tall fescue (*Festuca arundinacea*). However, those species are most commonly used in light shade or partial shade in Stillwater, and bermudagrass is by far the most common grass in the area. The bermudagrass cultivar or biotype used has to be more cold tolerant than most bermudagrasses to survive the winters in reasonably good condition. In this case, bermudagrass was not an option. Because the species grows so aggressively, it is a very tenacious weed in flower beds, and beds bordered by bermudagrass require considerable effort to be maintained bermudagrass free. For that reason, our customer insisted that we use a species other than bermudagrass. Although zoysiagrass, tall fescue and other species are less aggressive and a better choice near flower beds, they are rarely used because they are readily invaded by aggressive indigenous bermudagrass. Consequently, our most difficult decision in this project was choosing an appropriate turfgrass for the small turfgrass area in the back lawn. We needed a species that would not aggressively invade the planting beds but one that would resist bermudagrass encroachment.

14.2 Planning the Project

The back lawn of the residence was an area measuring 2556 square feet (237 m²) and was completely fenced with a 5-foot (1.5-m) high wood privacy fence. The area consisted of a wood deck attached to the back of the house and two small planting beds near the back fence (Fig. 14.1). A storage shed stood on the north side of the property supported on treated wood timbers placed over pea gravel about 8 inches (20 cm) deep. The lawn consisted of a cold tolerant common bermudagrass. Drainage on the site was poor. Runoff drained away from the west fence and away from the house to accumulate approximately in the middle of the lawn with nowhere to go. The bermudagrass was surprisingly healthy in spite of the puddling that followed any major rain event. That was only because the climate in Stillwater is reasonably dry at about 34 inches (86 cm) of precipitation per year. Rainfall events are often short but intense, sometimes with long periods (2 to 4 weeks) of no rain at all.

After thoroughly inspecting the site, recording measurements and observing deficiencies and potential problems, we determined a plan and completed a simple drawing (Fig. 14.2). The homeowners selected the materials to be used and the design for the pergola from several different options and construction began in the summer of 2006. We decided that the small section of lawn between the patio and the north fence would remain in bermudagrass because it did not border on any

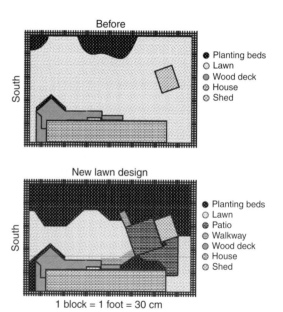

Fig. 14.2. A drawing of the site before construction (above) and a drawing of the proposed changes (below). In the new design the back lawn has been reduced from 2556 sf (237 m²) to 720 sf (67 m²). The small lawn will be easy to manage at near aesthetic perfection with minimal inputs and labor-intensive hand weeding practices. Drainage occurs under the borders of the planting beds and the walkway. Water from these drains and from lawn runoff collects under the patio and drains through a subsurface pipe along the north fence east into the sewers at the street.

Fig. 14.1. Photo of the homeowner's back lawn as it looked before reconstruction. The photographer is standing near the south fence facing north. Notice the shed on the north side, the small planting bed on the west and a small portion of the wood deck attached to the house supporting the hose caddy. A second planting bed in the southwest corner cannot be seen. The visible planting bed is in the approximate middle of the west fence line. The lawn is common bermudagrass (*Cynodon dactylon*). Runoff drains away from the house and away from the west fence to accumulate near the middle of the lawn.

planting beds except the area immediately behind the shed. That area was expected to remain in mulch deficient of any plant material. However, we needed to select a species other than bermudagrass for the center lawn that was now limited to a space of 720 square feet (67 m^2).

Facilitating drainage

Drainage was improved by supporting all of the new design elements on a bed of at least 12 inches (30 cm) of pea gravel that drained into a subsurface drain pipe. The drain pipe extended along the north fence line east and emptied into the sewers in the concrete street at the front of the house. Consequently, the large bed on the back fence drained into the gravel under its border then drained toward and under the patio. Water collecting in the middle of the lawn drained toward and under the patio and water draining from the bed near the house drained under the walkway toward the patio. The gravel under the shed also drained into the gravel under the patio and the patio gravel drained into the pipe and flowed toward the street. Since the reconstruction of the back lawn, drainage is no longer a problem.

We had some concern about using pea gravel as a base for the patio and walkway pavers. That is something that we had never attempted before and is not recommended. Normally a highly compactible fine gravel mixture is used under pavers to provide a solid base. That type of base, however, has attributes much like compacted soil has when compared with natural soil. Drainage from highly compactible gravel is faster than from compacted soil but the compactible gravel holds more water than other types of gravel and drains more slowly. Because the patio would be in the lowest part of the yard, we wanted it to be able to collect water underneath without heaving resulting from swelling and contracting during extremely wet or dry periods. We had to choose between attempting the pea gravel option, which we felt would be successful, or a more intricate and complicated drainage plan. We chose the pea gravel and have not been disappointed. Drainage in the lawn is excellent and there have been no problems with the patio or walkway. We placed a geotextile mat across the top of the gravel before applying the sand in which we set the pavers. The mat keeps the sand from migrating into the pea gravel.

We completed the drainage systems before moving on with the project so that we would not have to contend with standing water during the rest of the construction (Fig. 14.3). We used the same principles to determine how to drain the lawn that were presented in Chapter 8. We made no attempt to change the natural drainage of the site. We studied the situation and made decisions based on the simplest way to meet our objective. We designed the system to capture water and channel it to the lowest part of the site. We then determined a place for the water to go based on the natural drainage and the options available to us. The street being adjacent to the lowest part of the homeowner's property was the natural area in which to drain the water. The street also contained an elaborate sewer system to carry the water away. The only difficult part of the drainage construction was the ditch that had to be dug to the road. There was enough of a rise in the land between the back yard and the slope to the street that the water would not drain naturally. Consequently, the ditch was the only possibility for removing water from the lawn. Notice also that we considered drainage a top priority for the site. Without proper drainage, the backyard would never have been the site that the homeowners wished it to be and much of our effort would have been wasted. Drainage is always a top priority.

Fig. 14.3. The drainage system was the first part of the project completed. Here the borders for the planting beds have been installed over a deep gravel base. The excavation is complete for the patio, its borders are installed and the site has been backfilled with pea gravel. The pergola has also been constructed and work on the lawn is just beginning. Cultivation has also begun and the results are noticeable in a strip beside the large planting bed.

Once the drainage was complete, we were free to begin the renovation of the lawn. During the renovation, because of the way we planned the elevation of the beds, patio and walkway, we were able to further facilitate good drainage by smoothing the lawn surface to eliminate small collection pockets and by maintaining a slight but consistent slope to the patio. There was a small amount of soil redistribution during this project, but because the design improved upon existing patterns rather than changing them, no soil was added to or removed from the site. As we graded the lawn we were careful not to allow small collection pockets to develop as the soil settled back into place. Normally we would have rolled the surface to firm it and encourage settling, but in this case, it was not necessary. Because bermudagrass is such a difficult species to kill we had enough time between the initial cultivation of the site and seeding to allow the soil to settle sufficiently and to make fine adjustments to eliminate low spots.

The turfgrass species that we chose for the back lawn was Kentucky bluegrass (*Poa pratensis*). This is normally a poor choice for a lawn in Stillwater but, in this case, it met the objectives of the homeowner perfectly and provided what we believe was an excellent choice for the site. After 4 years in existence, the grass has exceeded our expectations. When making choices for turfgrass selection or for turfgrass maintenance, knowledge is very valuable. Consider all of your options, not just the popular ones. You might just know more than you think you know and you might just succeed where others fail. Let us consider the thought process that we used to select our grass.

Selecting the turfgrass

Kentucky bluegrass makes a beautiful ornamental grass, especially in a location like Stillwater where the warm-season grasses are dormant for 5 to 6 months. If irrigated and fertilized sufficiently, Kentucky bluegrass will remain green all year long in Stillwater and will be exceptionally pleasing during the springs and falls. Most Kentucky bluegrasses, however, do not tolerate the hot, sometimes dry, summers in Stillwater and except on sod farms, are simply not used. Because it spreads by rhizomes, Kentucky bluegrass is an excellent sod-forming turfgrass and is used by sod farms in mixtures with tall fescue to improve sod strength.

Because of its excellent heat tolerance and its ability to form a deep root system compared with other cool-season grasses, tall fescue is the most popular shade grass in the Turfgrass Transition Zone. Therefore, many sod growers have a consistent demand for tall fescue sod. However, because tall fescue is, for all practical purposes, a bunch-type grass, it does not form a good sod and generally falls apart during cutting or during installation. Sod growers in the area must support tall fescue sod either by using nylon mesh netting or by mixing it with Kentucky bluegrass. Most growers prefer the Kentucky bluegrass mix to nylon netting. A small amount, 5 to 10%, of Kentucky bluegrass is sufficient to provide good tall fescue sod handling characteristics and the bluegrass is expected to die during its first summer. However, in some cases, the bluegrass does not die and in others actually it spreads in competition with the tall fescue.

In Chapter 9 you learned about the NTEP (the National Turfgrass Evaluation Program) in the USA. Kentucky bluegrasses are tested in areas of the transition zone as well as in areas where they are better adapted for their ability to tolerate heat, drought and other conditions stressful for cool-season grasses. By studying the NTEP results from transition-zone locations, it was clear that a few recently developed Kentucky bluegrass cultivars tolerated transition-zone conditions very well. We selectively studied both the Kentucky bluegrass NTEP trials in Stillwater and those that had been performed close by in Wichita, Kansas, and found three cultivars that demonstrated good potential for use at our site. We ruled out other cool-season grasses except for tall fescue and all warm-season grasses except for zoysiagrass. After screening, we were left with three possible choices for our lawn, Kentucky bluegrass, tall fescue or zoysiagrass. Because of its persistence in tall fescue sod and the positive results of some cultivars in recent NTEP trials nearby, we considered Kentucky bluegrass to be an equally viable choice compared with zoysiagrass and tall fescue. A description of our selection process is presented later in this section.

As you may have noticed earlier, we designed the site so that the small ornamental lawn was completely enclosed by planting beds or structures except for the end bordered by the south fence only (Figure 14.2). For that reason, bermudagrass encroachment could not occur unless seed or sprigs were carried into the lawn by humans, birds or the wind. Although the neighbor's lawn to the south was bermudagrass we did not fear encroachment from that direction because the fence shaded a small area on its north side for nearly the entire day. Most bermudagrasses have no adaptation for shade and require full sun for more than half of each day to grow and spread. So we felt that it was unlikely that bermudagrass would encroach from the south neighbor's lawn, but that if it did, it could easily be removed by hand and a strip of mulch placed along the fence to prevent further encroachment. We used one of its natural enemies, shade, against the bermudagrass to restrict its encroachment. As you can see, the complete enclosure of the lawn by other structures also had purpose.

Once the attributes of the site and species adaptation to the transition-zone location had been considered to eliminate all but three candidates, it was time to consider the use and expectations of the site to choose the best grass. The lawn was to be primarily ornamental. Therefore, we wanted the best looking grass year-round. As zoysiagrass is dormant for nearly half of the year, it was eliminated, but not without discussion. Zoysiagrass has a very attractive gold color in dormancy and could have been a very nice choice. However, the homeowners preferred a green grass year-round to a gold one and were willing to mow early in the spring and late in the fall to maintain a green grass, another important consideration. Tall fescue is also a very attractive grass but it has a more coarse texture and does not get quite as dense as Kentucky bluegrass. In addition, it gets brown patch (*Rhizoctonia solani*) disease every summer in Stillwater and develops a brownish tint for a few weeks of the year because of the disease. If the brown patch infection is severe, it could require fungicide treatments, a possibility that we would like to eliminate. Kentucky bluegrass can get rust (*Puccinia* spp.) in the spring and fall in Stillwater that gives it a brownish tint but N fertilizer reduces the damage and pesticides are not necessary. It also gets a dangerous disease called summer patch (*Magnaporthe poae*) which could kill it, but our adapted cultivars appeared to have some resistance to the disease. Other diseases are possible, but dangerous infection is rare and the possibilities of infection were about the same for both species. Insect damage is also a possibility for both species, but is fairly rare and does not differ between the two grasses. Because of the likelihood of occasional fungicide applications (every 2 to 3 years) on brown patch, but mostly because Kentucky bluegrass has slightly more visual appeal, we chose bluegrass for the site. We would have preferred to seed a blend of the three cultivars that we identified as having positive potential, but a blend was not available at that time. Consequently, we seeded the most conveniently available cultivar of the three. In reality, the perfect plan is often not achievable. In this case, a second, single cultivar option worked out very well.

14.3 Turfgrass Establishment

Before the Kentucky bluegrass could be established, the bermudagrass on the main (center) lawn had to be removed. Simply removing the sod was not an option as the bermudagrass would readily grow back from rhizomes. The existing lawn had to be killed, and killing bermudagrass is difficult.

The first step in preparation for planting Kentucky bluegrass was to apply a full rate of glyphosate, a nonselective herbicide, to the existing bermudagrass lawn. Glyphosate is sold as Roundup and under several other trade names. It is nonselective, meaning that it kills anything green and is available for direct sale to homeowners in the USA; it is also a herbicide that is widely considered to be very safe to use. However, a single application of glyphosate or any other herbicide will not kill bermudagrass completely, and as a 0% threshold for bermudagrass encroachment was all that could be tolerated, we had to make sure that the existing stand was completely eliminated. For a second time in this project, we used the natural adaptation of bermudagrass against itself. Bermudagrass grows by both stolons and rhizomes. Although the stolons are usually killed by glyphosate, some of the rhizomes survive. Consequently, surviving rhizomes eventually produce new plants and, with time, can completely replenish the stand. For that reason, at least two applications of glyphosate, and often three, are needed to completely eliminate a bermudagrass stand. Knowledge of the pest, in this case a turfgrass, is required to control it.

Following the first glyphosate application, we continued a program to eliminate the existing bermudagrass completely. We could have waited until all of the bermudagrass was brown following the first glyphosate application, but that was not necessary. Once the initial glyphosate application was applied, we waited only long enough to observe the first signs of decline to begin cultivation. Because the site was small we used a simple garden cultivator to till the soil. If soil tests had revealed a need for a pH adjustment or a need for macronutrients we would have fertilized and tilled the nutrients, sulfur or lime into the soil at this point. In this case, however, the soil was in good condition and adjustments were not required. After tilling, we raked and prepared the site just as we would for a seed bed, but we did not apply seed. Instead we applied irrigation in the same manner that we would use on a new seed bed. The homeowner applied a very small amount of irrigation two to three times per day to keep the soil damp. Soon weeds began to germinate and bermudagrass began to sprout. A second application of glyphosate about 5 weeks after the first application killed the weeds and the bermudagrass.

Once the weeds had died completely following the second glyphosate application (about 2 weeks later) the site was raked a second time to fill small depressions that had occurred during soil settling, and the irrigation treatments were begun again. Following 3 weeks of irrigation treatments, few weeds had germinated and no bermudagrass plants were found. So glyphosate was applied a third time, this time as a spot spray. We waited 3 days to be sure that glyphosate absorption had occurred, and the seeding process began.

A final raking was applied to the site, the Kentucky bluegrass seed was scattered, and the site lightly raked again to partially cover the seed. Irrigation was applied as before, and about 2 weeks later, bluegrass germination began to occur. We advised the homeowner to continue irrigation three times daily for a week after germination, and then to start applying less frequently, but in a slightly greater amount. Irrigation was applied twice daily for the next week and once daily for two more weeks. By that time, the plants had developed reasonably good root systems so irrigation was applied on alternate days for 2 weeks and then twice a week for as long into the fall as mowing was required. After that irrigation was practiced only during very dry periods until spring.

We wanted the mowing height to be 3 inches for the first season to provide sufficient photosynthetic leaf area for the turf to develop a strong root system. Consequently, the first mowing occurred when the bluegrass reached a height of 4 inches (10 cm). The mower blade was sharpened in preparation for the first mowing to facilitate as clean a cut as possible and to minimize the possibility that the mower blade would grab young plants and pull them from the ground owing to their weak roots.

After mowing, the lawn was fertilized lightly and the fertilizer was irrigated in. After about three more weeks, the bluegrass was approaching full cover and we began a standard fertilizer maintenance program.

Assessment of the establishment procedures

Throughout the establishment process we did everything "by the book", so to speak, including time of establishment. It is usually best to establish cool-season grasses in the fall instead of the spring because there is less weed competition. We seeded our grass in mid September, which is a perfect time to establish a cool-season grass in the Stillwater area. However, we forgot to anticipate annual bluegrass (*Poa annua*) germination and competition with the Kentucky bluegrass. Because these two grasses are in the same genus they share almost the same niche. Consequently, annual bluegrass is a major competitor of Kentucky bluegrass. It seems that this particular lawn had a serious amount of annual bluegrass seed just waiting for the right conditions to germinate, and we provided those conditions for it. Out of curiosity, we checked the Kentucky bluegrass seed for annual bluegrass contamination and found absolutely none. The annual bluegrass was resident in the soil. We should have known.

Encroachment by weeds other than annual bluegrass was minor and the individual weeds were easily removed by hand. Once full cover had been reached that fall, however, annual bluegrass accounted for about 30% of the stand. Except for the visible presence of annual bluegrass the lawn was beautiful in the following spring. However, the annual bluegrass died during the summer, leaving some unsightly holes in the lawn (Fig. 14.4). In order to provide the Kentucky bluegrass with an opportunity to fill in over the fall, winter and spring without annual bluegrass competition we chose to apply a pre-emergence herbicide before annual bluegrass germination in the fall. As the following summer approached, the Kentucky bluegrass had filled in all but the small area where runoff drained under the patio, and full cover was achieved (Fig. 14.5). At present, fall pre-emergence applications are no longer necessary, and the Kentucky bluegrass has been quite reliable in successfully competing with the annual bluegrass. Only a small amount of annual bluegrass encroachment occurs each fall and this is easily hand weeded.

Fig. 14.4. The Kentucky bluegrass (*Poa pratensis*) lawn as it appeared in October 2007, one year after seeding. A pre-emergence herbicide has been applied to prevent the fall establishment of annual bluegrass (*Poa annua*). The open areas in the otherwise dense canopy are where annual bluegrass existed and died over the summer.

Fig. 14.5. The Kentucky bluegrass (*Poa pratensis*) lawn in June 2008. The only area that has yet to fill in is the area where runoff water drains under the patio. Fall pre-emergence herbicides are no longer necessary to control annual bluegrass (*Poa annua*), as the Kentucky bluegrass has proved to be an excellent competitor and little encroachment of annual bluegrass has occurred since full cover was achieved.

A lesson learned

Although our procedures for establishment were generally correct we could have saved ourselves some labor and a pesticide application by adjusting those procedures to minimize annual bluegrass encroachment. As you have been reminded throughout this text, establishment procedures and maintenance procedures need to be finely adjusted to the target site. In this case, we overlooked an important component of cool-season grass establishment in the area.

We know that annual bluegrass germinates as temperatures cool in the fall and the soil is sufficiently damp for seeds to imbibe water. Therefore, annual bluegrass germination is likely following the sufficient amount of rainfall that occurs after about mid September. In this case, we provided perfect conditions for germination of annual bluegrass as well as for our Kentucky bluegrass. The only reason that we had more Kentucky bluegrass than annual bluegrass germination is that there was more Kentucky bluegrass seed present. How could we have done this differently?

We know when annual bluegrass germinates. We also know approximately how long it will take for Kentucky bluegrass to germinate and establish. Consequently, because there is a reasonably mild fall season in the Stillwater area, we can establish cool-season grasses in October and often in November. In that case, we should have used the same procedures that we did to remove the possibility of bermudagrass encroachment. We should have treated the area like a seed bed in mid-September, allowed the annual bluegrass to germinate, removed that annual bluegrass and then seeded the Kentucky bluegrass. If we had followed that procedure, we would have had full Kentucky bluegrass cover in the first fall or early in the first spring with very minimal annual bluegrass encroachment. If we had thought things through a little better, we would have known how to proceed. Nonetheless, we learned from the mistake and we won't make that mistake again. In combination, knowledge and experience are great partners for learning and for the retention of what we learn.

14.4 Making a Management Plan

By its third year, our Kentucky bluegrass lawn was outstanding and had exceeded our expectations. It still looked great after surviving its third summer of 100°F (38°C) and higher temperatures (Fig. 14.6). As expected, the bluegrass remained green during the winter seasons, offering a nice contrast with the otherwise mostly brown and leafless landscape (Fig. 14.7). However, we did not achieve this success without a good management plan. It is especially difficult to manage a bermudagrass lawn in the front of a house and a Kentucky bluegrass lawn at the back. The management procedures are very different for the two grasses and good protocols for

Fig. 14.6. The Kentucky bluegrass (*Poa pratensis*) lawn as it looked near the end of the summer in August 2009. The turf has fully matured and it has developed a strong root system. Without such a root system it could not survive the extended 100 °F (38 °C) Oklahoma summers and still look like this.

Fig. 14.7. The lawn as it looked in December 2009. The grass is the only thing that is green in this photo. The rest of the ornamental plants have turned brown and the leaves fell from the trees long ago. The Kentucky bluegrass (*Poa pratensis*) adds a nice splash of green color to the otherwise brown landscape.

both have to be considered and executed. Even practices as simple as mowing become complicated enough to cause a decline in one grass or the other.

Mowing

Because Kentucky bluegrass is poorly adapted to the summer heat of Stillwater, we need to be careful not to mow the turf too low. Neither do we really want to mow the bermudagrass too high or it will look ragged and wavy. It is really no problem to adjust the mowing height on the mower from the front lawn to the back lawn, but what if the homeowner forgets to do it? It seems as if it would be a simple thing to remember but you have to think that someday the homeowner will mow the front lawn at 1.5 inches (38 mm), forget to raise the height to 3 inches (76 mm) in the back and scalp the bluegrass. That could result in severe damage to the back lawn. To prevent that from happening we recommended that the homeowner compromise and use a consistent mowing height of 2.5 inches (64 mm), the lowest height that we felt was comfortable for the Kentucky bluegrass. At that height, the bluegrass has remained in good condition and the bermudagrass looks reasonably good. To prove our point, the homeowner has already scalped the back lawn once while attempting to maintain at two mowing heights. Fortunately, it was not in the middle of the summer.

Irrigation

We did not design the backyard to include an automatic irrigation system. However, after irrigating all of the separate beds and lawns at the back, as well as those at the front, one at a time, for a year, an irrigation system was installed. The homeowners had always irrigated their bermudagrass lawn once a week when necessary at about 1 inch (25 mm) of irrigation, depending on the conditions. We recommended that the same amount of irrigation be applied to the Kentucky bluegrass lawn, but to apply it twice a week instead of once a week. We recommended that both grasses be irrigated twice a week at approximately 0.5 inches (13 mm) of precipitation each event to maintain consistency. We would not recommend that bermudagrass be irrigated more often than twice a week, but for bluegrass in Oklahoma, twice per week is a requirement. In cooler climates, Kentucky bluegrass will go dormant to protect itself from drought. It can remain dormant for several weeks under cool conditions and survive. However, we have found that when it goes dormant in our region, it can only survive for a short time, 1 or 2 weeks, before it dies. After experiencing a long, hot summer, there is very little root system surviving. Consequently, frequent irrigation is required to keep the bluegrass from wilting severely. As you recall, we irrigate to the depth of the root system on a cool-season grass in the summer. The grass is not healthy enough to grow deep roots to search for water at that time.

By installing an automatic irrigation system and using it wisely the homeowners were able to save water. It was no longer necessary to irrigate during the day when the temperatures were higher, the radiation was more intense and the wind was blowing. By setting the system to automatically irrigate before sunrise when it was cool and calm, less irrigation evaporated and less water was needed. The homeowners turn the irrigation system off following a rainfall and do not activate it again unless no precipitation occurs for several days. They also adjust to using less water during cloudy periods and to using more water during windy periods. They can also adjust by ET_c (crop evapotranspiration) if they desire, as the calculations for their area are available on the internet. They monitor their system to make sure that the heads are adjusted properly so that they are not irrigating the street, driveway or walkways and they observe and sometimes check their distribution patterns periodically. The proper installation and use of an automatic irrigation system can most definitely save water compared with the use of manually operated sprinklers. Most homeowners and commercial landowners, however, require education in the adjustment and use of their systems. You can help.

Fertilization and cultivation

Fertilization of a warm-season grass like bermudagrass and of a cool-season grass like Kentucky bluegrass differ substantially by season. Consequently, we had to develop different fertilizer programs for the front lawn and the back lawn. The bluegrass required less fertilizer than the bermudagrass and fewer applications. Bermudagrass is a nitrogen-loving plant. It grows and performs best when fertilized with substantial amounts of N. Rates of 0.75 to 1.00 pounds N per 1000 square feet per month (37–49 kg ha^{-1}) are common fertilization rates for bermudagrass in the Stillwater area. We begin fertilizing at greenup and stop about 6 weeks before dormancy is expected. Generally, fertilization would begin on approximately 1 April and end on 1 September, depending on annual conditions.

To improve efficient use of fertilizer we suggested that front lawn fertilization begin with 0.5 pounds N per 1000 square feet (24 kg ha^{-1}) at greenup followed by 1.0 pound N per 1000 square feet (49 kg ha^{-1}) in May and June. In July and August, fertilizer was to be applied at 0.50 to 0.75 pounds N per 1000 square feet (24–36 kg ha^{-1}) depending on recent temperatures. Even warm-season grasses use less fertilizer during hot summers. The front lawn was to receive less fertilizer when temperatures were extremely warm and the higher rate if the temperature had been more mild or close to normal. The final fertilization in September was to be applied at 1.0 pound N per 1000 square feet (49 kg ha^{-1}) and an optional application of 0.5 pounds N per 1000 square feet (24 kg ha^{-1}) and 1.0–2.0 pounds K per 1000 square feet (49–98 kg ha^{-1}) could be applied in October to help avoid winter kill if desired. Based on soil tests, monthly fertilizer is applied in a 17:3:5 ratio of N-P-K during normal monthly applications. The October fertilizer high in K has never been applied, and although some winter kill occurs each year, the lawn recovers rapidly. The original recommendations have proved to be acceptable and the front lawn is normally in very good condition. By customizing applications by season instead of applying a straight 1.0 pound N per 1000 square feet (49 kg ha^{-1}) every month the homeowner saves 1.0 pound N per 1000 square feet a year (49 kg ha^{-1}) with no loss of turfgrass quality.

The back lawn is fertilized differently from the front. The Kentucky bluegrass in the back is fertilized specifically to encourage root growth. Fertilizer is applied at 0.5 pounds N per 1000 square feet (24 kg ha^{-1}) in late February or early March following the first mowing, when the temperature has warmed and the grass has begun to grow. An application at that time when the temperature is still too cool for rapid shoot growth tends to encourage root growth. Fertilizer is not applied again until October when the air temperature cools and shoot growth begins to slow. The October application is followed by a November application and a December application to encourage root growth during the winter. Nitrogen is applied at 1.0 pound N per 1000 square feet (49 kg ha^{-1}) in October and November and at 0.5 pounds N per 1000 square feet in December. Air temperatures usually remain warm enough for active root growth in Stillwater until late December, and roots will continue to grow slowly over most of the winter as long as the grass remains green. Consequently, the fertilizer schedule encourages healthy deep roots as the bluegrass goes into summer.

By managing the fertilizer and irrigation properly, neither the front nor the back lawn requires aerification, vertical mowing or additional cultivation.

Pre-emergence herbicides are applied to the bermudagrass in the spring and fall, and yellow nutsedge (*Cyperus esculentus*) control and broadleaf weed control are usually applied once per year typically as a spot spray instead of a broadcast application. Yellow nutsedge is the major problem in the back lawn and has required a herbicide application, but weeds are generally removed by hand. By nature a Kentucky bluegrass canopy is extremely dense and severely limits weed encroachment. Once the annual bluegrass pressure was relieved, the yellow nutsedge was the only weed that required herbicide and it was applied as a spot spray rather than a broadcast spray. The bluegrass sometimes gets rust disease in the spring or fall when the weather remains cloudy and damp for several days, a condition that rarely occurs. The infections have not been severe and could probably be easily controlled with an additional fertilizer application if desired. Other predator problems front or back are insignificant. This would probably not be the case if the turf were managed at a high level like a golf course fairway or professional athletic field would be.

14.4 Chapter Summary

Hopefully, you have found this final chapter interesting as I described the thought process for the establishment and maintenance of a particularly unusual turfgrass for my region. Considering alternatives and thinking through problems is rewarding, not only in finding solutions but in practical and economic efficiency. Successful planning now often saves considerable time later.

In truth, I was very familiar with this particular job in all details because it was my lawn. My wife and I are the homeowners and my wife was my partner. Now we are working on the flower beds and I have become the partner.

References

Abernathy, S.D., White, R.H., Colbaugh, P.F., Engelke, M.C., Taylor, G.R. II and Hale, T.C. (2001) Dollar spot resistance among blends of creeping bentgrass cultivars. *Crop Science* 41, 806–809.

Abraham, E.M., Meyer, W.A., Bonos, S.A. and Huang, B. (2008) Differential responses of hybrid bluegrass and Kentucky bluegrass to drought and heat stress. *HortScience* 43, 2191–2195.

Ackerson, R.C. and Youngner, V.B. (1975) Responses of bermudagrass to salinity. *Agronomy Journal* 67, 678–680.

Adams, W.A. and Smith, J.N.G. (1993) Chemical properties of rootzones containing a black layer and some factors affecting sulphide production. *International Turfgrass Society Research Journal* 7, 540–545.

Adams, W.W. III and Demmig-Adams, B. (1992) Operation of the xanthophyll cycle in higher plants in response to diurnal changes in incident sunlight. *Planta* 186, 390–398.

Agnew, M.L. and Carrow, R.N. (1985) Soil compaction and moisture stress preconditioning in Kentucky bluegrass. I. Soil aeration, water use, and root responses. *Agronomy Journal* 77, 872–878.

Airfield Systems (2009) Home page. Available at: http://www.airfieldsystems.com (accessed 19 January 2010).

Aldous, D.E. and Kaufmann, J.E. (1979) Role of root temperature on shoot growth of two Kentucky bluegrass cultivars. *Agronomy Journal* 71, 545–547.

Aldous, D.E., Neylan, J.J. and Whykes, B. (2001) Improving surface quality in sports turf and reducing compaction using innovative aeration machinery. *Journal of Turfgrass Science* 77, 47–58.

Allen, R.G., Pereira, L.S., Raes, D. and Smith, M. (2004) *Crop Evapotranspiration: Guidelines for Computing Crop Water Requirements*, 2nd edn. FAO Irrigation and Drainage Paper No. 56, Food and Agriculture Organization, Rome.

Anderson, J.A., Taliaferro, C.M. and Martin, D.L. (2003) Longer exposure durations increase freeze damage to turf bermudagrasses. *Crop Science* 43, 973–977.

Annicchiarico, P., Russi, L., Piano, E. and Veronesi, F. (2006) Cultivar adaptation across Italian locations in four turfgrass species. *Crop Science* 46, 264–272.

Aronson, L.J., Gold, A.J. and Hull, R.J. (1987a) Cool-season turfgrass responses to drought stress. *Crop Science* 27, 1261–1266.

Aronson, L.J., Gold, A.J., Hull, R.J. and Cisar, J.L. (1987b) Evapotranspiration of cool-season turfgrasses in the humid northeast. *Agronomy Journal* 79, 901–905.

Arrieta, C., Busey, P. and Daroub, S.H. (2009) Goosegrass and bermudagrass competition under compaction. *Agronomy Journal* 101, 11–16.

Backman, P.A., Miltner, E.D., Stahnke, G.K. and Cook, T.W. (2001) Effects of cultural practices on earthworm casting on golf course fairways. *International Turfgrass Society Research Journal* 9, 823–827.

Bae, G. and Choi, G. (2008) Decoding of light signals by plant phytochromes and their interacting proteins. *Annual Review of Plant Biology* 59, 281–311.

Baker, S.W. (1994) The effect of the frequency of slit tine aeration on the quality of soccer and rugby pitches. *Journal of the Sports Turf Research Institute* 70, 44–54.

Baker, S.W., Mooney, S.J. and Cook, A. (1999a) The effects of sand type and rootzone amendments on golf green performance. I. Soil properties. *Journal of Turfgrass Science* 75, 2–17.

Baker, S.W., Mooney, S.J. and Cook, A. (1999b) The effects of sand type and rootzone amendments on golf green performance. II. Grass characteristics. *Journal of Turfgrass Science* 75, 18–26.

Baldwin, C.M., Liu, H., McCarty, L.B., Luo, H. and Toler, J.E. (2009) Nitrogen and plant growth regulator influence on 'Champion' bermudagrass putting green under reduced sunlight. *Agronomy Journal* 101, 75–81.

Balogh, J.C. and Walker, W.J. (eds) (1992) *Golf Course Management and Construction: Environmental Issues*. Lewis Publishers, Chelsea, Michigan.

Barbour, M.G., Burk, J.H., Pitts, W.D., Gilliam, F.S. and Schwartz, M.W. (1999) *Terrestrial Plant Ecology*, 3rd edn. Benjamin Cummings, Menlo Park, California.

Barden, J.A., Halfacre, R.G. and Parrish, D.J. (1987) *Plant Science*. McGraw-Hill, New York.

Beard, J.B. (1973) *Turfgrass Science and Culture*. Prentice-Hall, Englewood Cliffs, New Jersey.

Beard, J.B. and Beard, H.J. (2005) Turfgrass culture evolution at the St. Andrews Golf Links. *International Turfgrass Society Research Journal* 10, 70–77.

Beard, J.B. and Martin, D.P. (1970) Influence of water temperature on submersion tolerance of four grasses. *Agronomy Journal* 62, 257–259.

Beard, J.B. and Sifers, S.I. (1997) Genetic diversity in dehydration avoidance and drought resistance within the *Cynodon* and *Zoysia* species. *International Turfgrass Society Research Journal* 8, 603–610.

Beard, J.B., Croce, P., Mocioni, M., De Luca, A. and Volterrani, M. (2001) The comparative competitive ability of thirteen *Agrostis stolonifera* cultivars to *Poa annua*. *International Turfgrass Society Research Journal* 9, 828–831.

Bednarz, C.W., Nichols, R.L. and Brown, S.M. (2006) Plant density modifies within-canopy cotton fiber quality. *Crop Science* 46, 950–956.

Belesky, D.P. and Jung, G.A. (1982) Seasonal variation of water soluble and total zinc in cool-season grasses. *Agronomy Journal* 74, 1009–1012.

Bell, G.E. and Danneberger, T.K. (1999) Temporal shade on creeping bentgrass turf. *Crop Science* 39, 1142–1146.

Bell, G.E. and Moss, J.Q. (2008) Management practices that reduce runoff transport of nutrients and pesticides from turfgrass. In: Nett, M.T., Carroll, M.J., Horgan, B.P. and Petrovic, A.M. (eds) *The Fate of Nutrients and Pesticides in the Urban Environment*. American Chemical Society, Washington, D.C., pp. 133–150.

Bell, G.E. and Xiong, X. (2008) The history, role, and potential of optical sensing for practical turf management. In: Pessarakli, M. (ed.) *Handbook of Turfgrass Management and Physiology*. CRC Press, Boca Raton, Florida, pp. 641–660.

Bell, G.E., Danneberger, T.K. and McMahon, M.J. (2000) Spectral irradiance available for turfgrass growth in sun and shade. *Crop Science* 40, 189–195.

Bell, G.E., Howell, B.M., Johnson, G.V., Raun, W.R., Solie, J.B. and Stone, M.L. (2004) Optical sensing of turfgrass chlorophyll content and tissue nitrogen. *HortScience* 39, 1130–1132.

Berndt, W.L. (2008) Double exponential model describes decay of hybrid bermudagrass thatch. *Crop Science* 48, 2437–2446.

Berndt, W.L. and Vargas, J.M. Jr. (1992) Elemental sulfur lowers redox potential and produces sulfide in putting green sand. *HortScience* 27, 1188–1190.

Berry, J. and Björkman, O. (1980) Photosynthetic response and adaptation to temperature in higher plants. *Annual Review of Plant Physiology* 31, 491–543.

Bewick, T.A. (1996) Technological advancements in biological weed control with microorganisms: an introduction. *Weed Technology* 10, 600.

Bigelow, C.A., Bowman, D.C., Cassel, D.K. and Rufty, T.W. Jr. (2001) Creeping bentgrass response to inorganic soil amendments and mechanically induced subsurface drainage and aeration. *Crop Science* 41, 797–805.

Bigelow, C.A., Bowman, D.C. and Wollum, A.G. II (2002) Characterization of soil microbial population dynamics in newly constructed sand-based rootzones. *Crop Science* 42, 1611–1614.

Bigelow, C.A., Bowman, D.C. and Cassel, D.K. (2004) Physical properties of three sand size classes amended with inorganic materials or sphagnum peat moss for putting green rootzones. *Crop Science* 44, 900–907.

Black, C.C., Chen, T.M. and Brown, R.H. (1969) Biochemical basis for plant competition. *Weed Science* 17, 338–344.

Black, C.C., Campbell, W.H., Chen, T.M. and Dittrich, P. (1973) The monocotyledons: their evolution and comparative biology. III. Pathways of carbon metabolism related to net carbon dioxide assimilation by monocotyledons. *Quarterly Review of Biology* 48, 299–313.

Blackman, F.F. (1905) Optima limiting factors. *Annals of Botany* 19, 281–298.

Bonos, S.A., Wilson, M.M., Meyer, W.A. and Funk, C.R. (2005) Suppression of red thread in fine fescues through endophyte-mediated resistance. *Applied Turfgrass Science* doi:10.1094/ATS-2005-0725-01-RS.

Boulter, J.I., Trevors, J.T. and Boland, G.J. (2002) Microbial studies of compost: bacterial identification, and their potential for turfgrass pathogen suppression. *World Journal of Microbiology and Biotechnology* 18, 661–671.

Bowman, D.C. (2003) Daily vs. periodic nitrogen addition affects on growth and tissue nitrogen in perennial ryegrass turf. *Crop Science* 43, 631–638.

Brady, N.C. (1990) *The Nature and Properties of Soils*. Macmillan Publishing, New York.

Brede, A.D. (2004) Blending of Kentucky bluegrass cultivars of different quality performance levels. *Crop Science* 44, 561–566.

Brede, A.D. (2005) Necrotic ring spot and turfgrass quality as affected by Kentucky bluegrass cultivar and species mixtures. *International Turfgrass Society Research Journal* 10, 156–162.

Brede, A.D. and Duich, J.M. (1986) Plant interaction among *Poa annua*, *Poa pratensis*, and *Lolium perenne* turfgrasses. *Agronomy Journal* 78, 179–184.

Bremer, D.J., Su, K., Keeley, S.J. and Fry, J.D. (2006) Performance in the transition zone of two hybrid bluegrasses compared with Kentucky bluegrass and tall fescue. *Applied Turfgrass Science* doi:10.94/ATS-2006-0808-02-RS.

Brine, D.J. and Iqbal, M. (1983) Diffuse and global solar spectral irradiance under cloudless skies. *Solar Energy* 30, 447–453.

Brosnan, J.T., Ebdon, J.S. and Dest, W.M. (2005) Characteristics in diverse wear-tolerant genotypes of Kentucky bluegrass. *Crop Science* 45, 1917–1926.

Brown, R.H. (1978) A difference in N use efficiency in C_3 and C_4 plants and its implications in adaptation and evolution. *Crop Science* 18, 93–98.

Brown, T.L., LeMay, H.E. Jr. and Bursten, B.E. (1991) *Chemistry: The Central Science*, 5th edn. Prentice-Hall, Englewood Cliffs, New Jersey.

Bunnell, B.T., McCarty, L.B., Dodd, R.B. and Hill, H.S. (2002) Creeping bentgrass growth response to elevated soil carbon dioxide. *HortScience* 37, 367–370.

Busey, P. (2003) Cultural management of weeds in turfgrass: a review. *Crop Science* 43, 1899–1911.

Busey, P. and Snyder, G.H. (1993) Population outbreak of the southern chinch bug is regulated by fertilization. *International Turfgrass Society Research Journal* 7, 353–361.

Bushoven, J.R. and Hull, R.J. (2005) The role of nitrate in modulating growth and partitioning of nitrate assimilation between roots and leaves of perennial ryegrass (*Lolium perenne* L.). *International Turfgrass Society Research Journal* 10, 834–840.

Butler, J.D., Fults, J.L. and Sanks, G.D. (1974) Review of grasses for saline and alkali areas. In: Roberts, E.C. (ed.) *Proceedings of the Second International Turfgrass Conference*. ASA-CSSA-SSSA (American Society of Agronomy-Crop Science Society of America-Soil Science Society of America), Madison, Wisconsin, pp. 551–556.

Calhoun, R.N., Rinehart, G.J., Hathaway, A.D. and Buhler, K.K. (2005) Maximizing cultural practices to minimize weed pressure and extend herbicide treatment interval in a cool-season turfgrass mixture. *International Turfgrass Society Research Journal* 10, 1184–1188.

Canny, M.J. (1977) Flow and transport in plants. *Annual Review of Fluid Mechanics* 9, 275–296.

Carroll, M.J. and Petrovic, A.M. (1991) Nitrogen, potassium, and irrigation effects on water relations of Kentucky bluegrass leaves. *Crop Science* 31, 449–453.

Carroll, M.J., Slaughter, L.H. and Krouse, J.M. (1994) Turgor potential and osmotic constituents of Kentucky bluegrass leaves supplied with four levels of potassium. *Agronomy Journal* 86, 1079–1083.

Carrow, R.N. (1993) Canopy temperature irrigation scheduling indices for turfgrasses in humid climates. *International Turfgrass Society Research Journal* 7, 594–599.

Carrow, R.N. (1995) Drought resistance aspects of turfgrasses in the southeast: evapotranspiration and crop coefficients. *Crop Science* 35, 1685–1690.

Carrow, R.N. (1996) Drought resistance aspects of turfgrasses in the southeast: root–shoot responses. *Crop Science* 36, 687–694.

Carrow, R.N. (1997) Turfgrass response to slow-release nitrogen fertilizers. *Agronomy Journal* 89, 491–496.

Carrow, R.N. and Troll, J. (1977) Cutting height and nitrogen effects on improved perennial ryegrasses in monostand and polystand communities. *Agronomy Journal* 69, 5–10.

Carrow, R.N., Johnson, B.J. and Burns, R.E. (1987) Thatch and quality of Tifway bermudagrass turf in relation to fertility and cultivation. *Agronomy Journal* 79, 524–530.

Carrow, R.N., Johnson, B.J. and Landry, G.W. Jr. (1988) Centipedegrass response to foliar application of iron and nitrogen. *Agronomy Journal* 80, 746–750.

Carrow, R.N., Waddington, D.V. and Rieke, P.E. (2001) *Turfgrass Soil Fertility and Chemical Problems: Assessment and Management*. Ann Arbor Press, Chelsea, Michigan.

Carrow, R.N., Krum, J.M., Flitcroft, I. and Cline, V. (2010) Precision turfgrass management: challenges and field applications for mapping turfgrass soil and stress. *Precision Agriculture* 11, 115–134.

Castonguay, Y., Thibault, G., Rochette, P., Bertrand, A., Rochefort, S. and Dionne, J. (2009) Physiological responses of annual bluegrass and creeping bentgrass to contrasted levels of O_2 and CO_2 at low temperatures. *Crop Science* 49, 671–689.

Chappelle, E.W., Kim, M.S. and McMurtrey, J.E. III (1992) Ratio analysis of reflectance spectra (RARS): an algorithm for remote estimation of the concentrations of chlorophyll a, chlorophyll b, and carotenoids in soybean leaves. *Remote Sensing of Environment* 39, 239–247.

Chatterton, N.J., Harrison, P.A., Bennett, J.H. and Asay, K.H. (1989) Carbohydrate partitioning in 185 accessions of Gramineae grown under warm and cool temperatures. *Journal of Plant Physiology* 134, 169–179.

Chen, Y., Eggens, J.L. and Carey, K. (1997) Stress response of single and multiple cultivar populations of bent (*Agrostis* spp.). *International Turfgrass Society Research Journal* 8, 639–651.

Christians, N.E. (1993) Advances in plant nutrition and soil fertility. *International Turfgrass Society Research Journal* 7, 50–57.

Christians, N.E. (2007) *Fundamentals of Turfgrass Management*, 3rd edn. John Wiley and Sons, Hoboken, New Jersey.

Christians, N.E., Martin, D.P. and Wilkinson, J.F. (1979) Nitrogen, phosphorus, and potassium effects on quality and growth of Kentucky bluegrass and creeping bentgrass. *Agronomy Journal* 71, 564–567.

Christie, J.M. (2007) Phototropin blue-light receptors. *Annual Review of Plant Biology* 58, 21–45.

Cole, J.T., Baird, J.H., Basta, N.T., Huhnke, R.L., Storm, D.E., Johnson, G.V., Payton, M.E., Smolen, M.D., Martin, D.L. and Cole, J.C. (1997) Influence of buffers on pesticide and nutrient runoff from bermudagrass turf. *Journal of Environmental Quality* 26, 1589–1598.

Connell, J.H. and Slatyer, R.O. (1977) Mechanisms of succession in natural communities and their role in community stability and organization. *American Naturalist* 111, 1119–1144.

Couch, H.B. (1995) *Diseases of Turfgrasses*, 3rd edn. Krieger Publishing, Malabar, Florida.

Cox, C.J., McCarty, L.B., Toler, J.E., Lewis, S.A. and Martin, S.B. (2006) Suppressing sting nematodes with *Brassica* sp., poinsettia, and spotted spurge extracts. *Agronomy Journal* 98, 962–967.

Cullimore, D.R., Nilson, S., Taylor, S. and Nelson, K. (1990) Structure of a black plug layer in a turfgrass putting sand green. *Journal of Soil and Water Conservation* 45, 657–659.

Cyril, J., Powell, G.L., Duncan, R.R. and Baird, W.V. (2002) Changes in membrane polar lipid fatty acids of seashore paspalum in response to low temperature exposure. *Crop Science* 42, 2031–2037.

DaCosta, M. and Huang, B. (2006a) Minimum water requirements for creeping, colonial, and velvet bentgrasses under fairway conditions. *Crop Science* 46, 81–89.

DaCosta, M. and Huang, B. (2006b) Deficit irrigation effects on water use characteristics of bentgrass species. *Crop Science* 46, 1779–1786.

Danneberger, T.K. (1993) *Turfgrass Ecology and Management.* Franzak and Foster, Cleveland, Ohio.

Danneberger, T.K. and Vargas, J.M. Jr. (1984) Annual bluegrass seedhead emergence as predicted by degree-day accumulation. *Agronomy Journal* 76, 756–758.

Davis, D.L. and Gilbert, W.B. (1970) Winter hardiness and changes in soluble protein fractions of bermudagrass. *Crop Science* 10, 7–9.

Davis, W.B., Paul, J.L. and Bowman, D. (1990) The sand putting green: construction and management. *University of California Cooperative Extension Publication* No. 21448. Davis, California.

Dean, D.E., Devitt, D.A., Verchick, L.S. and Morris, R.L. (1996) Turfgrass quality, growth, and water use influenced by salinity and water stress. *Agronomy Journal* 88, 844–849.

Dellinger, A.E., Schmidt, J.P. and Beegle, D.B. (2008) Developing nitrogen fertilizer recommendations for corn using an active sensor. *Agronomy Journal* 100, 1546–1552.

Demmig-Adams, B. (1990) Carotenoids and photoprotection in plants: a role for the xanthophyll zeaxanthin. *Biochimica et Biophysica Acta* 1020, 1–24.

Dernoeden, P.H., Crahay, J.N. and Davis, D.B. (1991) Spring dead spot and bermudagrass quality as influenced by nitrogen source and potassium. *Crop Science* 31, 1674–1680.

Dernoeden, P.H., Carroll, M.J. and Krouse, J.M. (1993) Weed management and tall fescue quality as influenced by mowing, nitrogen, and herbicides. *Crop Science* 33, 1055–1061.

Dernoeden, P.H., Fidanza, M.A. and Krouse, J.M. (1998) Low maintenance performance of five *Festuca* species in monostands and mixtures. *Crop Science* 38, 434–439.

Devitt, D.A., Morris, R.L. and Bowman, D.C. (1992) Evapotranspiration, crop coefficients, and leaching fractions of irrigated desert turfgrass systems. *Agronomy Journal* 84, 717–723.

Devitt, D.A., Wright, L., Bowman, D.C., Morris, R.L. and Lockett, M. (2008) Nitrate-N concentrations in the soil solution below reuse irrigated golf course fairways. *HortScience* 43, 2196–2202.

DiMascio, J.A., Sweeney, P.M., Danneberger, T.K. and Kamalay, J.C. (1994) Analysis of heat shock response in perennial ryegrass using maize heat shock protein clones. *Crop Science* 34, 798–804.

Dionne, J., Castonguay, Y., Nadeau, P. and Desjardins, Y. (2001a) Freezing tolerance and carbohydrate changes during cold acclimation of green-type annual bluegrass (*Poa annua* L.) ecotypes. *Crop Science* 41, 443–451.

Dionne, J., Castonguay, Y., Nadeau, P. and Desjardins, Y. (2001b) Amino acid and protein changes during cold acclimation of green-type annual bluegrass (*Poa annua* L.) ecotypes. *Crop Science* 41, 1862–1870.

DiPaola, J.M. (1984) Syringing effects on the canopy temperatures of bentgrass greens. *Agronomy Journal* 76, 951–953.

DiPaola, J.M. and Beard, J.B. (1992) Physiological effects of temperature stress. In: Waddington, D.V., Carrow, R.N. and Shearman, R.C. (eds) *Turfgrass.* ASA-CSSA-ASSA (American Society of Agronomy-Crop Science Society of America-Soil Science Society of America), Madison, Wisconsin, pp. 231–268.

Doty, J.A., Braunworth, W.S., Jr., Tan, S., Lombard, P.B. and William, R.D. (1990) Evapotranspiration of cool-season grasses grown with minimal maintenance. *HortScience* 25, 529–531.

Duble, R.L. (1989) *Southern Turfgrasses, Their Management and Use.* TexScape, College Station, Texas.

Dudeck, A.E. and Peacock, C.H. (1985a) Effects of salinity on seashore paspalum turfgrasses. *Agronomy Journal* 77, 47–50.

Dudeck, A.E. and Peacock, C.H. (1985b) Salinity effects on perennial ryegrass germination. *HortScience* 20, 268–269.

Dudeck, A.E. and Peacock, C.H. (1992) Shade and turfgrass culture. In: Waddington, D.V., Carrow, R.N. and Shearman, R.C. (eds) *Turfgrass.* ASA-CSSA-ASSA (American Society of Agronomy-Crop Science Society of America-Soil Science Society of America), Madison, Wisconsin, pp. 269–284.

Dudeck, A.E. and Peacock, C.H. (1993) Salinity effects on growth and nutrient uptake of selected warm-season turf. *International Turfgrass Society Research Journal* 7, 680–686.

Dudeck, A.E., Singh, S., Giordano, C.E., Nell, T.A. and McConnell (1983) Effects of sodium chloride on *Cynodon* turfgrasses. *Agronomy Journal* 75, 927–930.

Dudeck, A.E., Peacock, C.H. and Wildmon, J.C. (1993) Physiological and growth responses of St. Augustinegrass cultivars to salinity. *HortScience* 28, 46–48.

Duell, R.W. and Markus, D.K. (1977) Guttation deposits on turfgrass. *Agronomy Journal* 69, 891–898.

Dunn, J.H. and Nelson, C.J. (1974) Chemical changes occurring in three bermudagrass turf cultivars in relation to cold hardiness. *Agronomy Journal* 66, 28–31.

Dunn, J.H., Sheffer, K.M. and Halisky, P.M. (1981) Thatch and quality of Meyer zoysia in relation to management. *Agronomy Journal* 73, 949–952.

Dunn, J.H., Minner, D.D., Fresenburg, B.F. and Bughrara, S.S. (1994) Bermudagrass and cool-season turfgrass mixtures: response to simulated traffic. *Agronomy Journal* 86, 10–16.

Dunn, J.H., Minner, D.D., Fresenburg, B.F., Bughrara, S.S. and Hohnstrater, C.H. (1995) Influence of core aerification, topdressing, and nitrogen on mat, roots, and quality of 'Meyer' zoysiagrass. *Agronomy Journal* 87, 891–894.

Dunn, J.H., Bughrara, S.S., Warmund, M.R. and Fresenburg, B.F. (1999) Low temperature tolerance of zoysiagrasses. *HortScience* 34, 96–99.

Dunn, J.H., Ervin, E.H. and Fresenburg, B.S. (2002) Turf performance of mixtures and blends of tall fescue, Kentucky bluegrass, and perennial ryegrass. *HortScience* 37, 214–217.

Ebdon, J.S. and Kopp, K.L. (2004) Relationships between water use efficiency, carbon isotope discrimination, and turf performance in genotypes of Kentucky bluegrass during drought. *Crop Science* 44, 1754–1762.

Ebdon, J.S., Petrovic, A.M. and White, R.A. (1999) Interaction of nitrogen, phosphorus, and potassium on evapotranspiration rate and growth of Kentucky bluegrass. *Crop Science* 39, 209–218.

Egler, F.E. (1954) Vegetation science concepts I: initial floristic composition, a factor in old-field vegetation development. *Vegetation* 4, 412–417.

Elford, E.M.A., Tardif, F.J., Robinson, D.E. and Lyons, E.M. (2008) Effect of perennial ryegrass overseeding on weed suppression and sward composition. *Weed Technology* 22, 231–239.

Elliott, M.L., Guertal, E.A. and Skipper, H.D. (2004) Rhizosphere bacterial population flux in golf course putting greens in the southeastern United States. *HortScience* 39, 1754–1758.

Ellram, A., Horgan, B. and Hulke, B. (2007) Mowing strategies and dew removal to minimize dollar spot on creeping bentgrass. *Crop Science* 47, 2129–2137.

Erickson, J.E., Cisar, J.L., Snyder, G.H., Park, D.M. and Williams, K.E. (2008) Does a mixed-species landscape reduce inorganic-nitrogen leaching compared to a conventional St Augustinegrass lawn? *Crop Science* 48, 1586–1594.

Fales, S.L. and Wakefield, R.C. (1981) Effects of turfgrass on establishment of woody plants. *Agronomy Journal* 73, 605–610.

Faust, M.B. and Christians, N.E. (2000) Copper reduces shoot growth and root development of creeping bentgrass. *Crop Science* 40, 498–502.

Feldhake, C.M., Danielson, R.E. and Butler, J.D. (1983) Turfgrass evapotranspiration. I. Factors influencing rate in urban environments. *Agronomy Journal* 75, 825–830.

Feldhake, C.M., Danielson, R.E. and Butler, J.D. (1984) Turfgrass evapotranspiration. II. Responses to deficit irrigation. *Agronomy Journal* 76, 85–89.

Fidanza, M.A., Dernoeden, P.H. and Zhang, M. (1996) Degree-days for predicting smooth crabgrass emergence in cool-season turfgrasses. *Crop Science* 36, 990–996.

Figliola, S.S., Camper, N.D. and Ridings, W.H. (1988) Potential biological control agents for goosegrass (*Eleusine indica*). *Weed Science* 36, 830–835.

Fitzpatrick, R.J.M. and Guillard, K. (2004) Kentucky bluegrass response to potassium and nitrogen fertilization. *Crop Science* 44, 1721–1728.

Frank, K.W., Gaussoin, R.E., Fry, J.D., Frost, M.D. and Baird, J.H. (2002) Nitrogen, phosphorus, and potassium effects on seeded buffalograss establishment. *HortScience* 37, 371–373.

French, C.S. (1961) Light pigments and photosynthesis. In: McElroy, W.D. and Blass, B. (eds) *A Symposium on Light and Life*. John Hopkins Press, Baltimore, Maryland.

Fry, J.D. and Butler, J.D. (1989) Responses of tall and hard fescue to deficit irrigation. *Crop Science* 29, 1536–1541.

Fry, J.D., Lang, N.S., Clifton, R.G.P. and Maier, F.P. (1993) Freezing tolerance and carbohydrate content of low-temperature-acclimated and nonacclimated centipedegrass. *Crop Science* 33, 1051–1055.

Fu, J. and Huang, B. (2003) Growth and physiological response of creeping bentgrass to elevated night temperature. *HortScience* 38, 299–301.

Fu, J., Fry, J.F. and Huang, B. (2004) Minimum water requirements of four turfgrasses in the transition zone. *HortScience* 39, 1740–1744.

Fu, J., Fry, J. and Huang, B. (2007) Tall fescue rooting as affected by deficit irrigation. *HortScience* 42, 688–691.

Furuya, M. (2004) An unforeseen voyage to the world of phytochromes. *Annual Review of Plant Biology* 55, 1–21.

Gardner, D.S. and Taylor, J.A. (2002) Change over time in quality and cover of various turfgrass species and cultivars maintained in shade. *HortTechnology* 12, 465–469.

Gates, D.M. (1966) Spectral distribution of solar radiation at the earth's surface. *Science* 151, 523–529.

Gatschet, M.J., Taliaferro, C.M., Porter, D.R., Anderson, M.P., Anderson, J.A. and Jackson, K.W. (1996) A cold-regulated protein from bermudagrass crowns is a chitinase. *Crop Science* 36, 712–718.

Gausman, H.W., Allen, W.A., Myers, V.I. and Cardenas, R. (1969) Reflectance and internal structure of cotton leaves, *Gossypium hirsutum* (L.). *Agronomy Journal* 61, 374–376.

Gaussoin, R.E. and Branham, B.E. (1989) Influence of cultural factors on species dominance in a mixed stand of annual bluegrass/creeping bentgrass. *Crop Science* 29, 480–484.

Gaussoin, R.E., Branham, B.E. and Flore, J.E. (2005) The influence of environmental variables on CO_2 exchange rates of three cool season turfgrasses. *International Turfgrass Society Research Journal* 10, 850–856.

Geiger, D.R. and Servaites, J.C. (1994) Diurnal regulation of photosynthetic carbon metabolism in C_3 plants. *Annual Review of Plant Physiology and Plant Molecular Biology* 45, 235–256.

Gething, M.J. (1997) *Guidebook to Molecular Chaperones and Protein-Folding Catalysts*. Sambrook and Tooze Publication at Oxford University Press, Oxford, UK.

Giancoli, D.C. (1998) *Physics: Principles with Applications*, 5th edn. Prentice Hall, Upper Saddle River, New Jersey.

Gibbs, R.J., Adams, W.A. and Baker, S.W. (1993) Changes in soil physical properties of different construction methods for soccer pitches under intensive use. *International Turfgrass Society Research Journal* 7, 413–421.

Giesler, L.J., Yuen, G.U. and Horst, G.L. (2000) Canopy microenvironments and applied bacteria population dynamics in shaded tall fescue. *Crop Science* 40, 1325–1332.

Gifford, R.M. and Evans, L.T. (1981) Photosynthesis, carbon partitioning and yield. *Annual Review of Plant Physiology* 32, 485–509.

Gilbert, W.B. and Davis, D.L. (1971) Influence of fertility ratios on winter hardiness of bermudagrass. *Agronomy Journal* 63, 591–593.

Goatley, J.M. Jr., Maddox, V., Lang, D.M. and Crouse, K.K. (1994) 'Tifgreen' bermudagrass response to late-season application of nitrogen and potassium. *Agronomy Journal* 86, 7–10.

Goatley, J.M. Jr., Maddox, V.L., Lang, D.L., Elmore, R.E. and Stewart, B.R. (2005) Temporary covers maintain fall bermudagrass quality, enhance spring greenup, and increase stem carbohydrate levels. *HortScience* 40, 227–231.

Golembiewski, R.C., Danneberger, T.K. and Sweeney, P.M. (2001) Lack of dollar spot (*Sclerotinia homeocarpa* F.T. Bennett) influence on changing the cultivar composition of a stand of creeping bentgrass. *International Turfgrass Society Research Journal* 9, 665–668.

Gore, J.P., Cox, R. and Davies, T.M. (1979) Wear tolerance of turfgrass mixtures. *Journal of the Sports Turf Research Institute* 55, 45–68.

Goss, R.L., Stanton, E.B. and Orton, S.P. (1979) Uptake of sulfur by bentgrass putting green turf. *Agronomy Journal* 71, 909–913.

Grant, R.H. (1997) Partitioning biologically active radiation in plant canopies. *International Journal of Biometeorology* 40, 26–40.

Greub, L.J., Drolsom, P.N. and Rohweder, D.A. (1985) Salt tolerance of grasses and legumes for roadside use. *Agronomy Journal* 77, 76–80.

Griffin, J.J., Reid, W.R. and Bremer, D.J. (2007) Turf species affects establishment and growth of redbud and pecan. *HortScience* 42, 267–271.

Grime, J.P. (1974) Vegetation classification by reference strategy. *Nature* 250, 26–31.

Grime, J.P. (1977) Evidence for the existence of three primary strategies in plants and its relevance to ecological and evolutionary theory. *American Naturalist* 111, 1169–1194.

Grossi, N., Lulli, F., Volterrani, M. and Miele, S. (2005) Timing of fall nitrogen application on tall fescue turf. *International Turfgrass Society Research Journal* 10, 462–465.

Guertal, E.A. and Shaw, J.N. (2004) Multispectral radiometer signatures for stress evaluation in compacted bermudagrass turf. *HortScience* 39, 403–407.

Guertal, E.A., van Santen, E. and Han, D.Y. (2005) Fan syringe application for cooling bentgrass greens. *Crop Science* 45, 245–250.

Hale, M.G. and Orcutt, D.M. (1987) *The Physiology of Plants Under Stress*. John Wiley, New York.

Hall, N.P. and Keys, A.J. (1983) Temperature dependence of the enzymic carboxylation and oxygenation of ribulose 1,5-bisphosphate in relation to effects of temperature on photosynthesis. *Plant Physiology* 72, 945–948.

Han, L., Song, G. and Zhang, S. (2008) Preliminary observations on physiological responses of three turfgrass species to traffic stress. *HortTechnology* 18, 139–143.

Harivandi, M.A., Butler, J.D. and Soltanpour, P.N. (1982) Salt influence on germination and seedling survival of six cool season turfgrass species. *Communications in Soil Science and Plant Analysis* 13, 519–529.

Harivandi, M.A., Butler, J.D. and Wu, L. (1992) Salinity and turfgrass culture. In: Waddington, D.V., Carrow, R.N. and Shearman, R.C. (eds) *Turfgrass*. ASA-CSSA-ASSA (American Society of Agronomy-Crop Science Society of America-Soil Science Society of America), Madison, Wisconsin, pp. 207–230.

Hayes, A.R., Mancino, C.F., Forden, W.Y., Kopec, D.M. and Pepper, I.L. (1990) Irrigation of turfgrass with secondary sewage effluent: II. Turf quality. *Agronomy Journal* 82, 943–946.

He, Y. and Huang, B. (2007) Protein changes during heat stress in three Kentucky bluegrass cultivars differing in heat tolerance. *Crop Science* 47, 2513–2520.

Heckman, J.R., Clarke, B.B. and Murphy, J.A. (2003) Optimizing manganese fertilization for the suppression of take-all patch disease on creeping bentgrass. *Crop Science* 43, 1395–1398.

Henry, G.M., Hart, S.E. and Murphy, J.A. (2005) Overseeding bentgrass species into existing stands of annual bluegrass. *HortScience* 40, 468–470.

Henry, G.M., Burton, M.G. and Yelverton, F.H. (2009) Heterogeneous distribution of weedy *Paspalum* species and edaphic variables in turfgrass. *HortScience* 44, 447–451.

Hesketh, J.D. (1963) Limitations to photosynthesis responsible for differences among species. *Crop Science* 3, 493–496.

Hodges, C.F. (1992a) Growth of *Agrostis palustris* in subsurface black-layered sand induced by cyanobacteria and sulfate-reducing bacteria. *Plant and Soil* 142, 91–96.

Hodges, C.F. (1992b) Pathogenicity of *Pythium torulosum* to roots of *Agrostis palustris* in black-layered sand produced by the interaction of the cyanobacteria species *Lyngbya*, *Phormidium*, and *Nostoc* with *Desulfovigrio desulfuricans*. *Canadian Journal of Botany* 70, 2193–2197.

Hollingsworth, B.S., Guertal, E.A. and Walker, R.H. (2005) Cultural management and nitrogen source effects on ultradwarf bermudagrass cultivars. *Crop Science* 45, 486–493.

Horgan, B.P. and Yelverton, F.H. (2001) Removal of perennial ryegrass from overseeded bermudagrass using cultural methods. *Crop Science* 41, 118–126.

Hoveland, C.S., Buchanan, G.A. and Harris, M.C. (1975) Response of weeds to soil phosphorus and potassium. *Weed Science* 24, 194–200.

Howieson, M.J. and Christians, N.E. (2008) Carbohydrate metabolism and efficiency of photosystem II in mown creeping bentgrass (*Agrostis stolonifera* L.). *HortScience* 43, 525–528.

Hsiang, T., Carey, K., Ge, B. and Eggens, J.L. (1997) Composition of mixtures of four turfgrass species four years after seeding under non-wear conditions. *International Turfgrass Society Research Journal* 8, 671–679.

Huang, B. and Gao, H. (2000) Growth and carbohydrate metabolism of creeping bentgrass cultivars in response to increasing temperatures. *Crop Science* 40, 1115–1120.

Huang, B. and Liu, X. (2003) Summer root decline: production and mortality for four cultivars of creeping bentgrass. *Crop Science* 43, 258–265.

Huang, B., Duncan, R.R. and Carrow, R.N. (1997) Root spatial distribution and activity of four turfgrass species in response to localised drought stress. *International Turfgrass Society Research Journal* 8, 681–690.

Huang, B., Liu, X. and Fry, J.D. (1998) Shoot physiological responses of two bentgrass cultivars to high temperature and poor soil aeration. *Crop Science* 38, 1219–1224.

Huff, D.R. (1998) The case for *Poa annua* on golf greens. *Golf Course Management* 66(10), 54–56.

Hull, R.J. (1976) A carbon-14 technique for measuring photosynthate distribution in field grown turf. *Agronomy Journal* 68, 99–102.

Hull, R.J. (1992) Energy relations and carbohydrate partitioning in turfgrass. In: Waddington, D.V., Carrow, R.N. and Shearman, R.C. (eds) *Turfgrass*. ASA-CSSA-ASSA (American Society of Agronomy-Crop Science Society of America-Soil Science Society of America), Madison, Wisconsin, pp. 175–206.

Hunter Irrigation Innovators (2009a) Residential Sprinkler System Design Handbook: A Step-By-Step Introduction to Design and Installation. Available at: http://www.hunterindustries.com/Resources/PDFs/Technical/Domestic/lit226w.pdf (accessed 7 October 2009).

Hunter Irrigation Innovators (2009b) The Handbook of Technical Irrigation Information: A Complete Reference Source for the Professional. Available at: http://www.hunterindustries.com/Resources/PDFs/Technical/Domestic/LIT194w.pdf (accessed 7 October 2009).

Hurto, K.A., Turgeon, A.J. and Spomer, L.A. (1980) Physical characteristics of thatch as a turfgrass growing medium. *Agronomy Journal* 72, 165–167.

Inguagiato, J.C., Murphy, J.A. and Clarke, B.B. (2008) Anthracnose severity on annual bluegrass influenced by nitrogen fertilization, growth regulators, and verticutting. *Crop Science* 48, 1595–1607.

Jaabak, G. (1993) Long-term improvement in drainage and aeration within impermeable golf greens. *International Turfgrass Society Research Journal* 7, 426–429.

Jiang, H., Fry, J. and Tisserat, N. (1998) Assessing irrigation management for its effects on disease and weed levels in perennial ryegrass. *Crop Science* 38, 440–445.

Jiang, Y. and Huang, B. (2000a) Osmotic adjustment and root growth associated with drought preconditioning-enhanced heat tolerance in Kentucky bluegrass. *Crop Science* 41, 1168–1173.

Jiang, Y. and Huang, B. (2000b) Effects of drought or heat stress alone and in combination on Kentucky bluegrass. *Crop Science* 40, 1358–1362.

Jiang, Y. and Huang, B. (2001) Drought and heat stress injury to cool-season turfgrasses in relation to antioxidant metabolism and lipid peroxidation. *Crop Science* 41, 436–442.

Jiang, Y. and Wang, K. (2006) Growth, physiological, and anatomical responses of creeping bentgrass cultivars to different depths of waterlogging. *Crop Science* 46, 2420–2426.

Jiang, Y., Liu, H. and Cline, V. (2009) Correlations of leaf relative water content, canopy temperature, and spectral reflectance in perennial ryegrass under water deficit conditions. *HortScience* 44, 459–462.

Jo, Y. and Smitely, D.R. (2006) Impact of soil moisture and mowing height on *Ataenius spretulus* (Coleoptera: Scarabaeidae) selection of golf course turf habitat in choice tests. *HortScience* 41, 459–462.

Johnson, B.J. (1994) Biological control of annual bluegrass with *Xanthomonas campestris* pv. *Poa annua*. *HortScience* 29, 659–662.

Johnson, B.J., Carrow, R.N. and Burns, R.E. (1987) Bermudagrass turf response to mowing practices and fertilizer. *Agronomy Journal* 79, 677–680.

Johnson, B.J., Carrow, R.N. and Burns, R.E. (1988) Centipedegrass decline and recovery as affected by fertilizer and cultural treatments. *Agronomy Journal* 80, 479–456.

Johnson, P.G. (2003) Mixtures of buffalograss and fine fescue or streambank wheatgrass as a low-maintenance turf. *HortScience* 38, 1214–1217.

Jones, J.R. Jr. (1980) Turf analysis. *Golf Course Management* 48(1), 29–32.

Jordan, J.E., White, R.H., Thomas, J.C., Hale, T.C. and Victor, D.M. (2005) Irrigation frequency effects on turgor pressure of creeping bentgrass and soil air composition. *HortScience* 40, 232–236.

Juska, F.V. and Hanson, A.A. (1961) Effects of interval and height of mowing on growth of Merion and common Kentucky bluegrass (*Poa pratensis* L.). *Agronomy Journal* 53, 385–388.

Juska, F.V. and Murray, J.J. (1974) Performance of bermudagrasses in the transition zone as affected by potassium and nitrogen. In: *Proceedings of the Second International Turfgrass Research Conference*, Blacksburg, Virginia, pp. 149–154.

Kaminski, J.E. and Dernoeden, P.H. (2005) Nitrogen source impact on dead spot (*Ophiosphaerella agrostis*) recovery in creeping bentgrass. *International Turfgrass Society Research Journal* 10, 214–223.

Kaminski, J.E. and Dernoeden, P.H. (2007) Seasonal *Poa annua* L. seedling emergence patterns in Maryland. *Crop Science* 47, 775–781.

Kandel, E.R., Schwartz, J.H. and Jessell, T.M. (1991) *Principles of Neural Science,* 3rd edn. Elsevier, New York.

Kandel, H.J., Porter, P.M., Johnson, B.L., Henson, R.A., Hanson, B.K., Weisberg, S. and LeGare, D.G. (2004) Plant population influences niger seed yield in the northern great plains. *Crop Science* 44, 190–197.

Karcher, D.E. and Rieke, P.E. (2005) Water injection cultivation of a sand-topdressed putting green. *International Turfgrass Society Research Journal* 10, 1094–1098.

Karcher, D.E., Richardson, M.D., Landreth, U.W. and McCalla, J.H. Jr. (2005a) Recovery of bermudagrass varieties from injury. *Applied Turfgrass Science* doi:10.1094/ATS-2005-0117-01-RS.

Karcher, D.E., Richardson, M.D., Landreth, U.W. and McCalla, J.H. Jr. (2005b) Recovery of zoysiagrass varieties from injury. *Applied Turfgrass Science* doi:10.1094/ATS-2005-0728-01-RS.

Karnok, K.J., Rowland, E.J. and Tan, K.H. (1993) High pH treatments and the alleviation of soil hydrophobicity on golf greens. *Agronomy Journal* 85, 983–986.

Kendrick, D.L. and Danneberger, T.K. (2002) Lack of competitive success of an intraseeded creeping bentgrass cultivar into an established putting green. *Crop Science* 42, 1615–1620.

Kneebone, W.R. and Pepper, W.R. (1982) Consumptive water use by sub-irrigated turfgrasses under desert conditions. *Agronomy Journal* 74, 419–423.

Kneebone, W.R. and Pepper, W.R. (1984) Luxury water use by bermudagrass turf. *Agronomy Journal* 76, 999–1002.

Kneebone, W.R., Kopec, D.M. and Mancino, C.F. (1992) Water requirements and irrigation. In: Waddington, D.V., Carrow, R.N. and Shearman, R.C. (eds) *Turfgrass.* ASA-CSSA-ASSA (American Society of Agronomy-Crop Science Society of America-Soil Science Society of America), Madison, Wisconsin, pp. 441–472.

Knipling, E.B. (1970) Physical and physiological basis for the reflectance of visible and near-infrared radiation from vegetation. *Remote Sensing of Environment* 1, 155–159.

Koh, K., Bell, G.E., Martin, D.L. and Walker, N.R. (2003) Shade and airflow restriction effects on creeping bentgrass golf greens. *Crop Science* 43, 2182–2188.

Kopp, K.L. and Guillard, K. (2002) Clipping management and nitrogen fertilization of turfgrass: growth, nitrogen utilization, and quality. *Crop Science* 42, 1225–1231.

Koski, A.J., Street, J.R. and Danneberger, T.K. (1988) Prediction of Kentucky bluegrass root growth using degree-day accumulation. *Crop Science* 28, 848–850.

Kowalewski, A.R., Buhler, D.D., Lang, N.S., Nair, M.G. and Rogers, J.N. III (2009) Mulched maple and oak leaves associated with a reduction in common dandelion populations in established Kentucky bluegrass. *HortTechnology* 19, 297–304.

Krans, J.V. and Beard, J.B. (1985) Effects of clipping on growth and physiology of 'Merion' Kentucky bluegrass. *Agronomy Journal* 25, 17–20.

Krenzer, E.G. Jr. and Moss, D.N. (1969) Carbon dioxide compensation in grasses. *Crop Science* 9, 619–621.

Krömer, S. (1995) Respiration during photosynthesis. *Annual Review of Plant Physiology and Plant Molecular Biology* 46, 45–70.

Krum, J.M., Carrow, R.N. and Karnok, K. (2010) Spatial mapping of complex turfgrass sites: site-specific management units and protocols. *Crop Science* 50, 301–315.

Lalonde, S., Wipf, D. and Frommer, W.B. (2004) Transport mechanisms for organic forms of carbon and nitrogen between source and sink. *Annual Review of Plant Biology* 55, 341–372.

Landschoot, P.J. and McNitt, A.S. (1997) Effect of nitrogen fertilizers on suppression of dollar spot disease of *Agrostis stolonifera* L. *International Turfgrass Society Research Journal* 8, 905–911.

Ledeboer, F.B. and Skogley, C.R. (1967) Investigations into the nature of thatch and methods for its decomposition. *Agronomy Journal* 59, 320–323.

Lee, C.W., Jackson, M.B., Duysen, M.E., Freeman, T.P. and Self, J.R. (1996) Induced micronutrient toxicity in 'Touchdown' Kentucky bluegrass. *Crop Science* 36, 705–712.

Lee, G., Duncan, R.R. and Carrow, R.N. (2004a) Salinity tolerance of seashore paspalum ecotypes: shoot growth responses and criteria. *HortScience* 39, 1138–1142.

Lee, G., Carrow, R.N. and Duncan, R.R. (2004b) Salinity tolerance of selected seashore paspalums and bermudagrasses: root and verdure responses and criteria. *HortScience* 39, 1143–1147.

Leibig, J. (1840) *Organic Chemistry and its Application to Vegetable Physiology and Agriculture*. T.B. Peterson, Philadelphia, Pennsylvania.

Leinauer, B.N., Karcher, D., Barrick, T., Ikemura, Y., Hubble, H. and Makk, J. (2007) Water repellency varies with depth and season in sandy rootzones treated with ten wetting agents. *Applied Turfgrass Science* doi:10/1094-ATS-2007-0221-01-RS.

Levitt, J. (1980) *Responses of Plants to Environmental Stresses*, Vol. 1, 2nd edn. Academic Press, New York.

Li, D., Minner, D.D., Christians, N.E. and Logsdon, S. (2005) Evaluating the impact of variable root zone depth on the hydraulic properties of sand-based turf systems. *International Turfgrass Society Research Journal* 10, 1100–1107.

Lickfeldt, D.W., Voigt, T.B., Branhan, B.E. and Fermanian, T.W. (2001) Evaluation of allelopathy in cool season turfgrass species. *International Turfgrass Society Research Journal* 9, 1013–1018.

Lickfeldt, D.W., Voigt, T.B. and Hamblin, A.M. (2002a) Composition and characteristics of blended Kentucky bluegrass stands. *HortScience* 37, 1124–1126.

Lickfeldt, D.W., Voigt, T.B. and Hamblin, A.M. (2002b) Cultivar composition and spatial patterns in Kentucky bluegrass blends. *Crop Science* 42, 842–847.

Lin, C. and Shalitin, D. (2003) Cryptochrome structure and signal transduction. *Annual Review of Plant Biology* 54, 469–496.

Liu, H. and Hull, R.J. (2006) Comparing cultivars of three cool-season turfgrasses for nitrogen recovery in clippings. *HortScience* 41, 827–831.

Liu, X. and Huang, B. (2000) Heat stress injury in relation to membrane lipid peroxidation in creeping bentgrass. *Crop Science* 40, 503–510.

Liu, X. and Huang, B. (2001) Seasonal changes and cultivar difference in turf quality, photosynthesis, and respiration of creeping bentgrass. *HortScience* 36, 1131–1135.

Liu, X. and Huang, B. (2002a) Cytokinin effects on creeping bentgrass response to heat stress: II. Leaf senescence and antioxidant metabolism. *Crop Science* 42, 466–472.

Liu, X. and Huang, B. (2002b) Mowing effects on root production, growth, and mortality of creeping bentgrass. *Crop Science* 42, 1241–1250.

Liu, X., Huang, B. and Banowetz, G. (2002) Cytokinin effects on creeping bentgrass responses to heat stress: I. Shoot and root growth. *Crop Science* 42, 457–465.

Lockett, A.M., Devitt, D.A. and Morris, R.L. (2008) Impact of reuse water on golf course soil and turfgrass parameters monitored over a 4.5-year period. *HortScience* 43, 2210–2218.

Long, S.P., Humphries, S. and Falkowski, P.G. (1994) Photoinhibition of photosynthesis in nature. *Annual Review of Plant Physiology and Plant Molecular Biology* 45, 633–662.

Lunt, O.R., Youngner, V.B. and Oertli, J.J. (1961) Salinity tolerance of five turfgrass varieties. *Agronomy Journal* 53, 247–249.

Lush, W.M. (1988a) Biology of *Poa annua* in a temperate zone golf putting green (*Agrostis stolonifera/Poa annua*) I. The above-ground population. *Journal of Applied Ecology* 25, 977–988.

Lush, W.M. (1988b) Biology of *Poa annua* in a temperate zone golf putting green (*Agrostis stolonifera/Poa annua*) II. The seed bank. *Journal of Applied Ecology* 25, 989–997.

Lush, W.M. (1990) Turf growth and performance evaluation based on turf biomass and tiller density. *Agronomy Journal* 82, 505–511.

Lush, W.M. and Rogers, M.E. (1992) Cutting height and the biomass and tiller density of *Lolium perenne* amenity turfs. *Journal of Applied Ecology* 29, 611–618.

Lyons, E.M., Snyder, R.H. and Lynch, J.P. (2008) Regulation of root distribution and depth by phosphorus localization in *Agrostis stolonifera*. *HortScience* 43, 2203–2209.

Lyons, E.M., Jordan, K.S. and Carey, K. (2009) Use of wetting agents to relieve hydrophobicity in sand root-zone putting greens in a temperate climate zone. *International Turfgrass Society Research Journal* 11, 1131–1138.

Maas, S.J. and Dunlap, J.R. (1989) Reflectance, transmittance, and absorptance of light by normal, etiolated, and albino corn leaves. *Agronomy Journal* 81, 105–110.

MacArthur, R.H. and Wilson, E.O. (1967) *The Theory of Island Biogeography*. Princeton University Press, Princeton, New Jersey.

Maddox, V.L., Goatley, J.M. Jr., Philley, H.W., Stewart, B. and Wells, D.W. (2007) Maximizing 'Cimarron' little bluestem establishment as secondary rough for a golf course. *Applied Turfgrass Science* doi:10.1094/ATS-2007-0802-01-RS.

Madison, J.H. (1962) Mowing of turfgrass. III. The effect of rest on seaside bentgrass turf mowed daily. *Agronomy Journal* 54, 252–253.

Madison, J.H. and Hagan, R.M. (1962) Extraction of soil moisture by Merion bluegrass (*Poa pratensis* L. 'Merion') turf, as affected by irrigation frequency, mowing height, and other cultural operations. *Agronomy Journal* 54, 157–160.

Mancino, C.F. and Pepper, I.L. (1992) Irrigation of turfgrass with secondary sewage effluent: soil quality. *Agronomy Journal* 84, 650–654.

Marcum, K.B. (1998) Cell membrane thermostability and whole-plant heat tolerance of Kentucky bluegrass. *Crop Science* 38, 1214–1218.

Marcum, K.B. (1999) Salinity tolerance mechanisms of grasses in the subfamily Chloridoideae. *Crop Science* 39, 1153–1160.

Marcum, K.B. (2001) Salinity tolerance of 35 bentgrass cultivars. *HortScience* 36, 374–376.

Marcum, K.B. (2008) Relative salinity tolerance of turfgrass species and cultivars. In: Pessarakli, M. (ed.) *Handbook of Turfgrass Management and Physiology.* CRC Press, Boca Raton, Florida, pp. 389–406.

Marcum, K.B. and Murdoch, C.L. (1990) Growth responses, ion relations, and osmotic adaptations of eleven C_4 turfgrasses to salinity. *Agronomy Journal* 82, 892–896.

Marcum, K.B. and Murdoch, C.L. (1994) Salinity tolerance mechanisms of six C_4 turfgrasses. *Journal of the American Society for Horticultural Science* 119, 779–784.

Marcum, K.B. and Pessarakli, M. (2006) Salinity tolerance and salt gland excretion efficiency of bermudagrass turf cultivars. *Crop Science* 46, 2571–2574.

Martin, D.L. and Wehner, D.J. (1987) Influence of prestress environment on annual bluegrass heat tolerance. *Crop Science* 27, 579–585.

Martin, D.L., Bell, G.E., Baird, J.H., Taliaferro, C.M., Tisserat, N.A., Kuzmic, R.M., Dobson, D.D. and Anderson, J.A. (2001) Spring dead spot resistance and quality of seeded bermudagrasses under different mowing heights. *Crop Science* 41, 451–456.

Matsuoka, M., Furbank, R.T., Fukayama, H. and Miyao, M. (2001) *Annual Review of Plant Physiology and Plant Molecular Biology* 52, 297–314.

McCarty, L.B. (2001) *Best Golf Course Management Practices.* Prentice Hall, Upper Saddle River, New Jersey.

McCarty, L.B. and Miller, G.L. (2002) *Managing Bermudagrass Turf: Selection, Construction, Cultural Practices and Pest Management Strategies.* Ann Arbor Press, Chelsea, Michigan.

McCarty, L.B. and Tudker, B.J. (2005) Prospects for managing turf weeds without protective chemicals. *International Turfgrass Society Research Journal* 10, 34–41.

McCarty, L.B., Gregg, M.F. and Toler, J.E. (2007) Thatch and mat management in an established creeping bentgrass golf green. *Agronomy Journal* 99, 1530–1537.

McClellan, T.A., Shearman, R.C., Gaussoin, R.E., Mamo, M., Wortmann, C.S., Horst, G.L. and Marx, D.B. (2007) Nutrient and chemical characterization of aging golf course putting greens: establishment and rootzone mixture treatment effects. *Crop Science* 47, 193–199.

McClellan, T.A., Gaussoin, R.E., Shearman, R.C., Wortmann, C.S., Mamo, M., Horst, G.L. and Marx, D.B. (2009) Nutrient and chemical properties of aging golf course putting greens as impacted by soil depth and mat development. *HortScience* 44, 452–458.

McClendon, J.H. and McMillen, G.G. (1982) The control of leaf morphology and the tolerance of shade by woody plants. *Botanical Gazette* 143, 79–83.

McDade, M.C. and Christians, N.E. (2001) Corn gluten hydrolysate for crabgrass (*Digitaria* spp.) control in turf. *International Turfgrass Research Journal* 9, 1026–1029.

McElroy, J.S., Kopsell, D.A., Sorochan, J.C. and Sams, C.E. (2006) Response of creeping bentgrass carotenoid composition to high and low irradiance. *Crop Science* 46, 2606–2612.

McGuan, M.J., Danneberger, T.K. and Gardner, D.S. (2004) Regional differences in the relative competitive ability of annual bluegrass (*Poa annua* L.). *HortScience* 39, 1736–1739.

McInnes, K. and Thomas, J. (2008) A comparison of water drainage and storage in putting greens built using Airfield Systems and USGA methods of construction. 2008 USGA Turfgrass and Environmental Research Summary. Available at: http://turf.lib.msu.edu/ressum/2008/2.pdf (accessed 12 August 2010).

McKell, C.M., Youngner, V.B., Nudge, F.J. and Chatterton, N.J. (1969) Carbohydrate accumulation of coastal bermudagrass and Kentucky bluegrass in relation to temperature regimes. *Crop Science* 9, 534–537.

Mehall, B.J., Hull, R.J. and Skogley, C.R. (1984) Turf quality of Kentucky bluegrass cultivars and energy relations. *Agronomy Journal* 76, 47–50.

Merriam-Webster Online (2009) Available at: http://www.merriam-webster.com/ (accessed 5 April 2009).

Miller, G.L. (1999) Potassium application reduces calcium and magnesium levels in bermudagrass leaf tissue and soil. *HortScience* 34, 265–268.

Miltner, E.D., Stahnke, G.K. and Rinehart, G.J. (2005) Mowing height, nitrogen rate, and organic and synthetic fertilizer effects on perennial ryegrass quality and pest occurrence. *International Turfgrass Society Research Journal* 10, 982–988.

Minner, D.D. and Valverde, F.J. (2005a) The effect of traffic intensity and periodicity on *Poa pratensis* L. performance. *International Turfgrass Society Research Journal* 10, 387–392.

Minner, D.D. and Valverde, F.J. (2005b) Performance of established cool-season grass species under simulated traffic. *International Turfgrass Society Research Journal* 10, 393–397.

Minner, D.D., Dunn, J.H., Bughrara, S.S. and Fresenburg, B.F. (1997) Effect of topdressing with 'Profile' porous ceramic clay on putting green quality, incidence of dry spot and hydraulic conductivity. *International Turfgrass Society Research Journal* 8, 1240–1249.

Moges, S.M., Raun, W.R., Mullen, R.W., Freeman, K.W., Johnson, G.V. and Solie, J.B. (2004) Evaluation of green, red, and near infrared bands for predicting winter wheat biomass, nitrogen uptake, and final grain yield. *Journal of Plant Nutrition* 27, 1431–1441.

Monteith, J.L. and Unsworth, M.H. (1990) *Principles of Environmental Physics.* Edward Arnold, London.

Moore, R. and Black, C.C. Jr. (1979) Nitrogen assimilation pathways in leaf mesophyll and bundle sheath cells of C_4 photosynthesis plants formulated from comparative studies with *Digitaria sanguinalis* (L.) Scop. *Plant Physiology* 64, 309–313.

Moore, R.W., Christians, N.E. and Agnew, M.L. (1996) Response of three Kentucky bluegrass cultivars to sprayable nitrogen fertilizer programs. *Crop Science* 36, 1296–1301.

Morgan, W.C., Letey, J. and Stolzy, L.H. (1965) Turfgrass renovation by deep aerification. *Agronomy Journal* 57, 494–496.

Moss, D.N. and Smith, L.H. (1972) A simple classroom demonstration of differences in photosynthetic capacity among species. *Journal of Agronomic Education* 1, 16–17.

Münch, E. (1927) Versuche über den Saftkreislauf (Experiments on the circulation of sap). *Ber. Deutsch. Bot. Ges.* 45, 340–356.

Münch, E. (1930) *Die Stoffbewegungen in der Pflanze (Translocation in Plants)*. Fischer, Jena, Germany.

Murphy, J.A., Rieke, P.E. and Erickson, A.E. (1993a) Core cultivation of a putting green with hollow and solid tines. *Agronomy Journal* 85, 1–9.

Murphy, J.W., Field, T.R.O. and Hickey, M.J. (1993b) Age development in sand-based turf. *International Turfgrass Society Research Journal* 7, 464–468.

Murray, J.J. and Juska, F.V. (1977) Effect of management practices on thatch accumulation, turf quality, and leaf spot damage in common Kentucky bluegrass. *Agronomy Journal* 69, 365–369.

Nagy, F. and Schäfer, E. (2002) Phytochromes control photomorphogenesis by differentially regulated, interacting signaling pathways in higher plants. *Annual Review of Plant Biology* 53, 329–355.

NASA/Cool Cosmos EPO (Education and Public Outreach) (2010) Infrared Astronomy: Near, Mid and Far Infrared. Available at: http://coolcosmos.ipac.caltech.edu/cosmic_classroom/ir_tutorial/irregions.html (accessed 27 July 2010).

Nektarios, P.A., Economou, G. and Avgoulas, C. (2005) Allelopathic effects of *Pinus halepensis* needles on turfgrasses and biosensor plants. *HortScience* 40, 246–250.

Nielsen, A.P. and Wakefield, R.C. (1978) Competitive effects of turfgrass on the growth of ornamental shrubs. *Agronomy Journal* 70, 39–42.

Niemczyk, H.D. and Shetlar, D.J. (2000) *Destructive Turf Insects*. H.D.N. Books, Wooster, Ohio.

Niyogi, K.K. (1999) Photoprotection revisited: genetic and molecular approaches. *Annual Review of Plant Physiology and Plant Molecular Biology* 50, 333–359.

O'Neil, K.J. and Carrow, R.N. (1982) Kentucky bluegrass growth and water use under different soil compaction and irrigation regimes. *Agronomy Journal* 74, 934–936.

O'Neil, K.J. and Carrow, R.N. (1983) Perennial ryegrass growth, water use, and soil aeration status under soil compaction. *Agronomy Journal* 75, 177–180.

Oparka, K.J. and Cruz, S.S. (2000) The great escape: phloem transport and unloading of macromolecules. *Annual Review of Plant Physiology and Plant Molecular Biology* 51, 323–347.

Oparka, K.J. and Turgeon, R. (1999) Sieve elements and companion cells – traffic control centers of the phloem. *The Plant Cell* 11, 739–750.

Park, D.M., Cisar, J.L., Snyder, G.H., Erickson, J.E., Daroub, S.H. and Williams, K.E. (2005) Comparison of actual and predicted water budgets from two contrasting residential landscapes in south Florida. *International Turfgrass Society Research Journal* 10, 115–120.

Patton, A.J., Cunningham, S.M., Volenec, J.J. and Reicher, Z.J. (2007a) Differences in freeze tolerance of zoysiagrasses: I. Role of proteins. *Crop Science* 47, 2162–2169.

Patton, A.J., Cunningham, S.M., Volenec, J.J. and Reicher, Z.J. (2007b) Differences in freeze tolerance of zoysiagrasses: II. Carbohydrate and proline accumulation. *Crop Science* 47, 2170–2181.

Peacock, C.H. and Dudeck, A.E. (1984) Physiological response of St. Augustinegrass to irrigation scheduling. *Agronomy Journal* 76, 275–279.

Peacock, C.H. and Dudeck, A.E. (1985) A comparative study of turfgrass physiological responses to salinity. In: *Proceedings of the Fifth International Research Conference*. INRA Publications, Paris, pp. 821–830.

Peacock, C.H., Bruneau, A.H. and Dipaola, J.M. (1997) Response of the *Cynodon* cultivar 'Tifgreen' to potassium fertilization. *International Turfgrass Society Research Journal* 8, 1308–1313.

Perdomo, P., Murphy, J.A. and Berkowitz, G.A. (1996) Physiological changes associated with performance of Kentucky bluegrass cultivars during summer stress. *HortScience* 31, 1182–1186.

Petri, A.N. and Petrovic, A.M. (2001) Cation exchange capacity impacts on shoot growth and nutrient recovery in sand based creeping bentgrass greens. *International Turfgrass Society Research Journal* 9, 422–427.

Petrovic, A.M. (1990) The fate of nitrogenous fertilizers applied to turfgrass. *Journal of Environmental Quality* 19, 1–14.

Petrovic, A.M. (1993) Leaching: current status of research. *International Turfgrass Society Research Journal* 7, 139–147.

Petrovic, A.M., Soldat, D., Gruttadaurio, J. and Barlow, J. (2005) Turfgrass growth and quality related to soil and tissue nutrient content. *International Turfgrass Society Research Journal* 10, 989–997.

Picasso, V.D., Brummer, E.C., Liebman, M.L., Dixon, P.M. and Wilsey, B.J. (2008) Crop species diversity affects productivity and weed suppression in perennial polycultures under two management strategies. *Crop Science* 48, 331–342.

Plaxton, W.C. (1996) The organization and regulation of plant glycolysis. *Annual Review of Plant Physiology and Plant Molecular Biology* 47, 185–214.

Pollock, C.J. and Cairns, A.J. (1991) Fructan metabolism in grasses and cereals. *Annual Review of Plant Physiology and Plant Molecular Biology* 42, 77–101.

Potter, D.A., Bridges, B.L. and Gordon, F.C. (1985) Effect of N fertilization on earthworm and microarthropod populations in Kentucky bluegrass turf. *Agronomy Journal* 77, 367–372.

Potter, D.A., Patterson, C.G. and Redmond, C.T. (1992) Influence of turfgrass species and tall fescue endophyte on feeding ecology of Japanese beetle and southern masked chafer grubs (Coleoptera: Scarabaeidae). *Journal of Economic Entomology* 85, 900–909.

Praemassing, W., Reinders, A. and Franken, H. (2009) Physical properties of different turfgrass soils influenced by aeration treatments. *International Turfgrass Society Research Journal* 11, 1139–1152.

Prettyman, G.W. and McCoy, E.L. (2003) Profile layering, root zone permeability, and slope affect [effect] on soil water content during putting green drainage. *Crop Science* 43, 985–994.

Puhalla, J., Krans, J. and Goatley, M. (1999) *Sports Fields: A Manual for Design, Construction and Maintenance*. Ann Arbor Press, Chelsea, Michigan.

Qian, Y.L. and Engelke, M.C. (1999) Performance of five turfgrasses under linear gradient irrigation. *HortScience* 34, 893–896.

Qian, Y.L., Fry, J.D. and Upham, W.S. (1997) Rooting and drought avoidance of warm-season turfgrasses and tall fescue in Kansas. *Crop Science* 37, 905–910.

Qian, Y.L., Wilhelm, S.J. and Marcum, K.B. (2001) Comparative responses of two Kentucky bluegrass cultivars to salinity stress. *Crop Science* 41, 1895–1900.

Reeves, S.A. Jr. and McBee, G.G. (1972) Nutritional influences on cold hardiness of St. Augustinegrass (*Stenotaphrum secundatum*). *Agronomy Journal* 64, 447–450.

Reeves, S.A. Jr., McBee, G.G. and Bloodworth, M.E. (1970) Effect of N, P, and K tissue levels and late fall fertilization on the cold hardiness of Tifgreen bermudagrass (*Cynodon dactylon* × *C. transvaalensis*). *Agronomy Journal* 62, 659–662.

Richardson, M.D. (2002) Turf quality and freezing tolerance of 'Tifway' bermudagrass as affected by late-season nitrogen and trinexapac-ethyl. *Crop Science* 42, 1621–1626.

Richardson, M.D., Hoveland, C.S. and Bacon, C.W. (1993) Photosynthesis and stomatal conductance of symbiotic and nonsymbiotic tall fescue. *Crop Science* 33, 145–149.

Richie, W.E., Green, R.L., Klein, G.J. and Hartin, J.S. (2002) Tall fescue performance influenced by irrigation scheduling, cultivar, and mowing height. *Crop Science* 42, 2011–2017.

Richmond, D.S. and Shetlar, D.J. (1999) Larval survival and movement of bluegrass webworm in mixed stands of endophytic perennial ryegrass and Kentucky bluegrass. *Journal of Economic Entomology* 92, 1329–1334.

Richmond, D.S. and Shetlar, D.J. (2000) Hairy chinch bug (Hemiptera: Lygaeidae) damage, population density, and movement in relation to the incidence of perennial ryegrass infected by *Neotyphodium* endophytes. *Journal of Economic Entomology* 93, 1167–1172.

Richmond, D.S., Niemczyk, H.D. and Shetlar, D.J. (2000) Overseeding endophytic perennial ryegrass into stands of Kentucky bluegrass to manage bluegrass billbug (Coleoptera: Curculionidae). *Journal of Economic Entomology* 93, 1662–1668.

Riddle, G.E., Burpee, L.L. and Boland, G.J. (1991) Virulence of *Sclerotinia sclerotiorum* and *S. minor* on dandelion (*Taraxacum officinale*). *Weed Science* 39, 109–118.

Rochefort, S., Desjardins, Y., Shetlar, D.J. and Brodeur, J. (2007) Establishment and survival of endophyte-infected and uninfected tall fescue and perennial ryegrass overseeded into existing Kentucky bluegrass lawns in northeastern North America. *HortScience* 42, 682–687.

Rodriguez, I.R., McCarty, L.B. and Toler, J.E. (2005a) Effects of misting and subsurface air movement on bentgrass putting greens. *Agronomy Journal* 67, 1438–1442.

Rodriguez, I.R., McCarty, L.B., Toler, J.E. and Dodd, R.B. (2005b) Soil CO_2 concentration effects on creeping bentgrass grown under various soil moisture and temperature conditions. *HortScience* 40, 839–841.

Rogers, R.A., Dunn, J.H. and Brown, N.F. (1976) Ultrastructural characterization of the storage organs of zoysia and bermudagrass. *Crop Science* 16, 639–642.

Rogers, R.A., Dunn, J.H. and Nelson, C.J. (1977) Photosynthesis and cold hardening in zoysia and bermudagrass. *Crop Science* 17, 727–732.

Rose-Fricker, C., Fraser, M. and Meyer, W.A. (1997) Competitive abilities and performance of cool season turfgrass species in mixtures in the Willamette Valley of Oregon, USA, under high and low maintenance turf conditions. *International Turfgrass Society Research Journal* 8, 1330–1335.

Rouse, J.W., Haas, R.H., Schell, J.A. and Deering, D.W. (1973) Monitoring vegetation systems in the great plains with ERTS. In: *Proceedings of the 3rd ERTS Symposium*, NASA SP-351. NASA, Washington, D.C., pp. 309–317.

Sack, L. and Holbrook, N.M. (2006) Leaf hydraulics. *Annual Review of Plant Biology* 57, 361–381.

Saha, S.K., Trenholm, L.E. and Unruh, J.G. (2005) Effect of fertilizer source on water use of St. Augustinegrass and ornamental plants. *HortScience* 40, 2164–2166.

Salisbury, F.B. and Ross, C.W. (1992) *Plant Physiology*, 4th edn. Wadsworth Publishing, Belmont, California.

Samaranayake, H., Lawson, T.J. and Murphy, J.A. (2008) Traffic stress effects on bentgrass putting green and fairway turf. *Crop Science* 48, 1193–1202.

Sartain, J.B. (1985) Effect of acidity and N source on growth of thatch accumulation of Tifgreen bermudagrass and on soil nutrient retention. *Agronomy Journal* 77, 33–36.

Sartain, J.B. (2002) Tifway bermudagrass response to potassium fertilization. *Crop Science* 42, 507–512.

Sartain, J.B. and Volk, B.G. (1984) Influence of selected white-rot fungi and topdressings on the composition of thatch components of four turfgrasses. *Agronomy Journal* 76, 359–362.

Sass, J.F. and Horgan, B.P. (2006) Irrigation scheduling on sand-based creeping bentgrass: evaluating evapotranspiration estimation, capacitance sensors, and deficit irrigation in the upper Midwest. *Applied Turfgrass Science* doi:10/1094/ATS-2006-0330-01-RS.

Schmidt, R.E. and Snyder, V. (1984) Effects of N, temperature, and moisture stress on the growth and physiology of creeping bentgrass and response to chelated iron. *Agronomy Journal* 76, 590–594.

Schulze, E.D., Beck, E. and Müller-Hohenstein, K. (2002) *Plant Ecology*. Springer, Berlin/Heidelberg.

Shanahan, J.F., Schepers, J.S., Francis, D.D., Varvel, G.E., Wilhelm, W.W., Tringe, M.T., Schlemmer, M.R. and Major, D.J. (2001) Use of remote-sensing imagery to estimate corn grain yield. *Agronomy Journal* 93, 583–589.

Sharkey, T.D. and Raschke, K. (1981) Separation and measurement of direct and indirect effects of light on stomata. *Plant Physiology* 68, 33–40.

Shearman, R.C. and Beard, J.B. (1975a) Turfgrass wear tolerance mechanisms: I. Wear tolerance of seven turfgrass species and quantitative methods for determining turfgrass wear injury. *Agronomy Journal* 67, 208–211.

Shearman, R.C. and Beard, J.B. (1975b) Turfgrass wear tolerance mechanisms: II. Effects of cell wall constituents on turfgrass wear tolerance. *Agronomy Journal* 67, 211–215.

Shearman, R.C., Kinbacher, E.J., Riordan, T.P. and Steinegger, D.H. (1980) Thatch accumulation in Kentucky bluegrass as influenced by cultivar, mowing, and nitrogen. *HortScience* 15, 312–313.

Shearman, R.C., Erusha, K.S. and Wit, L.A. (2005) Irrigation and potassium effects on *Poa pratensis* L. fairway turf. *International Turfgrass Society Research Journal* 10, 998–1004.

Shimazaki, K., Doi, M., Assmann, S.M. and Kinoshita, T. (2007) Light regulation of stomatal movement. *Annual Review of Plant Biology* 58, 219–247.

Shuman, L.M. (2002) Phosphorus and nitrate nitrogen in runoff following fertilizer application to turfgrass. *Journal of Environmental Quality* 31, 1710–1715.

Sifers, S.I. and Beard, J.B. (1993) Comparative inter- and intra-specific leaf firing resistance to supraoptimal air and soil temps in cool-season turfgrass genotypes. *International Turfgrass Society Research Journal* 7, 621–628.

Smith, H. (1982) Light quality, photoperception, and plant strategy. *Annual Review of Plant Physiology* 33, 481–518.

Smith, M.W., Wolf, M.E., Cheary, B.S. and Carroll, B.L. (2001) Allelopathy of bermudagrass, tall fescue, redroot pigweed, and cutleaf evening primrose on pecan. *HortScience* 36, 1047–1048.

Smith, R.L. and Smith, T.M. (2001) *Ecology and Field Biology*, 6th edn. Benjamin Cummings, New York.

Snyder, G.H. and Cisar, J.L. (2000) Nitrogen/potassium fertilization ratios for bermudagrass turf. *Crop Science* 40, 1719–1723.

Snyder, G.H. and Cisar, J.L. (2005) Potassium fertilization responses as affected by sodium. *International Turfgrass Society Research Journal* 10, 428–435.

Snyder, G.H., Burt, E.O. and Gascho, G.J. (1979) Correcting pH-induced manganese deficiency in bermudagrass turf. *Agronomy Journal* 71, 603–608.

Soldat, D.J. and Petrovic, A.M. (2008) The fate and transport of phosphorus in turfgrass ecosystems. *Crop Science* 48, 2051–2065.

Soper, D.Z., Dunn, J.H., Minner, D.D. and Sleper, D.A. (1988) Effects of clipping disposal, nitrogen, and growth retardants on thatch and tiller density in zoysiagrass. *Crop Science* 28, 325–328.

Sorochan, J.C., Rogers, J.N. III, Stier, J.C. and Karcher, D.E. (2001) Fertility and simulated traffic effects on Kentucky bluegrass/supina bluegrass mixtures. *International Turfgrass Society Research Journal* 9, 941–946.

Spak, D.R., Dipaola, J.M. and Anderson, C.E. (1993) Tall fescue sward dynamics: I. Seasonal patterns of turf shoot development. *Crop Science* 33, 300–304.

Stanford, R.L., White, R.H., Krausz, J.P., Thomas, J.C., Colbaugh, P. and Abernathy, S.D. (2005) Temperature, nitrogen and light effects on hybrid bermudagrass growth and development. *Crop Science* 45, 2491–2496.

Starr, C. (2000) *Biology: Concepts and Applications*, 4th edn. Brooks/Cole, Pacific Grove, California.

Starr, J.L. and DeRoo, H.C. (1981) The fate of nitrogen fertilizer applied to turfgrass. *Crop Science* 21, 531–536.

Steinke, K., Chalmers, D.R., Thomas, J.C. and White, R.H. (2009) Summer drought effects on warm-season turfgrass canopy temperatures. *Applied Turfgrass Science* doi:101094/ATS-2009-0303-01-RS.

Stern, K.R. (1991) *Plant Biology*, 5th edn. Wm. C. Brown Publishers, Dubuque, Iowa.

Steudle, E. (2001) The cohesion–tension mechanism and the acquisition of water by plant roots. *Annual Review of Plant Physiology and Plant Molecular Biology* 52, 847–875.

Stiff, M.L. and Powell, J.B. (1974) Stem anatomy of turfgrass. *Crop Science* 14, 181–186.

Tan, Z.G. and Qian, Y.L. (2003) Light intensity affects gibberellic acid content in Kentucky bluegrass. *HortScience* 38, 113–116.

Teuton, T.C., Sorochan, J.C., Main, C.L., Samples, T.J., Parham, J.M. and Mueller, T.C. (2007) Hybrid bluegrass, Kentucky bluegrass and tall fescue response to nitrogen fertilization in the transition zone. *HortScience* 42, 369–372.

Thomas, J.C., White, R.H., Vorheis, J.R., Harris, H.G. and Diehl, K. (2006) Environmental impact of irrigating turf with type I recycled water. *Agronomy Journal* 98, 951–961.

Thompson, S.R., Brandenburg, R.L. and Arends, J.J. (2005) Conidial viability and pathogenicity of *Beauveria bassiana* (Balsamo) Buillemin for mole cricket (Orthoptera: Bryllotalpidae) control in turfgrass. *International Turfgrass Society Research Journal* 10, 784 792.

Throssell, C.S., Carrow, R.N. and Milliken, G.A. (1987) Canopy temperature based irrigation scheduling indices for Kentucky bluegrass turf. *Crop Science* 27, 126–131.

Throssel, C.S., Lyman, G.T., Johnson, M.E. and Stacey, G.A. (2009) Golf course environmental profile measures water use, source, cost, quality, and management and conservation strategies. *Applied Turfgrass Science* doi:10.1094/ATS-2009-0129-01-RS.

Tilman, D. (1988) *Plant Strategies and the Dynamic and Structure of Plant Communities*. Princeton University Press, Princeton, New Jersey.

Tilman, E.A., Tilman, D., Crawley, M.J. and Johnston, A.E. (1999) Biological weed control via nutrient competition: potassium limitation of dandelions. *Ecological Applications* 9, 103–111.

Tisserat, N. and Fry, J. (1997) Cultural practices to reduce spring dead spot (*Ophiosphaerella herpotricha*) severity in *Cynodon dactylon*. *International Turfgrass Society Research Journal* 8, 931–936.

Titko, S. III, Street, J.R. and Logan, T.J. (1987) Volatilization of ammonia from granular and dissolved urea applied to turfgrass. *Agronomy Journal* 79, 535–540.

Toler, J.E., Higingbottom, J.K. and McCarty, L.B. (2007) Influence of fertility and mowing height on performance of established centipedegrass. *HortScience* 42, 678–681.

Tomaso-Peterson, M. and Perry, D.H. (2007) The role of biofungicides and organic fertilizer in the management of dollar spot in bermudagrass. *Applied Turfgrass Science* doi:10.1094/ATS-2007-0911-01-RS.

Tompkins, D.K., Ross, J.B. and Moroz, D.L. (2000) Dehardening of annual bluegrass and creeping bentgrass during late winter and early spring. *Agronomy Journal* 92, 5–9.

Tompkins, D.K., Ross, J.B. and Moroz, D.L. (2004) Effects of ice cover on annual bluegrass and creeping bentgrass putting greens. *Crop Science* 44, 2175–2179.

Torello, W.A. and Rice, L.A. (1986) Effects of NaCl stress on proline and cation accumulation in salt sensitive and tolerant turfgrasses. *Plant and Soil* 93, 241–247.

Torello, W.A. and Symington, A.G. (1984) Screening of turfgrass species and cultivars for NaCl tolerance. *Plant and Soil* 82, 155–161.

Torello, W.A., Wehner, D.J. and Turgeon, A.J. (1983) Ammonia volatilization form fertilized turfgrass stands. *Agronomy Journal* 75, 454–456.

Toro (2010) Hydroject® 3010. Available at: http://www.toro.com/grounds/cultivation/aerator/small/3010.html (accessed 2 February 2010).

Treadway, L.P., Soika, M.D. and Clarke, B.B. (2001) Red thread development in perennial ryegrass in response to nitrogen, phosphorous, and potassium fertilizer applications. *International Turfgrass Society Research Journal* 9, 715–722.

Trenholm, L.E., Duncan, R.R. and Carrow, R.N. (1999) Wear tolerance, shoot performance, and spectral reflectance of seashore paspalum and bermudagrass. *Crop Science* 39, 1147–1152.

Tucker, B.J., McCarty, L.B., Liu, H., Wells, C.E. and Rieck, J.R. (2006) Mowing height, nitrogen rate, and biostimulant influence root development of field-grown 'TifEagle' bermudagrass. *HortScience* 41, 805–807.

Tucker, K.A., Karnok, K.J., Radcliffe, D.E., Landry, G. Jr., Roncadori, R.W. and Tan, K.H. (1990) Localized dry spots as caused by hydrophobic sands on bentgrass greens. *Agronomy Journal* 82, 549–555.

Turgeon, A.J. (2008) *Turfgrass Management*, 8th edn. Pearson Education, Upper Saddle River, New Jersey.

Turgeon, A.J. and Lester, G. (1976) Xanthophyll levels in turfgrass clippings. *Agronomy Journal* 68, 946–948.

Turner, T.R. and Hummel, N.W. Jr. (1992) Nutritional requirements and fertilization. In: Waddington, D.V., Carrow, R.N. and Shearman, R.C. (eds) *Turfgrass*. ASA-CSSA-SSSA (American Society of Agronomy-Crop Science Society of America-Soil Science Society of America), Madison, Wisconsin, pp. 385–440.

Tworkoski, T.J. and Glenn, D.M. (2001) Yield, shoot and root growth, and physiological responses of mature peach trees to grass competition. *HortScience* 36, 1214–1218.

Tyystjärvi, E. (2008) Photoinhibition of photosystem II and photodamage of the oxygen evolving manganese cluster. *Coordination Chemistry Reviews* 252, 361–376.

U.S. Congress (1987) Water Quality Act of 1987. 100th United States Congress, Washington, D.C.

USGA Green Section Staff (1993) USGA recommendations for a method of putting green construction: the 1993 revision. *USGA Green Section Record* 31 (2) 1–3. Available at: http://turf.lib.msu.edu/1990s/1993/index.htm (accessed 6 August 2010).

USGA Green Section Staff (2004) USGA Recommendations for a Method of Putting Green Construction: 2004 revision. Available at: http://www.usga.org/course_care/articles/construction/greens/Green-Section-Recommendations-For-A-Method-Of-Putting-Green-Construction/ (accessed 19 January 2010).

USSL (United States Salinity Laboratory Staff) (1954) Saline and Alkali Soils: Diagnosis and Improvement.

Handbook No. 60. available at: http://www.ars.usda.gov/Services/docs.htm?docid=10158&page=2 (accessed 1 February 2010).

Vargas, J.M. Jr. (1994) *Management of Turfgrass Diseases*, 2nd edn. CRC Press, Boca Raton, Florida.

Vargas, J.M., Danneberger, R.K. and Jones, A.L. (1993) Effects of temperature, leaf wetness duration and inoculum concentration on infection of annual bluegrass by *Colletotrichum graminicola*. *International Turfgrass Society Research Journal* 7, 324–328.

Vazquez, J.C. and Buss, E.A. (2006) Southern chinch bug feeding impact on St. Augustinegrass growth under different irrigation regimes. *Applied Turfgrass Science* doi:10.1094/ATS-2006-0711-01-RS.

Voet, D.V. and Voet, J.G. (1990) *Biochemistry*. John Wiley, New York.

Voigt, T.B., Fermanian, T.W. and Haley, J.E. (2001) Influence of mowing and nitrogen fertility on tall fescue turf. *International Turfgrass Society Research Journal* 9, 953–956.

Volk, G.M. and Horn, G.C. (1975) Response curves of various turfgrasses to application of several controlled-release nitrogen sources. *Agronomy Journal* 67, 201–204.

Volkmar, K.M., Hu, Y. and Steppuhn, H. (1998) Physiological responses of plants to salinity: a review. *Canadian Journal of Plant Science* 78, 19–27.

Waddington, D.V. (1992) Soils, soil mixtures, and soil amendments. In: Waddington, D.V., Carrow, R.N. and Shearman, R.C. (eds) *Turfgrass*. ASA-CSSA-ASSA (American Society of Agronomy-Crop Science Society of America-Soil Science Society of America), Madison, Wisconsin, pp. 331–384.

Waddington, D.V. and Duich, J.M. (1976) Evaluation of slow-release nitrogen fertilizers on Pennpar creeping bentgrass. *Agronomy Journal* 68, 812–815.

Waddington, D.V., Carrow, R.N. and Shearman, R.C. (eds) (1992) *Turfgrass*. ASA-CSSA-SSSA (American Society of Agronomy-Crop Science Society of America-Soil Science Society of America), Madison, Wisconsin.

Walker, N.R., Zhang, H. and Martin, D.L. (2005) Potential management approaches for the sting nematode in bermudagrass sod production. *International Turfgrass Society Research Journal* 10, 793–796.

Waltz, F.C. Jr. and McCarty, L.B. (2005) Field evaluation of soil amendments used in rootzone mixes for golf course putting greens. *International Turfgrass Society Research Journal* 10, 1150–1158.

Wang, Y., Crocker, R.L., Wilson, L.T., Smart, G., Wei, X., Nailon, W.T. Jr. and Cobb, P.P. (2001) Effect of nematode and fungal treatments on nontarget turfgrass-inhabiting arthropod and nematode populations. *Environmental Entomology* 30, 196–203.

Wang, Z. and Huang, B. (2004) Physiological recovery of Kentucky bluegrass from simultaneous drought and heat stress. *Crop Science* 44, 1729–1736.

Wang, Z., Huang, B., Bonos, S.A. and Meyer, W.A. (2004) Abscisic acid accumulation in relation to drought tolerance in Kentucky bluegrass. *HortScience* 39, 1133–1137.

Watschke, T.L., Schmidt, R.E., Carson, E.W. and Blaser, R.E. (1972) Some metabolic phenomena of Kentucky bluegrass under high temperature. *Crop Science* 12, 87–90.

Watschke, T.L., Schmidt, R.E., Carson, E.W. and Blaser, R.E. (1973) Temperature influence on the physiology of selected cool season turfgrasses and bermudagrass. *Agronomy Journal* 65, 591–594.

Webster, D.E. and Ebdon, J.S. (2005) Effects of nitrogen and potassium fertilization on perennial ryegrass cold tolerance during deacclimation in late winter and early spring. *HortScience* 40, 842–849.

Wehner, D.J. and Haley, J.E. (1993) Effects of late fall fertilization on turfgrass as influenced by application timing and N source. *International Turfgrass Society Research Journal* 7, 580–586.

Wehner, D.J., Haley, J.E. and Martin, D.L. (1988) Late fall fertilization of Kentucky bluegrass. *Agronomy Journal* 80, 466–471.

Wherley, B.G., Gardner, D.S. and Metzger, J.D. (2005) Tall fescue photomorphogenesis as influenced by changes in the spectral composition and light intensity. *Crop Science* 45, 562–568.

Whitcomb, C.E. (1972) Influence of tree root competition on growth response of four cool season turfgrasses. *Agronomy Journal* 64, 355–359.

White, J. and Harper, J.L. (1970) Correlated changes in plant size and number in plant populations. *Journal of Ecology* 58, 467–485.

White, R.H. and Dickens, R. (1984) Thatch accumulation in bermudagrass as influenced by cultural practices. *Agronomy Journal* 76, 19–22.

White, R.H., Engelke, M.C., Morton, J., Johnson-Cicalese, J.M. and Ruemmele, B.A. (1992) *Acremonium* endophyte effects on tall fescue drought tolerance. *Crop Science* 32, 1392–1396.

Wiecko, G., Carrow, R.N. and Karnok, K.J. (1993) Turfgrass cultivation methods: influence on soil physical, root/shoot, and water relationships. *International Turfgrass Society Research Journal* 7, 451–457.

Wilkinson, J.F. (1977) Effect of IBDU and UF rate, date, and frequency of application on Merion Kentucky bluegrass. *Agronomy Journal* 69, 657–661.

Wilkinson, J.F. and Beard, J.B. (1975) Anatomical responses of 'Merion' Kentucky bluegrass and 'Pennlawn' red fescue to reduced light conditions. *Crop Science* 15, 189–194.

Wilkinson, J.F. and Miller, R.H. (1978) Investigation and treatment of localized dry spots on sand golf greens. *Agronomy Journal* 70, 299–304.

Wilkinson, J.F., Beard, J.B. and Krans, J.V. (1975) Photosynthetic-respiratory responses of 'Merion' Kentucky bluegrass and 'Pennlawn' red fescue at reduced light intensities. *Crop Science* 15, 165–168.

Williams, D. and McCarty, B. (2005) Cultural practices for golf courses. In: McCarty, L.B. (ed.) *Best Golf Course Management Practices*, 2nd edn. Prentice Hall, Upper Saddle River, New Jersey, pp. 423–455.

Williams, D.W., Powell, A.J. Jr., Vincelli, P. and Dougherty, C.R. (1996) Dollar spot on bentgrass influenced by displacement of leaf surface moisture, nitrogen, and clipping removal. *Crop Science* 36, 1304–1309.

Williams, D.W., Burrus, P.B. and Vincelli, P. (2001) Severity of gray leaf spot in perennial ryegrass as influenced by mowing height and nitrogen level. *Crop Science* 41, 1207–1211.

Williamson, R.C. and Potter, D.A. (2001) Survival and development of black cutworm (Lepidoptera: Noctuidae) larvae on creeping bentgrass cultivars. *International Turfgrass Society Research Journal* 9, 810–813.

Woods, M.W., Ketterings, Q.M. and Rossi, F.S. (2005) Measuring the effects of potassium application on calcium and magnesium availability in a calcareous sand. *International Turfgrass Society Research Journal* 10, 1015–1020.

Wright, J.P. and Fisher, D.B. (1980) Direct measurement of sieve tube turgor pressure using severed aphid stylets. *Plant Physiology* 65, 1133–1135.

Wu, L. and Lin, H. (1993) Salt concentration effects on buffalograss germplasm seed germination and seedling establishment. *International Turfgrass Society Research Journal* 7 823–828.

Xiong, X., Bell, G.E., Smith, M.W. and Martin, B. (2006) Comparison of the USGA and airfield sand systems for sports turf construction. *Applied Turfgrass Science* doi:10.1094/ATS-2006-0531-02-RS.

Xiong, X., Bell, G.E., Solie, J.B., Smith, M.W. and Martin, B. (2007) Bermudagrass seasonal responses to nitrogen fertilization and irrigation detected using optical sensing. *Crop Science* 47, 1603–1610.

Xu, Q. and Huang, B. (2000) Growth and physiological responses of creeping bentgrass to changes in air and soil temperatures. *Crop Science* 40, 1363–1368.

Xu, Q. and Huang, B. (2001) Morphological and physiological characteristics associated with heat tolerance in creeping bentgrass. *Crop Science* 41, 127–133.

Xu, Q. and Huang, B. (2003) Seasonal changes in carbohydrate accumulation for two creeping bentgrass cultivars. *Crop Science* 43, 266–271.

Xu, Q. and Huang, B. (2004) Antioxidant metabolism associated with summer leaf senescence and turf quality decline for creeping bentgrass. *Crop Science* 44, 553–560.

Xu, Q., Huang, B. and Wang, Z. (2003) Differential effects of lower day and night soil temperatures on shoot and root growth of creeping bentgrass. *HortScience* 38, 449–454.

Xu, X. and Mancino, C.F. (2001) Annual bluegrass and creeping bentgrass response to varying levels of iron. *HortScience* 36, 371–373.

Yoda, K., Kira, T., Ogawa, H. and Hozumi, K. (1963) Self-thinning in overcrowded pure stands under cultivated and natural conditions. *Journal of Biology at Osaka University* 14, 107–129.

York, C.A. (1993) A questionnaire survey of dry patch on golf courses in the United Kingdom. *Journal of the Sports Turf Research Institute* 69, 20–30.

York, C.A. and Baldwin, N.A. (1992) Dry patch on golf greens: a review. *Journal of the Sports Turf Research Institute* 68, 7–19.

Youngner, V.B. and Lunt, O.R. (1967) Salinity effects on roots and tops of bermudagrass. *Grass and Forage Science* 1967, 257–259.

Youngner, V.B. and Nudge, F.J. (1976) Soil temperature, air temperature and defoliation effects on growth and nonstructural carbohydrates of Kentucky bluegrass. *Agronomy Journal* 68, 257–260.

Yust, A.K., Wehner, D.J. and Fermanian, T.W. (1984) Foliar application of N and Fe to Kentucky bluegrass. *Agronomy Journal* 76, 934–938.

Zeiger, P. and Hepler, P.K. (1977) Light and stomatal function: blue light stimulates swelling of guard cell protoplasts. *Science* 196, 887–889.

Zhang, M., Nyborg, M. and Malhi, S.S. (1998) Comparison of controlled-release nitrogen fertilizers on turfgrass in a moderate temperature area. *HortScience* 33, 1203–1206.

Zhang, X., Ervin, E.H. and LaBranche, A.J. (2006) Metabolic defense responses of seeded bermudagrass during acclimation to freezing stress. *Crop Science* 46, 2598–2605.

Zimmerman, U., Wagner, H.J., Schneider, H., Rokitta, M., Haase, A. and Bentrup, F.W. (2000) Water ascent in plants: the ongoing debate. *Trends in Plant Science* 5, 145–146.

Zscheile, F.P. and Comar, C.L. (1941) Influence of preparative procedure on the purity of chlorophyll components as shown by absorption spectra. *Botanical Gazette* 102, 463–481.

Zuk, A.J. and Fry, J.D. (2006) Inhibition of 'Zenith' zoysiagrass seedling emergence and growth by perennial ryegrass leaves and roots. *HortScience* 41, 818–821.

Index

Page numbers in **bold** refer to illustrations and tables

abscisic acid (ABA) 55, **57**
absorbance 77, 78, 79
acclimation 137, 139, 142–143
adaptation 24–**25**, 33, 34, 123–124, 166, 176
adenosine diphosphate (ADP) 18, 19, 48, **49**
adenosine triphosphate (ATP)
    glycolysis role 45
    production 22, 47, 57
    proton pump fuelling 42
    solute pumping role 51
    synthesis 18, 48, **49**, **50**
    see also energy
aerification (aeration)
    aggressive 148
    explained 9, 72, 147
    methods 71, 72, 147–150, 151, 152
    requirement reasons 161
    seasonal 138
    surface disruption calculation formula 149
    timing 147–148
    see also air; compaction
aggregates 70, **92**–93, 147
air
    injection 150
    movement
        improvement 21, 23
        insufficiency 86
        maintenance 32
        restriction 20, 21, 50–51, 82–83, 87
    temperature **132**
    see also aerification; boundary layer
Airfield System construction method 152, **153**
alkali 157–158
allelopathy 178–179, 182, 184
amino acids 48, **49**
amendments 153–154
    see also fertilizers; nutrition
amylase 38
anatomy 30, **42**
anions 89–90, 158
anoxia 142
antibiotics, resistance 185
antioxidants 133
Any Valley Country Club 196–197
apoplast 42, 51, **52**

bacteria 150, 151, 185
    see also pathogens
basins 125
bentgrass (*Agrostis* spp.) **25**, 59, 72
    creeping bentgrass (*A. stolonifera*)
        color measurement 7
        differential adaptation 176
        ecological population 165
        performance observation 6
        root mass **86**, **122**
        seasonal spectral reflectance response **131**
        shade effect 84–85
        spring health appearance **41**
bermudagrass (*Cynodon dactylon*)
    aggressiveness trait 35
    color measurement 7
    construction project use 201
    control 205
    dew formation **119**
    dormancy 34
    fertilization 209
    management practices 64
    performance observation 6
    seasonal spectral reflectance response **131**
    species **25**
best looking 3, 4
best management practices 2, 3
bicarbonate 31–**32**, 33
biocides 195–196
biology 2, 3, 4, 5, 7, 10
biomass 63, 70, **169**, **170**, 172, **173**, 174
black layer 150–**151**, 161
blends 171, 176, **194**
    see also mixtures
bluegrass (*Poa* spp.) **25**
    annual bluegrass (*P. annua*) 40, **41**, 63, 136–137, 182, 207
    Kentucky bluegrass (*P. pratensis*) 34, 165, 204, 205, **207**–**208**, 209
boron 159
boundary layer 23, 27, 50–51
breeding programs 195
brown patch disease **63**, 205
brushing 137
buffalograss (*Buchloe dactyloides*) **25**, 33, 34, 78, **131**

buildings 21, **81**
    see also air, movement, restriction; shade
bunkers 124

C-S-R model, three-strategy model 179
calcium (Ca) 97, 104, 106, 159
calcium sulfate (Gypsum) 95
Calvin cycle (dark reaction)
    bundle sheath cell role 30–31, 35, **53**
    carbon dioxide importance 23
    compounds 45, 46–47
    fuelling 16, 17, 22–23, 31
    process **19–20**
    Rubisco interaction 25–26
canopies 77, **79**, 86, 139
    see also shade; trees
capillary action 10, 54, 72
carbohydrates
    conversion to sucrose 142
    efficient use encouragement 37
    energy provision 57
    partitioning 43–44
    production 136
    storage 37, 132, 143
    see also starch; sugars
carbon 22, 26, 31–33, 35, 93–94, 136
carbon dioxide (CO_2)
    availability 10, 57, 120
    binding **26**, 35
    compensation point 37
    concentrations 31, 32–33, 35
    conversion to sugar 18
    deficiency 20
    fixing 37
    forms in water 31
    loss 33
    production 9
    seasonal requirement 17
    water ratio 29–30
carotenoids 16, 137
carpetgrass (*Axonopus* spp.) **25**
carrying capacity **165**, **168**, 170, 172, **173**, 187
cation exchange capacity (CEC) 70, 89–**90**, 94, 155
cations, components 158
centipedegrass (*Eremochla ophiuroides*) **25**
chaperonins (heat shock proteins) 134
checklists 66, **67**
chlorine excretion 158
chlorophyll
    degradation 17
    energy carrier 16–17, 31
    light energy receptor 16
    photon absorption 15, 22
    photon color affinity 15, 77
    photooxidation 141

    production 16, 105
    synthesis **16**, 137, 138
chloroplast 26, **27**, 39
cholesterol precursors 48, **49**
citric acid (Krebs) cycle 48, **49**
clay **89**, 90, 91, 92
climate 98, 166
cohesion-tension theory 55
cold 33, 130, 139–144, 176
    see also temperature
colligative properties 38, 52, 139, 140, 143
    see also osmosis
colloids 90, 94
color
    enzyme affinity **15**, 77
    health indicator 77
    measurement 7
    photons 15
    photosynthetically active radiation 22
    turf 77, **78**, 101, **102**, 130
    see also greenness; light
communities, ecological 166–167
compaction (compression)
    causes 72, 90–91, 92, **154**
    explained 9, 51, 91, 147
    layer **149**
    remedies 60, 72, 93, 94, 161
    water drainage problems 126
    see also aerification
competition
    control 63
    inhibition 191
    interspecific 176, 177–179, 191
    intraspecific 6, 82, 169–170, 182
    management 175–186, 187–199
    maturity advantage 171
    pests 187–199
    root 84, 86
    species selection role 169–170, 171
    strategies 179–183
    see also trees; vegetation control; weeds
compression *see* compaction
conduction 134–135
conductivity, electrical 157, 158
construction 2–3, 152–154, 157, 201–210
convection 134–135
cooling 21, 57, 119, 134, 139
    see also conduction; convection; transpiration
crabgrass (*Digitaria* spp.) 63
crop coefficient (K_c) 116, 127
crowns 19, 143
cryptochrome 80
cultivars 68, 144, 171
cultivation 183, 209–210
cutworm 197, 198, 199
cytokinins (plant hormone) 133

dallisgrass (*Paspalum dilatatum*) **183**
dark reaction *see* Calvin cycle
day length response 137–138
de-acclimation 143
decision-making 7, 11, 65, 95, 201–210
dehydration 62, 140-1, 141–142, 143, 144
denaturation 51, 134
density **169**, 172, **173**, 174, 178
deoxyribonucleic acid (DNA) 47, 106
dew removal 197
disease
    cause **63**
    control 193–194, 197, 205
    damage 196
    hot spots recording 192
    pressure 192
    research requirement 197
    resistance 190
    spread 171
    susceptibility 86, 138, 190, 193
    see also pathogens
disturbances 168–169, 174, 179
    see also drought; ice; mowing; traffic; weather; wind speed
ditches, sub-surface **126**
diversity 171, 176
dollar spot (*Sclerotinia homeocarpa*) 185, 193, **197**
dominance 166, 171, 174, 176–177, 182
dormancy 34, 51, 180, **181**
drainage 107–108, 124–127, 128, 152–153
drought
    damage **180**
    heat stress relationship 136
    physiological 158, 161
    preconditioning 120
    protection 57
    tolerance 34, 35, 116, 123, 124

ecology 2, 3, 4, 6, 163–174, 175
ecophysiology 165
ecosystems 6, 167–168
electron transport system **47**, 48–51
elimination 170
    see also self-thinning
endophytes 195
    see also fungi
energy
    cells 48, 106
    components 37–38
    conversion 8, 12
    extraction 44–45
    loss 17, 51, 57
    molecules 102
    photons 22
    redistribution 81–82
    release 57

requirement 40
    sequestration processes 20
    sinks 19, 30, 37, 38–40, 57
        see also root systems; seeds
    sources 19, 30, 38–40, 44, 57
        see also leaves; sucrose
    storage 8, 40
    transfer 9
    transport 57
    see also adenosine triphosphate; glycolysis; heat; light; photosynthesis
environment 2–3, 95, 103, 105, 180
enzymes 7, **15**, **26**, 31, 38
    see also Rubisco
equilibrium, dynamic 28–29, 134, **135**, **165**, 173
equipment 65, **67**, **71**, 112–115, 127, **137**
    see also mowers
establishment 205–207
evaporation 9, 10, 62, 119, 120
    see also evapotranspiration
evapotranspiration (ET)
    calculation 116, 120, 127–128, 209
    factors 118
    irrigation frequency factor 121
    percentage replacement 116
    process 115–**117**
    rate 119, 120
    reference standard 116–117, 127
exchangeable sodium percentage (ESP) 157, 158

Facilitation-Tolerance-Inhibition Model 179
fans 21, 139
fatty acids 48, **49**, 133, 141
feeding 95, 156
    see also fertilizers; nutrition
fermentation 45
fertilizers
    acidifying 95, 194
    applications frequency 103
    color response 101, **102**
    decision-making 95
    degradation 100
    effects 70, 71
    formulations 97, 144
    guidelines 98
    management 88–89
    rates 209–210
    recommendations 95, 98
    release rates 100–102, 156
    seasonal 143, 209
    stability promotion 172
    timing 99, 138, 144, 209–210
    see also nitrogen; nutrients; nutrition; phosphorus; potassium; sulfur
fescue (*Fescue* spp.) 25, 34, **63**, **131**, 201
fluid systems 51–**52**, **54**, 197

frost line 144
fructans 8, 37, 143
fungi 62, 185, 195, 196, 197
    see also pathogens
fungicides 176, 185, 198, 205

genetics 7, 76, 174, 176, 194–195
germination 180, **181**, 207
glucose 8, 23, 37, 38, **49**
glycolysis 44–**46**, **47**, 48
glyphosate (Roundup) 152, 205–206
golf courses 2, **117**, 124, 196
    see also putting greens
grasses
    cool-season (C_3) 24–35, 85, 130–131,
        135–137, 138–139
        see also bentgrass; bluegrass;
            fescue; ryegrass
    warm-season (C_4) 24–35, 130–131,
        133–135, 141, 144
        see also bermudagrass; buffalograss;
            centipedegrass; St. Augustinegrass;
            zoysiagrasses
gravel 153, 203
grazing 59
greenness 130, 177
    see also color
growth 12–13, 101, 169, 170, 172
    see also root systems; shoot growth
gypsum (calcium sulfate) 95, 160

habitat management 178
hail damage **178**
healing 61–62, 86
    see also injury
health **41**, 77, 84, 130, 192, 199
heat
    dissipation need 9
    generation 136
    production 17
    shock proteins 134
    sources 76
    stress 130, 136, 138
    tolerance 133, 176
    water energy source 29
    see also temperature
heme synthesis pathway 105
henbit (*Lamium amplexicaule*) **182**
herbicides 40, 138, 175, 183–185, 205, 210
human activity 59, 168–169
    see also traffic
human expectations 175–176
humidity 55, 100, 105, 118–119
hydathode 54
hydrogen peroxide 152

hydrophobicity 69, 152
hydroponics 89

ice damage 142, 144
immaturity 172
information gathering 11, **191**
Initial Floristic Competition model 179
injury 60–61, **62**, 66
    see also healing; mowing
inoculation 68
insecticides 176, 198
insects 138, 190, 192–193, 194, 196
inspections 124–125, 198
integrated pest management
        (IPM) 10–11, 189–190, 199
    see also pesticides
ion exclusion 158
iron (Fe) **16**, 95, 104, 105, 144
irradiance 76–81
irrigation
    concerns, salt-affected sites 159–160
    design 108–115, **157**
    efficiency 115
    effluent water use 115, 123
    frequency 84, 120–122, 206
    management 108–115, 117–118, 154–155
    recommendations 208–209
    seasonal 121, 122
    spray heads 112–115, 127
    system faults 115
    wetting agents applications 152
    withholding 122
    zones 111, 112, **113**, 127
    see also water
isobutylidine diurea (IBDU) 86, 100

Japanese lawngrass (*Zoysia japonica*)
    see zoysiagrasses

knowledge 191–192, 201–202
    see also information
Kranz anatomy 30–33, 35
Krebs cycle (citric acid cycle) 48, **49**

laboratories 96, 97, 157, 159–160
landscape 66, **69**, 168–169
law of limiting factors 166, 170–171, 172
law of minimum 165
law of self-thinning **170**, 172
lawns 3, **202**
layering problems 138
leaching 101, 155, 156, **157**, 158, 160, 161
leaf fire measurement 123

leaf spot 190
leaves
    energy sinks/sources 38, **39**
    salt glands 158
    tip damage **62**, **63**, **68**
    vascular system **30**
    visual quality 123
light
    adaptation 33, 34
    bands 22, 75–76, 79
    colors **14**, 22, 76, 78, 79, 80
    day length response 137–138
    distribution **14**, 79
    energy **13**, **17**, 22
    interception reduction 137
    management 93
    penetration restriction **79**
    performance effect 35
    quality 79, **80**, 81, 85
        see also shade
    reaction (z-scheme) **16**–17, 22, 31, 35
    reflected 76–77, 78, 79
    saturation **27**–**28**
    spectrum 13
    transmission 77, 79
    wavelengths **13**, **14**, 22, 76
    see also photosynthesis
lime 95
lipids 133, 141
loam **90**, 93, 154–155
localized dry spots (LDS) 151–152, 161

macromolecules 41–42
    see also proteins; ribonucleic acid
macronutrients 88–89, 97–104
    see also nitrogen; phosphorus; potassium
macropores **92**, 93, 94
magnesium (Mg) **16**, 97, 104, 106, 159
malate 31, 48, **49**
management practices 58–73, 130–131, 144, 145
    see also aerification; irrigation; mowing; topdressing
manganese (Mn) 104–105
mapping **117**, 128, 192
mat 70
maturity 172, 174, 183
membrane, cell 133, 141, 143
metabolism 7, 119–120, 139–140, 141
    see also photosynthesis; physiology; respiration
metals 89
    see also micronutrients; sodium
microbes 62, 68, 70, 188
micronutrients 89, 95, 104
    see also calcium ; magnesium; sulfur
microorganisms 54, 62, 190
micropores **92**, 93, 94

minerals 89, **92**
mistakes, management 168
mitochondria 26, **27**, 47
mixtures 171, 176–177, **194**
moisture data collection 128
molecules, binding opportunity 15
monitoring 115, 117, 138, 192, 194
morphology **80**
mowers **65**, 71, **137**, 197
    see also equipment
mowing
    damage 59, 60–66, 208
    effect 63, 64, **82**, 166, **167**
    facilitation **69**
    frequency 65
    height
        adjustment 87, 174
        comparison **82**
        effect 94, 138
        low, avoidance 172, **173**
        variation 206
    injury shock period 81–82
    management 64–66
    methods 137
    reasons 71, 169, 172
    specialized 60
    timing 197
    tolerance 73, 137
mutation, genetic 76

NADPH (reduced form of nicotinamide adenine dinucleotide phosphate) 17–18, 22, 45, 47, 48, **49**
nanometer (nm) 13
National Turfgrass Evaluation Program (NTEP) 204
nematodes control 195–196
niche 167, 170, 171, 174, 177, 206
nitrogen (N)
    detrimental 85
    fertilization 97–102, 106, 183, 186, 209
    leaching 155, 156
    loss propensity 100
    overapplication hazard 95
    potassium interaction 105
    recommendations 34–35
    release rates 100–102
    soil content reports 95
    sources 97–98, 102
    use efficiency 34–35
    see also fertilizers
normalized difference vegetation index (NDVI) 6, **131**
nutrients
    application 155–156, 161
    availability limiting 95
    deficiencies 97, 155, 156
    interactions 105

nutrients (*continued*)
    loss 155
    management 93
    overapplication hazard 95
    rankings 88–89
    retention 91, 155
    soil content reports 95, 97, **98**
    solubility 93
    tissue concentrations 158–159
    see also oxygen; solutes
nutrition 88–106, 156
    see also fertilizers; nutrients
nutsedge 210

observation 6–7, 117, 192–193
organelles 26, **27**, 37
osmosis 38, 43, 52, 94, 140, 161
overseeding 176–177
oxaloacetate 31, 48, **49**
oxidation, chemical 18
oxygen
    air percentage 25
    assimilation equation 26
    consumption 8
    electron transport role 50
    encouragement 106
    lack 142
    nutrition role 94
    in water 29–30
oxygen-evolving complex **18**

paints 15, 16, 78–79
pathogens
    as biological herbicides 184, 185
    control 195
    population growth 6, 192
    prey adaptation 190
    specificity 171
    spread delay 197
    see also bacteria; disease; fungi
Penman-Monteith model 116
pentose phosphate pathway 46–47, 57
perennial ryegrass (*Lolium perenne*) **25**
performance factors 6–7, 29–30, 33, 35
peroxisome 26, **27**
personnel management 66
    see also training
pesticides 175, 176, 177–178, 196–**198**
    see also integrated pest management
pests 187–199
Pfr state 80, **81**
pH treatments 152
phloem 30, 40–43, **44**, 51, **53**, 57
phosphates 45, 47, 48
    see also NADPH

phosphoenolpyruvate (PEP) carboxylase 31, **32**, 33, 45
phosphoglycerate (PGA) 20, 22, 23, 26, **27**, 31
phosphoglycolate (PGL) 26
phosphorus (P)
    cold-protection qualities interference 143–144
    fertilizer 102–103, 106, 144, 186
    overapplication hazard 95, 183
    recommendations 103
    soil reports 97
photochrome 80–**81**, **82**
    see also pigments
photographs 192, **202**
photoinhibition 144
photon flux 13, **14**, 15, 22, **28**, 79
photooxidation 141
photorespiration 25–30, 35, 44, 57, 99, 136
photosynthates 37, **39**
photosynthesis
    components 13–22
        see also Calvin cycle; carbon dioxide; light; water
    decline 132, 133, 141
    defined 7, 12
    disruption 62
    encouragement 12–13, 105
    energy provision 57, 199
    improvement 75–76
    light role 13–16, **78**, 105
    limiting factors 133
    maintenance 32–33
    material removal 60, 62
    pathways 22, 24, 33, 35
    potential 63
    process 9, 12–23
    promotion factors 172, 174
    rate 5, 40, 137
    reduction 20
    requirements 8, 33, **78**, 86, 93, 178
    sinks 38, 40
    sources 37, **39**
    stalling 10
    types 8
    wind speed effect 120
    see also Calvin cycle; light; z-scheme
photosynthetically active radiation (PAR) 13, **14**, 22
photosystems **17**, 22, 141
physiology 2, 3, 4, 5–6, 7, 80
    see also, metabolism; photosynthesis; respiration; transpiration; temperature
phytochrome 79, 80–81
    see also pigments
pigments 16, 77, 79, 80–**81**
    see also carotenoids
pipes **110**, 111–112, 125, **126**, 127
planning 130, 199, 207
plants
    characteristics 33–35, 59, **60**, 62
    classifications 179–180

hormones 133
needs diagnosis 1–11
populations 34, 165, **168**, 171, 174
soil structure facilitators 92
see also grasses; leaves; vascular system
potassium (K)
application rates 103–104
DNA/RNA synthesis 106
fertilization timing 138
leaching 156
membrane-regulated exchange 143
nitrogen interaction 105
osmosis management 38
overapplication hazard 183
pumps 56
purposes 105
sodium uptake interference 159
soil reports 97
stress relief qualities 103
utilization sufficiency 143–144
precipitation 115, 127
see also irrigation; rainfall
predators
characteristics 191
control 190–191, 193, 194, 195–196
management 188–195
/prey relationships 168, 183–184, 187–**188**
resistance 190, **191**, 194–195
proteins 18–19, 40, 51, 134, 142–143
protons 18–19, 22, 42, 48, 51, 56
protoplasm 51, 52
pumps 18–19, 42, 51, 56
putting greens 2–3, **69**, 82, **83**, 152
see also golf courses
pyruvate 31, 45, **46**, **49**

radiance **13**, 15, 17
see also light
radiation 13, **14**, 22, 76, 118, 120
rainfall 99, 107, 108
see also precipitation
random model **188**
reaction centres *see* photosystems
record keeping 117, 194
recuperative ability 73
reduction, chemical 18
reflectance 15, 77, 79, **117**, **131**
reflection 76–77
reproduction 171
research 4–5
Resource Ratio Model 179, 180
respiration
defined 7
encouragement 105
equation 8–9
explained 37

growing medium performance 150, 161
oxygen role 5, 94, 105–106
plant/animal differences 45
poor, symptoms 5
process 37–51, 57
promotion 172, 174
seasonal 40
soil compaction role 72
types 44, 48–51, 57
water uptake energy provision 70
see also photorespiration
Rhizoctonia solani 63
ribonucleic acid (RNA) 40–41, 47, 106
ribulose bisphosphate carboxylase oxygenase
see Rubisco
ribulose bisphosphate (RuBP) 20, 22, 26
root systems
building 120–121
competition 84, 86
encouragement 87, 131
growing medium effect 150, 158
growth 87, 121–122, 131
mass **122**
pressure 53–54
state of health 99
temperature effect 85–86
water absorption 53–54
water uptake 57
Rubisco (ribulose bisphosphate carboxylase oxygenase)
deficiencies 33
role 20, 22, 25–**26**, 27, 29, 31, 35
runoff 99, 100
rust 194, 205
ryegrass (*Lolium* spp.) **25**

safety 66
St. Augustinegrass (*Stenophrum secundatum*) **25**, 34, 124
salt 89, 94–95, 156–161
sand systems
advantages **92**, 161
characteristics 14–17, 91, 152, **153**
construction 152–154, 157
core aerification **148**
design 152–154
drainage 124–125
fertilization 155
particle size **89**, **90**, 153, **154**
problems 150–152
sustainability 154
water inputs 161
scales, balance, equilibrium **135**
science 3–4
seashore paspalum (*Paspalum vaginatum* Swartz) **25**, 160
seasons 40, **41**, 130, 144, 145, 172

seeds **181**, 182, 183, **184**
selection
    competitive 169–170, 171, 182
    decisions 182, 185
    information 144
    knowledge importance 204–205
    traits 73
self-thinning **170**, 172, **173**
sensors **117**, 130
shade
    adaptation 82
    differences **81**
    effect 81, 83, **84–86**
    elimination 93–94
    light penetration 77
    management 82, 83–85, 87
    problems 10
    responses **80**, 81
    stress, measurement 44
    tolerance 85, 86
shoot growth 63, 81, 85, 86, 99
Siemen 157, 158
silt **89**, 91, **92**
single-tactic systems 10–11, 190–191, 199
sodium 94, 157, 158–159
soil
    acidity 70
    aeration 50–51, 91, 147–150
    attributes 89
    classifications **91**, 157–158, 161
    fertile, appearance 69–70
    moisture 121, 143
    pH 95, 152, 194
    problems 147
    profile, sand putting green **69**
    saturation 9, 97, **98**, 124, 127, 155
    sodicity 157–158, 159–160, 161, 204
    structure **92–93**, 159
    temperature **132**, 143
    test reports 95–97
    texture 90–92
    water-holding capacity 91
    see also aerification; clay; compaction;
        sand systems; silt
solar spectrum categories 75–76
solutes 9, 19, 37, **38**, 51, 140
    see also potassium
species
    list **25**
    number limitation 166, **167**
    spreading methods 169, 171
    see also grasses; plants; selection
spectral sensing 128, 130, **131**
sports fields 124
spraying 6, 178
    see also irrigation

sprinklers 112, **114–115**
    see also irrigation
stability 167–168, 171, 174
starch 8, 37–38, **39**, 143
stems 19, 40
stolons 38
stomates **38**, 55–57
stress
    human interventions 59
    measurement 44
    relief qualities, potassium 103
    resistant strategists 180
    seasonal 27, 29–30
    susceptibility 63
    temperature 130, 131–136, 139–141
    water deficit 121
    see also cold; compaction; drought; heat;
        mowing; shade; soil,
        saturation
stroma 18
succulence 72
sucrose 23, 37, **38**, 40, 42, **43**
    see also sugars
sugars 18, 19, 37, 40, 46–47
    see also glucose; sucrose
sulfur 95, 104, 106, 150, 151
summer **28**, 29, 99, 122, 138–139
summer patch disease 205
survival 59, 61
sustainability 168, 174
symplast 42, 51, **52**

techniques journal 192
temperature
    chemical reaction speed effect 70
    climatic 130
    daily 131, **132**
    increase 29
    influence 98–99, 100, 102
    management practices adjustment
        need 19–20, 118, 144
    physiological effects 35, 99, 119,
        131–132, 133, 136
    reduction, canopies 139
    tolerance 33–34, 133–134
thatch 66–72, 92, 138, 153
thykaloid 18, 22, 42
tillers (vegetative growth) 169, 172
tissue **42**, 97, 158–159
topdressing 60, 70, 71–72, 139, 152
toxins 45, 159, 178–179, 184
tracheids 52, **53**
traffic
    control 94, 138–139
    damage 69, 72, 73, 138, 168

problems 59–60
sustainability effect 154
tolerance 58, 86
see also compaction
training 65–66
transpiration
   activity 5
   components 51–57
   control 62
   cooling role 9, 51
   defined 7
   disruption 62
   encouragement 105, 172, 174
   explained 37
   irrigation adjustment factor 121
   likened to perspiration 133
   process 9–10
   pumping mechanisms 51
   reduction 20, 63, 83
   regulation 55
   water uptake effect 9–10, 70
   wind speed effect 120
   see also evaoptranspiration
trees 21, 23, 84, 86, 178
turf/turfgrasses, functional characteristics 59, **60**, 62
turfgrass, *defined* 58–59
Turfgrass Transition Zone, United States of America 201, 204
turgor (hydrostatic) pressure 38, 121, 123

ubiquinol (UQH$_2$) **47, 48, 49**
United States Golf Association (USGA) 2–3, 152, **153**, 154, 157
United States Salinity Laboratory (USSL) 157, 159–160
University of California Sand Putting Green Construction and Management method 152, **153**, 154

vacuoles 37, **38**, 51, **52**
   see also organelles
vapor pressure 28–29
vascular system 30, **30**, 51, 52–**53**, 55, 140
   see also phloem
vegetation control 21, 23
   see also shade; trees
vessels 52, **53**
viruses 190
volatilization 100

water
   adhesion 10, 52, 55
   amounts 128
   availability 110
   cellular/intracellular 139–**140**
   cohesion 10, 52, 55
   effluent 115, 123
   energy source 29
   expansion 140
   flow friction factor **110**, 111–**112**
   frequency 155
   loss **62**, 65
   management 107–128
   molecules motion 29
   movement 9–10, 52–53
   oxidation **50**
   pH 31, **32**
   potential 10, 42–43, 52, 55, 158
   pressure **110**, 127
   sources 107, 109–110
   splitting 18, 22
   transport regulation 55
   uptake 9–10, 57, 70, 158
   use 34, 35, 37, 118–124
   viscosity 141–142
   window 109, 127
   see also irrigation; soil, saturation
wear 72–73, 177
   see also traffic
weather 98
   see also climate; temperature; wind speed
weeds
   accumulation (soil bank) 181
   categories 63, 136–137, 179–180, **181**, **182**
   competition management 40, 63, 179–183
   control 63, 64, 179–186, 204–205, 206, 210
   eradication 177–178
   identification 180
   strategists 179–180
   see also bluegrass (*Poa* spp.), annual
wilt 121, 123, 155
wind speed 21, 118, 120

xanthophylls 16, 77
xeriscape **61**
xylem 30, 51, 52–**53**, 55, 140

z-scheme (light reaction) **16**–17, 22, 31, 35
   see also electron transport system
zoysiagrasses 25, 33, 34, **117**, **119**, 124, 201